P9-CQL-249

THE FEMINIST DOLLAR

The Wise Woman's Buying Guide

THE FEMINIST DOLLAR

The Wise Woman's Buying Guide

PHYLLIS A. KATZ, Ph. D.
and
MARGARET KATZ

PLENUM TRADE • NEW YORK AND LONDON

Library of Congress Cataloging-in-Publication Data

Katz, Phyllis A.
The feminist dollar : the wise woman's buying guide / Phyllis A. Katz and Margaret Katz.
p. cm.
Includes bibliographical references and index.
ISBN 0-306-45562-5 (hardbound). -- ISBN 0-306-45563-3 (pbk.)
1. Consumer protection--United States. 2. Consumer education--United States. 3. Women consumers--United States. 4. Feminism--United States. 5. Corporations--United States--Rankings. 6. Sex discrimination against women--United States. 7. Shopping--United States. I. Katz, Margaret. II. Title.
HC110.C63K37 1997
381.3'082--dc21 96-51694
CIP

The authors have made every effort to present the most recent information available to them at the time of their writing; they cannot be responsible for subsequent changes in information.

ISBN 0-306-45562-5 (Hardbound)
ISBN 0-306-45563-3 (Paperback)

Plenum Press is a Division of Plenum Publishing Corporation
233 Spring Street, New York, N.Y. 10013-1578
http://www.plenum.com

10 9 8 7 6 5 4 3 2 1

Printed in the United States of America

Foreword

Some years ago, Phyllis Katz was lunching at a restaurant with her colleagues when one of the women ordered South African lobster tails. The others in the group were appalled; Nelson Mandela was still in prison, apartheid was still the law of the land—how could this woman patronize the homeland of institutionalized racism? Shouldn't people of conscience register their opposition by boycotting all South African products?

Good point, thought Dr. Katz, but why just apartheid? Why not boycott countries that treat *women* badly, countries where institutionalized sexism is the law of the land? The thought came and went, as worthy thoughts often do, but fortunately for us, it resurfaced again not long afterward when Dr. Katz was vacationing in Ecuador. She read in the local newspaper that Ecuadorean women could be sent to prison for up to 12 years for having an abortion. "I found myself wondering why I was spending so much money in a country with so little regard for its female citizens," she said. "Although my involvement with women's issues goes back more than two decades, my travel arrangements were made with no knowledge of how women were treated at my destination."

Because she cares about how women are treated at *every* destination—whether it's a country, state, corporation, or neighborhood business in the downtown mall—Dr. Katz started collecting mountains of data and distilling them into the FEM scores that made *The Feminist Dollar* possible. And as a result of her exhaustive research, you now hold in your hands the most important new shopping manual since *Consumer Reports*, the most innovative travel book ever since the *Guide Michelin*, and the most practical feminist handbook since *Our Bodies, Ourselves*.

How can *The Feminist Dollar* help you? Let me count the ways—or at least three of many hundreds of ways:

- Say you live in Denver and you're shopping for a new car. You visit various automobile dealerships in town to check out the best deals and you can't help but notice that only two of the eight establishments employ female salespeople. You question the manager of one of the all-male operations, and he says his dealership hires only men because "women are too sensitive, couldn't handle the pressure, aren't aggressive enough, and don't want to work long hours." If this deliberate exclusionary policy operates so blatantly on the floor of the shop, it's safe to assume that women do no better at the management level. Why spend thousands of your hard-earned dollars in a place that doesn't support women? Check out the automobile FEM scores first.
- Say you're the person in your company who's responsible for setting up conference arrangements. It's your job to choose the hotel or conference center where 150 of the firm's managers and sales staff will meet twice each year. It's also up to you to book round-trip air travel for them to get there. Obviously, you will search out the best travel deal and the highest quality accommodations that your budget will allow. But let's assume that, when all the bids are in, you end up with five packages that are

comparable in every respect. Why not give your company's business to the hotel chain and airline that practice the best policies toward women? And why not book the conference in a state whose leaders have put improving women's status high on their agenda? In other words, check out their FEM scores first.

- Say you're a family of six and you use a lot of over-the-counter and prescription drugs. Say you also care deeply about maintaining women's reproductive freedom; you practiced family planning and you want every woman in the United States to have that right. But you've heard that certain drug companies have caved in to the threats and boycotts of anti-choice groups; they have been pulling money away from contraceptive research or have refused to market RU-486, the new French "morning-after pill." Why would you want to make such a drug company even one dollar richer? Why not buy your brand name drugs *after* you know the manufacturer's FEM score?

In *The Feminist Dollar*, Phyllis Katz and Margaret Katz provide readers with the basic facts about gender fairness and equity as it is—or is *not*—practiced by corporations and governments. Once armed with that information, it's up to us to decide what we want to do with our purchasing power and whether we care enough about social and economic justice to vote with our feet. The authors make a persuasive case for us to walk away from commercial interests that discriminate, diminish, or otherwise turn their backs on women. Their book actually makes it easy for you to speak in "pocketbook prose," to register your opposition to sexist attitudes and behaviors in the marketplace here and abroad, and to put your money where it can reward those who respect the dignity and potential of women.

From now on, every home, office, or library reference shelf will be incomplete without *The Feminist Dollar*. Keep it near your Yellow Pages. Consult it often. Open it before you open your checkbook. And don't leave home without knowing what the FEM score is at your destination.

Letty Cottin Pogrebin
Founding Editor, Ms. Magazine

Preface

This book has an ambitious mission: to help you choose products in a way that empowers women. Spending behavior is being used more and more as a source of power. For example, you can choose an investment plan that supports certain progressive companies or find a credit card that gives a percentage of dollars spent to your favorite charity. Either way, you are using your spending decisions to further a cause. This book is intended to add even more to your power as a consumer, especially if your agenda includes helping women succeed and achieve equality in the work force.

You will find here a guide to almost 400 companies that make and market the products you buy and use every day, as well as a guide to some of the states and countries to which you might travel. It is meant to be used for all your purchases. Take it grocery shopping; consult it before buying a car, a computer, or an airline ticket; talk about it with friends and colleagues. Use the "Buying Guide," a summarized version of the ratings that appears in a special highlighted section beginning on page 282.

In rating these companies, we have tried to gather information as objectively as possible, but there were a number of judgment calls we had to make. Each company was given multiple chances to provide and correct information. When that didn't happen, we searched for the data elsewhere, from government and private sources. We have tried to maintain consistency, fairness, and objectivity in compiling numbers. Because information came from such disparate sources, there were decisions that had to be made on a continual basis. Whenever we could, we let the companies guide us. For example, we originally set out to find the percentage of women among each company's top 25 executives. But some companies had only 7 executives, whereas others had 127. To accommodate these differences, we let the companies guide us, by counting the executives considered important enough to name in their annual reports.

Another judgment we were required to make was how to treat missing information. Not wanting to penalize private companies or nongovernment contractors (getting information for those companies proved a far more difficult task), we chose to give industry average (with a maximum of 10) scores in categories with missing information. Companies that are missing more than one category are marked with an asterisk throughout the book.

Yet another judgment we made was to choose what to highlight about companies in the text. We chose the things that stood out and the things we, as female consumers, might want to know when making a purchasing decision.

The state of corporate America is not static; mergers, spin-offs, and acquisitions make these companies moving targets. We have tried to keep up with these changes to the extent possible but cannot be responsible for what has changed since we went to press.

Overall, we have striven to make these ratings as objective, explicit, and scientific as possible (for more specific information of principles of research and methodology, see Appendix One). We feel they will be useful. By exercising our power as consumers, we can put pressure on companies to hire, promote, and make working life easier for women and mothers in this country. Let your dollars count for women!

Acknowledgments

Tracking down, compiling, and analyzing the mountains of information that went into the ratings for this book was a mammoth task. Fortunately, some extremely able individuals believed in the project sufficiently to assist in this quest at its various stages. For the early stages of the project, we are most grateful to Carol Hoffeld for her wisdom and encouragement; to Susan Cohen, our agent, for her savvy and persuasive skills; and to Elizabeth Daly, Jennifer Rovich, Isabel Friedman, Earnestine Simmons, and Marilyn Gelman for their research assistance. Additionally, steady and excellent help was provided throughout the project by Linda Dirnberger, whose diligence and continuing optimism in contacting corporations was wonderful. In the middle and later stages of the project, we were particularly fortunate to have the help of Christine Berkowitz and Genevieve Austin, both excellent researchers with fine editorial skills. Editorial help was given at various stages by Henry Ferris, Geoff Shandler, and Erika Goldman, and we appreciate their time and level of expertise.

The project could not have been completed without the organizational genius of Stephanie Roth-Nelson. A fine author in her own right [her own book, titled *S.E.E.K.* (*Self-Esteem Enhancement Kit*), is a journal-based workbook for teenagers], she gave her talents unstintingly for more than 2 years. Our gratitude goes beyond words. We would also like particularly to thank Dr. Marty Barrett, our statistician, both for her wonderful skills and for her patience and willingness to redo analyses as our data were being organized. Sylvia Grove, an irrepressible and irreplaceable administrative assistant, was invaluable throughout. We are most appreciative of her ideas and attention to detail, as well as her excellent secretarial skills.

We are also most fortunate to have a cooperative and skilled family, and we would like to express our gratitude to Aron Katz, a helpful husband and father; to Martin Katz (son and brother), who so willingly shared his legal and editorial expertise; and to Marcelina Rivera (daughter- and sister-in-law), who helped analyze early drafts with her marvelous combination of legal knowledge, ebullience, and feminist focus.

In addition, a number of individuals and organizations were generous in sharing information with us. For the corporate sections, we are particularly grateful to Alice Tepper Marlin and Steve Dyott at the Council of Economic Priorities; to Steve Lydenberg of Kinder, Lydenberg, Domini & Co., Inc.; to Bill Matthews of The Foundation Center; to Rochelle Sharpe of *The Wall Street Journal*; to Vivian Todini at Catalyst; to Pamela Rosenau and Shirley Mabe for their advice regarding financial institutions; and to the many corporate personnel who took the time to complete our questionnaire. We also appreciate the willingness of Annie Blackwell, Robert Greaux, and Joe Kennedy of the U. S. Department of Labor to process our multiple requests, if not always the speed of their response. For information relevant to states, we appreciate the help of Laurie Wood with the National Center for Women and Family Law and Susan Williams with Women Executives in State Government. Finally, gathering information about countries was assisted enormously by the help of International Planned Parenthood and Tim Wirth's office at the State Department. Although the ideas expressed in this volume are solely our own, it is most gratifying to be involved in the broad network of individuals and organizations who are helping women in so many ways. A list of these excellent organizations (with their addresses) is contained in Appendix Three for the reader's use.

Several very talented and busy people kindly took the time to serve on an Advisory Committee for this project. We are grateful to all of them: Dr. Barbara Bergmann, Dr. Elise Boulding, Dr. Leonard Chusmir, Dr. Faye Crosby, Dr. Barbara Gutek, Dr. Ellen Langer, Lettie Pogrebin, and Pamela Rosenau.

Contents

PART ONE WOMEN AND CONSUMERISM

Chapter One Women's Power: Making Good Use of Your Economic Clout 3

Chapter Two La Crème de la FEMME: Our Scoring Guide 15

PART TWO WOMEN AS DOMESTIC SPENDERS

Chapter Three The Kitchen CEO 31

Chapter Four Beyond the Grocery Store: Product Ratings for Household Goods, Stores, Clothes, and Children's Products 58

Chapter Five — Household-Based Consumption: Ratings of Other Products and Services Often Purchased by Women 95

PART THREE — WOMEN AS BUSINESS SPENDERS

Chapter Six — Making Female-Friendly Consumerism Work from Work: Rating Products and Services Related to the Office 137

PART FOUR — WOMEN AS TRAVEL CONSUMERS

Chapter Seven — The States of Women's Status 163

Chapter Eight — Globe-Trotting Women 219

Endnotes 274

Buying Guide — Product Names and Companies by Category 282

Appendix One — Methodology, Scoring, and References 347

Appendix Two — Category Ratings of Companies 369

Appendix Three — Organizations that Help Women 384

Appendix Four — Governors' Names, Addresses, and Phone Numbers 392

About the Authors 396

Index 397

PART ONE

Women and Consumerism

Chapter One

Women's Power

Making Good Use of Your Economic Clout

With money in your pocket, you are wise, and you are handsome, and you sing well too.

Jewish Proverb

The consumer is not a moron—she is your wife.

David Ogilvy

ARE WOMEN POOR OR POWERFUL?

There are two ideas concerning women's economic status that seem, at first glance, to contradict one another:

1. Women are poorer than men and think of themselves as powerless.
2. U.S. women are potentially the most economically powerful group in the world in terms of their purchasing power.

How can we reconcile this paradox?

We're sure most of you are familiar with the first notion. You know how hard it is to make ends meet, particularly if you are a single mother with no access to the father's income for household purchases. A disproportionate number of you, including those who work, live below the poverty line. You know firsthand the meaning of the dismal and often-repeated statistics of how much more men earn than women. You earned only 72 percent of what men earned in comparable positions in 1994[1] (the last year for which complete published numbers were available): a median income of $22,205 for women versus $30,854 for men. Another way of viewing this is that you would be likely to receive the equivalent of a *39 percent* raise if you were a man. The wage gap is even wider for women of color: African-American women earn only 63 cents of the white man's dollar and Hispanic women earn an abysmal 55 cents. Of those earning the minimum wage, 64 percent are women.[2] Despite your educational attainments, most of you still earn considerably less than men because the average woman with a college degree earns just slightly more than the average man with a high school degree.[3] Despite equal employment legislation and the vastly increased numbers of women entering the work force over the past two decades, you are still very likely to be subject to discrimination in hiring, pay, and promotion or to sexual harassment. Anita Hill's testimony at Justice Clarence Thomas's Supreme Court confirmation hearings struck a chord because so many of you have had similar experiences. Gender discrimination contributes heavily to women's low economic status and a general sense of powerlessness.

You are probably much less familiar with the second notion, which is that women are economically powerful. Information about women's power as consumers does not get much coverage. You may even have trouble believing it. But *U.S. women actually have more purchasing power than the total economic output of any other country*, including Japan! To put this power into perspective, and to understand its full extent, a few economic terms and statistics need to be examined.

One of the basic measures economists use to compare the economic output and progress of different countries is called the *gross domestic product* (or GDP). This refers to the *total output of goods and services produced within a country and available for consumption in a given year*. In industrialized countries, the GDP is measured in billions of dollars. It is composed of three major parts: spending by consumers, spending by the government, and spending by businesses for investment purposes. In 1995, for example, the GDP for the United States was $6150.3 billion[4]—a very large number. It is twice as large as that of our closest competitor, Japan. It looks even larger when it is written out numerically: $6,150,300,000,000.

Consumer spending makes up the biggest part, somewhat more than two-thirds. Government spending accounts for almost one-fifth of the pie, and investment spending by businesses on such items as capital improvements, equipment, and inventory accounts for the rest (see pie chart).

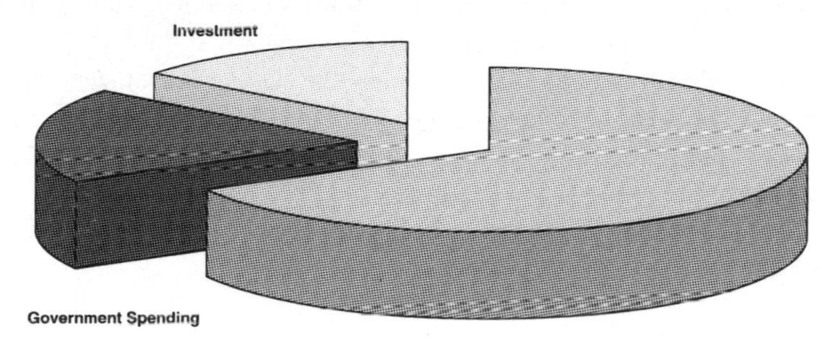

U.S. GDP (in billions of $)

The consumer portion is the one of major interest to us. Susan Faludi estimates in her book *Backlash* that *80 percent of all consumer spending is done by women.*[5] Economists have suggested that this number may be even higher when the sheer number of purchasing decisions are considered, rather than the total amount of money spent. Since women are usually the major decision-makers for most consumer purchases made by their households, their economic power is not limited to their own incomes. In terms of the 1995 pie, then, we estimate women's spending power in the consumer category to be $3300 billion—more than three times their own earning capacity.[6]

The amount of consumption women control, however, may be even larger when one considers that female employees also make many spending decisions in the other two parts of the money pie: the government and investment pieces. It's a bit harder to quantify these amounts, however. The government category includes all defense expenditures, an area in which women probably have little say. Similarly, women may have little input in investment decisions concerning new plants and major pieces of industrial machinery. But they often make decisions about purchases of office furniture, equipment, and supplies. Although it is harder to estimate precisely what these amounts might be, we can probably safely assume that at least 5 percent of the spending decisions in the government and investment sectors are made by women. Even this very low estimate brings the total amount of consuming power of U.S. women to approximately $3400 billion. Note that this is more than half of the total pie.

It is difficult for most people to grasp the significance of such large numbers. Let us compare the spending power of U.S. women with the total economic power (GDP) realized by other industrialized countries in the same year.

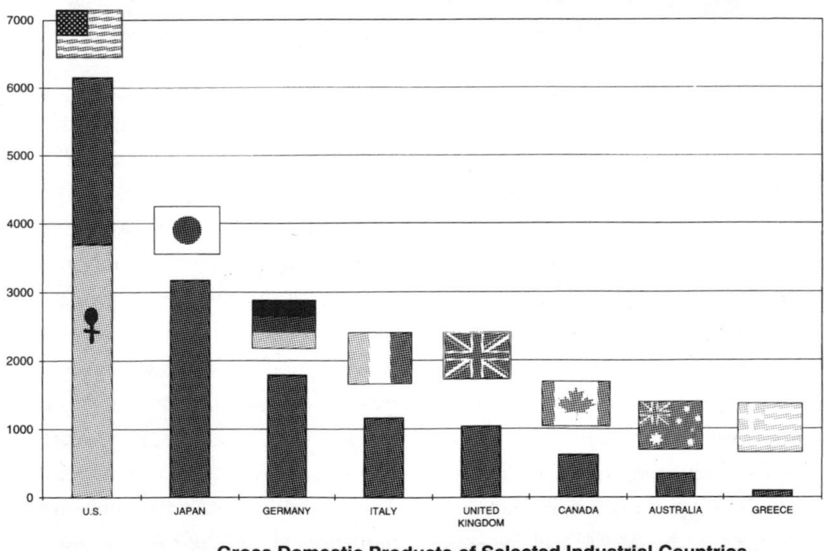

Gross Domestic Products of Selected Industrial Countries for 1995 (in Billions of Dollars)

In the bar graph, it can be seen that women in the United States have purchasing power *somewhat greater than the total economic output of Japan*, twice as much as Germany, three times as much as Italy or the

United Kingdom, and six times as much as Canada or Australia. This is an enormous degree of economic clout. But this power has not been used to benefit women. If used systematically, women could exercise vastly more influence than they do now.

How, then, can we resolve this seeming paradox between the first notion, women's feelings of *individual* lack of financial status and power, and the second notion, their very real *collective* potential for economic power? In fact, both are true—but in different contexts. It all depends on what comparison you wish to make. With respect to a man in the United States, a woman earns considerably less. As we have seen, however, the purchasing power of U.S. women as a group exceeds the total economic output of any other country in the world. Women can, therefore, wield considerably more economic influence than they now do. Three things are needed to accomplish this:

1 *Recognition of women's economic strength*
2 *The will to use it*
3 *The relevant information*

This book was written to help you in all three areas. After reading it, you will know how to use your consumer power to support companies and communities that treat women well and to shun those that are highly discriminatory. By doing this, you can communicate that *gender equality pays*. Even women who don't label themselves as feminists have been frustrated by discrimination at work: being come on to by a male boss, making less money than male colleagues, being the only woman in the boardroom or, more likely, feeling like the boardroom is—and will always be—off-limits. You have the power to force companies to treat women more fairly. You can also exert pressure on states and foreign countries to take women's issues more seriously. In short, you can impact the many institutions that are highly dependent on female consumers and, by doing so, increase women's earning power and improve their work environments.

Because our male-dominated legislatures and courts have not been very effective to date, the "pocketbook prose" approach we are suggesting has great appeal. In our work with women, we hear more and more about how fed up they are with the way things are. Many are searching for ways to improve their economic condition and their quality of life. Here's one very important thing you can do: think about how you spend the resources available to you, and make sure that you consider the effect of these decisions on women.

WHAT DETERMINES HOW WOMEN MAKE SPENDING DECISIONS?

People often pay little attention to why they buy particular products, but their purchasing patterns are studied very carefully by market researchers. Price is obviously one important factor in buying choices. Particular features of the product also play a role. Some consumers rely on objective ratings of quality made by organizations such as *Consumer Reports*. Most consumers, however, buy things as they need them and can afford them. Familiarity also plays a major role. Whether we admit it or not, we are all very much influenced by advertising and marketing techniques.

Corporations spend huge amounts of money to keep their products in the public eye and to appeal to motivations that often have little to do with the product. Volumes have been written on how best to influence purchasing decisions and increase sales. Because of their collective economic power, women are a prime target for advertisers, and because many commonly purchased products do not differ very much from each other in either quality or price, advertisers and marketing experts focus largely on irrelevant cues such as name brand identification, packaging attractiveness, or shelf location.

People do occasionally base their purchasing behavior on ideological grounds. Nestlé products, for example, were avoided by many consumers during the early 1980s because of their hard-sell campaign of infant formula to third-world countries and its disastrous health consequences. Poor women in those countries often could not afford to use the formulas consistently or in correct quantities. Thus, they were often watered down so much that infants received little nourishment from them. Also, the unsanitary water in many underdeveloped countries was dangerous to infants; breast-feeding would have been a much healthier alternative. As Nestlé no longer markets infant formula to developing areas of the world, the boycott ultimately had some impact.

People's political sensitivities also lead them, from time to time, to avoid buying products from countries engaged in objectionable practices.

For many years, South Africa's apartheid policy was a reason for many to boycott that country's exports. But what about boycotting products from countries that treated women badly? Were people interested in trying to find out, for example, whether the gasoline bought for their car came from Saudi Arabia, where women don't have the right to vote or to drive a car? And what about actually traveling abroad?

There are books that serve as guides to women traveling alone, but they focus mostly on issues of harassment and safety, and sometimes offer peculiar advice. A recent travel guide directed toward women,[7] for example, suggests, somewhat astonishingly, that women should suspend their feminist beliefs when they travel and accept the prevailing cultural practices. But what would the effects be? If women remain silent and continue their economic support of countries that abrogate women's rights or endorse violence toward, or mistreatment of, women, the traveler's compliance gives legitimacy to such practices. Such extreme cultural relativism has not been consistently applied. It should be recalled that South African groups defended apartheid by arguing that racial segregation represented their specific religious and cultural beliefs. Yet most Americans placed little credence in that view and were willing to impose their own moral beliefs in an attempt to change an unacceptable status quo. Why should we not be willing to do the same when the victims are female?

This book began, then, with the realization that very little information was available to consumers about how women are treated in the different states and countries they visit and at the companies that furnish the products and services they buy. If the relevant information became easily available (as it is in this book), what would the effects be if our readers consistently used this information?

Imagine scenarios in which countries heavily dependent on tourism suddenly experienced a substantial drop in the number of female—and accompanying male—tourists because of the country's disregard for women's reproductive and other rights. What would they do? What would happen if companies noted a precipitous decline in their sales because they had not hired, promoted, or paid female employees on the same basis as their male counterparts? Or because they had not provided family-friendly benefits? It is hard to believe that the application of such economic pressure would not prompt some rethinking about the political and economic practices that female tourists and consumers find offensive.

HOW POCKETBOOK PROSE CAN INFLUENCE THE TREATMENT OF WOMEN

Pocketbook prose can speak loudly. Interestingly, it has often worked to the detriment of women's interests. The mere threat by right-wing conservative groups to boycott drug companies that market birth control devices, for example, has made national and international corporations quiver. Because of this fear, there have been serious declines in research on contraception and a longtime lack of access for U.S. women to drugs used in other countries (such as RU-486) that give women more freedom in their reproductive choices.[8]

But the fact is there are far more female consumers than right-wing conservatives—or any other political group—in the world, and the time has come for women to use their own economic power to further their own goals.

Although women have not often used their economic power in this way, there are occasional success stories. For example, former Governor Cecil Andrus of Idaho vetoed an anti-choice bill passed by the state legislature, at least in part because of a threat by pro-choice women to boycott Idaho potatoes. Andrus was quoted as saying that he vetoed the bill because a diminution of sales in a major cash crop would be extremely serious for his state.

Monitoring of company practices for the purposes of buying or not buying has been done by other groups, most notably ethnic minority members and those concerned with environmental protections. The Council of Economic Priorities issues reports and publications that rate companies along many dimensions of social responsibility. Some companies are relatively good in most areas, including their treatment of women. For example, Hewlett–Packard was one of the companies to receive the Council of Economic Priorities 1996 Corporate Conscience Awards for its commitment to avoiding layoffs. Interestingly, Hewlett–Packard also ranks in our sample as a company that treats its female employees well. There are other companies, however, that do not score well on all dimensions. Some may have stronger records with regard to environmental concerns or general community involvement than, say, with women's issues. This is certainly the case for corporate giving, where only 5 percent typically goes to groups primarily concerned with girls and women. Thus, although we recognize that our readers may be rightfully concerned about many legitimate causes, we believe that it is time for women to focus their economic attention on the issues that most affect their own lives.

Voting with one's pocketbook can be *empowering* for women. It is a very effective way of expressing one's agenda. Pocketbook prose is a universal language, and will quickly be understood by the upper-level, mostly male managers trying to lure female dollars. *If economic decisions are made on the basis of female-relevant factors, women's treatment in the workplace will improve.* In the following chapters, we will describe how readers can become more effective consumers according to our newly developed FEMME system. FEMME is an acronym for Feminist Evaluation Method for Measuring Equality. The unit of measurement within this system is accordingly called a FEM (Feminist Evaluation Measure), and can range from 0 to 100. Comparative FEM ratings are presented for three types of entities: large companies that sell products and services to women, states where women spend business and travel dollars, and countries where women travel and buy products. Armed with such information, you can spend your consumer and business dollars in places that are supportive of their female employees and citizens.

The cynics among you may still be wondering whether such enlightened feminist consumerism can actually be effective. The general situation for women in corporate settings is pretty dismal. *Working Woman* magazine reported in 1995 that only *one* of the Fortune 500 companies has a female chief executive officer. Other sources report that although a majority of these companies now have one woman on their boards, women still occupy less than 10 percent of the seats available. The number of women at the upper-management levels of these successful companies is negligible, not only in the United States (97 percent of senior managers are men),[9] but throughout the world. Reading such dismal statistics, you may ask yourself if it really matters. Why spend time and effort choosing the lesser of evils if there's not much difference?

Making choices matters for a number of reasons. For starters, female-friendly consumerism can influence progress. The progress of women up the corporate ladder is not self-propelling. Despite the fact that women are making up more and more equal parts of university and graduate school classes, they are not automatically finding themselves promoted into upper- or even middle-management jobs. From the time the authors began examining data to the actual numbers used for scores, 3 years later, the averages for women's standings over the corporate sample did *not* drift upward. Instead, it would seem, some companies are making decisions to expand and improve perspective by promoting women and enabling women with families to work; others are not. As mentioned, there are still important differences among companies with regard to policies that matter to women. So by noticing and responding economically to these differences, the currently dismal averages can be improved, because companies care enormously about female consumer dollars. Finally, when corporations realize that female consumers are monitoring their behavior, they may think harder about decisions usually made behind closed doors.

There has been progress made over the long term, even if the picture often seems mixed. Although very few women have gotten through the glass ceiling, more women in the United States are entering management at lower and intermediate levels. In the mid-1960s, only 16 percent of managers were women (up from 4 percent in 1900), compared with 26 percent in 1980, and 39 percent in 1990. This figure continues to increase.[10] Women have also made recent inroads into corporate boards of directors. Yet they still exert little influence, and 75 percent of working women earn less than $25,000 a year—a phenomenon that some have referred to as the "sticky floor."[11]

There are important differences between companies that show up more prominently on some dimensions than on others. Whereas most corporations may have very few upper-level female managers, distinctions can still be made in other areas that significantly impact women's advancement. Some companies, for example, may offer more middle-management opportunities to women, may provide excellent child care and family leave options, or may contribute generously to organizations that benefit women. In contrast, other companies may offer women few opportunities for advancement, may lobby against government-sanctioned family leave policies, and may focus their corporate gifts on the American Enterprise Institute rather than Planned Parenthood. Even small but above-average steps that help women should be encouraged because such encouragement will surely create larger steps. Not paying attention to such differences means abrogating our responsibilities and squandering our potentially enormous influence.

Thus, although the overall situation may not be so great, companies still vary. Some are clearly above their industry averages in ways that are important to women, whereas others are way below average. Paying attention and applying pressure can help improve these overall averages, particularly if companies are made aware of the reasons for changes in consumer behavior.

Up until now, it has been hard to pay attention to these aspects of companies because the information needed has been either unavailable or widely dispersed, but this book makes this effort easy.

Here are a few concrete examples. If you shop for food—and women clearly comprise the majority of shoppers at any grocery store—Kroger, Food Lion, and Safeway may be among your choices. Kroger, which operates King Soopers, Dillon Supermarkets, and a host of convenience stores in addition to Kroger's, offers good benefits, including a year of job-guaranteed maternity leave and some access to job sharing, flex time, and a work-at-home option. Twenty percent of Kroger's board are women. Food Lion and Safeway, on the other hand, both have all-male boards and fewer benefits. Wouldn't you rather spend your grocery money to support an organization that treats its female employees well? If you are making travel plans, American Airlines has women represented at all of the top echelons of the company. It offers excellent benefits and gives at least an average amount to women's groups as part of its charitable donations. If American flies where you're going, isn't it just as easy to fly American over Continental, TWA, or Northwest, and reward the company that offers the best support to its female employees?

If products and places are comparable in price and quality, why not use your consumer dollars to purchase from companies most supportive of gender equality? It doesn't usually cost more to support women. All it takes is concern and information.

THE FEMINIST DOLLAR

Why did we choose to call this book *The Feminist Dollar*? The term *feminism* has many connotations for different people. Despite the fact that some see it as a dirty word, we believe its basic tenets still hold the most promise for helping women in all areas of their lives.

When we use the word *feminism*, we see it as involving two basic concepts: *fairness* and *choice*. The first implies that people be treated fairly and be given equal opportunity, whether they are male or female. This does not mean that all women must participate in paid employment. It only means that when they do—and most women in this country now do[12]—the playing field should be a level one.

The second basic element is freedom of choice. This does not proscribe a particular choice; it means just what it says. Freedom to choose means that women themselves can decide how they wish to allocate their time, their effort, and their training. When gender discrimination exists, freedom of choice is greatly diminished. This in turn has negative consequences for individuals, society, and business; arbitrary limitation of choices can diminish well-being, productivity, and potential profits.

But what if you're uncomfortable with labeling yourself a feminist? You should still proceed with reading and using this book. Recent surveys have shown that an overwhelming majority of women and men are very concerned with such issues as pay equity, equal opportunity for advancement, and how to juggle family and work responsibilities, even though they might not do so under the ideological banner of feminism. Even in the narrowest sense, these are women's issues, not simply feminist issues. In a broader sense, these are issues of concern to all of us, regardless of gender, since we all have mothers and most of us have daughters, daughters-in-law, or sisters. Thus, the issues addressed by this book hit close to home for everyone.

Women in the work force are on the front line. Their work has never been found to be inferior to men's. They have been equally educated, and are supposedly protected by laws. Yet they still bring home much less pay and have fewer opportunities for advancement than their male colleagues. If they are single parents, their children suffer disproportionately. If they are married and working, their total family income is still much lower than it would be in the absence of gender discrimination, so husbands of working wives should be equally concerned. Quite simply, gender discrimination hurts *everyone*, and we owe it to past, present, and future generations to change the current system by all means at our disposal. Exerting economic leverage can be an important tool in this process. Let's learn, then, how to make wise economic choices that will benefit ourselves, our mothers, our sisters, our wives, our daughters—and our economy, as well. Viva la FEMME!

Chapter Two

La Crème de la FEMME

Our Scoring Guide

We haven't come a long way, we've come a short way. If we had come a long way, no one would be calling us "baby."

Elizabeth Janeway

When money talks, there are few interruptions.

Herbert V. Prochnow

It is as children that we are first taught to be consumers, and for young girls perhaps the most important influence is blonde and less than a foot tall, namely, Barbie. Dolls are an important part of introducing little girls to their future roles as consumers, and their importance lies not in the doll itself, but in the huge array of outfits and props that can be bought for her. This is also true for other toys directed at girls: infant dolls are marketed with diapers, clothes, and carriages; dollhouses are incomplete without elaborate miniature furniture and appliances; play kitchens must be equipped with pots, pans, and cake mixes. In fact, nearly *every* traditional play activity for little girls is packaged to include lessons in incipient household consumerism. Although boys are also encouraged toward consumerism, the carryover to adult life is not as broad: buying laser guns may only have later relevance for those who work for the Pentagon.

Learning that it is fun to buy things, however, is only part of the story. It is easy to be a consumer in the United States because there are so many choices. But rarely do we learn how to be a dynamic, value-oriented consumer, creating power through purchasing. We almost never receive lessons as children or adults that show us how to use our enormous power to further our own interests.

WHY DO WE NEED A FEMOMETER?

Considerable media attention has been given to the general issue of gender equality, particularly in the business world. There is certainly sufficient information available to conclude that some corporations provide better opportunities for their female employees than others, but this is like saying that Mexico is warmer than Alaska. Without a thermometer, we can't go beyond simple comparisons. Similarly, without a more detailed system to measure female-friendly practices—a femometer—we can't make more refined judgments about companies.

The FEMME system evaluates companies and political entities along a number of dimensions that reflect how positive or negative their climate is for women. The final measure, the FEM rating, is a numerical score that can range from 0, which would be the worst for women, to 100, which would be the best. These FEM scores will be used to rate companies, states, and countries. The following sections describe how the scores were derived.

CORPORATE FEMS

There are many factors that reflect how conducive a corporate climate is to women's dignity and achievement. Some factors directly assess the present levels of achievement of women within each company; other factors reflect how female-friendly the working atmosphere is. We have chosen five factors to use in our ratings of corporations.

Factor 1. Management Opportunities: The Case of Rosie the Riveter

Rosie worked for Rickety Ratchet Company, riveting ratchet parts. She made a reasonably good salary. In her 10 years at Rickety, she had had four different male supervisors, all of whom made more money and understood less about ratchet assembly than she did. She decided that she could do a better job than Joe Blow, her current supervisor, and requested a transfer from her riveting position to an entry-level management training program. Even though she was willing to take a pay cut, her request was denied. The reason given was that they couldn't spare her in her present job. She suspected, however, that they simply didn't want any women in the program. As a consequence, she was riveted to her place in the organization. Rosie was not alone. Rickety had a uniformly poor record for hiring or promoting women into entry-level management positions.

The first factor to be measured, then, is how willing a corporation is to give people like Rosie a chance to become a manager. It is based on three measures: the percentage of female managers, the percentage of women in professional positions, and the presence or absence of active recruitment and training programs for women.[1]

Now, the average number of women in management for Rosie's industry, miscellaneous fabricated steel products, is not very good to start with. According to the most recent report by the Equal Employment Opportunity Commission,[2] only 4.5 percent of all female employees in this industry are in managerial positions; the rest of the managers are men. Thus, a company with these numbers would earn an average score. Our rating scale ranged from 0 to 20 points, so the average score would be 10, by definition. Rosie's hypothetical company, however, had *no* female managers or administrators, so they would get a lower score for this measure.[3] A percentage a little worse than the average could get a 5; those a bit above the 4.5 percent average might get a 12; those considerably above average might get a 15. This rating scale is used in the same way for the remaining four factors.

Factor 2. The Glass Ceiling: The Case of Gloria the Go-Getter

Gloria has worked for Eternity Insurance Company for 8 years. She is young, talented, and ambitious. She has received praise, raises, and promotions for her work and moved quickly into a middle-management position. She currently earns $50,000 a year; the CEO of Eternity makes $870,000. She is, therefore, waiting for her big break—a promotion to a top manager—but this doesn't seem likely as all of the vice presidents are relatively young men.

Gloria happens to work in an industry with a reasonably good record of women in total managerial positions, 43.7 percent, according to the Equal Employment Opportunity Commission, but this figure is based mostly on entry- and middle-level positions. The rates for women and men at Gloria's office are quite different. Twenty-seven percent of all male employees are classified as managers, compared with 11.7 percent of the female employees. Thus, her chances of getting through that glass ceiling, particularly at Eternity, are not great, despite the industry's good record. Women comprise only 5 percent of senior management in all companies, and less than that in Fortune 500 companies.[4] Additionally, only a small

percentage of those on corporate boards of directors have been women.[5] To evaluate the glass ceiling factor, we compare companies as to how many women serve as senior managers and officers relative to the averages of their industries. We also look at whether women have made it into the company's true top echelons: the highest-paid officials category and the board of directors.

Factor 3. Support of Women's Organizations: The Case of Cindy Lou, Mother of Two

Cindy Lou's long-term, live-in boyfriend split when he found out she was pregnant. Although he had a steady job, he said he wasn't ready for the commitment of a child. (Wouldn't he have been surprised if he had stuck around to find out it was twins!)

Cindy was lucky enough to live near her family, and moved back home. She relied on inexpensive prenatal care at Planned Parenthood, and is now the proud mother of two healthy twins, Boo and McGoo. Thanks to corporate scholarships awarded by a local junior college program, Cindy Lou also earned a degree in word processing, and now works at Eggyolk Software, where she earns enough to pay the rent on a small apartment and buy food. She is saving for a new car. Without the subsidized healthcare she received, or the scholarship money, she would probably be on welfare.

Companies in this country give an average of 5.2 percent of their charitable donations to women's causes, such as Planned Parenthood or scholarships for women.[6] Cindy Lou couldn't have gotten back on her feet without those corporate donations. She likes working for a company that has shown above-average charitable giving to women and likes that her company reaches down and helps others like her make it into jobs. Eggyolk Software gives 12 percent of its charitable donations to women's groups, which is well above average. In addition to scholarships and healthcare, some of the groups benefiting from such donations include breast cancer research organizations, Big Sister programs, women's shelters, and the YWCA. Eggyolk, because it gives more than twice the average to women's groups, would receive higher than the average score of 10 for this category, though how much higher would depend on the statistical spread of other companies in the software industry.

Factor 4. Benefits that Benefit Women: The Case of Chubby Charlene

Charlene was a senior account executive at Tinsel Town Toys, a national company. Her kindly manager left the company, and his replacement turned out to be quite sexist. He thought there were too many women in his department, and believed pregnant women should not interact with senior-level clients. When he found out about Charlene's pregnancy, he stopped assigning her to important accounts. She put up with the unofficial demotion because she assumed all would return after the baby was born. She planned to take 3 months off, and knew she'd be in top form when she returned to work. However, her company's policy was to offer 6 weeks of disability leave for childbirth; it was a small office and therefore was not required to offer the 12 weeks mandated by the Family Medical and Leave Act. Any granting of extended family leave or a part-time return was discretionary in this company. Her boss refused to grant either to her, so she quit.

In a recent talk by former U.S. Representative Pat Schroeder of Colorado, she mentioned an interesting survey her office engaged in with companies that *believed* their policies were very family friendly. Her staff asked female employees what would happen if their children were ill and they could not come to work. What excuse would they give to their bosses for their absence? *More than 95 percent of those interviewed said they would rather say they couldn't come in because their car broke down than because their child was sick!* Apparently, broken cars are more acceptable to employers than sick children.

Women thrive best in environments where there are real options available that are responsive to their needs and the choice is left up to them, rather than others. The Family Medical and Leave Act of 1993 guarantees parents 12 weeks of job-guaranteed leave, but this is leave without pay, and most women can afford only the standard 6 weeks of paid disability most companies offer. However, there is a lot that companies can do—and a few have—to ease the transition of new moms and dads returning from family leave. Unlike the first two factors we looked at when constructing our ratings, which measure opportunity for advancement directly, the quality of personnel benefits has a more indirect effect on women. With more women of childbearing age entering the work force, there has been an increasing need for options like maternity and family leave, child care, flex time, job sharing, and part-time return from leave as well as education and counseling programs on balancing family and work. Although such benefits may also be useful to men, the reality is that most women in the work force have primary responsibility for their families, and thus these policies have more impact on women. In a recent survey conducted by the Women's Bureau,[7] 60 percent of the respondents said that balancing work and family was a major concern—especially for the 75 percent of working women with children under 6 years old.

Evaluation of this factor in the FEMME system is based on the availability and variety of such options as extended maternity leave, family leave, child care (either on-site or financial assistance provided elsewhere), flex time, and other benefits that help to balance work with families. Companies that clearly exceed the norm—such as providing long job-guaranteed leaves, formal or nondiscretionary scheduling options, or even rooms for breast-feeding—score way above average in this category.

Factor 5. The Femlin Factor: The Case of Anita Uphill, The Aspiring Attorney

Anita was an upper-level associate at the law firm of Humperdinck and Pumpernickel. The company has excellent pay and benefits and little disparity in pay between male and female associates. There is also an active policy of recruiting female and minority graduates out of law school so that the firm has almost as many female as male associates. Anita really wanted to become a partner in this firm. The problem was that she often worked for Mr. Humperdinck, the senior partner, who kept asking her out and telling her dirty jokes, which she did not find funny.

She was in a quandary. On the one hand, as a lawyer, she was well aware of her legal right to be free of such harassment; on the other hand, she knew that if she antagonized Mr. Humperdinck by initiating a complaint, she would probably not become a partner (even though this would have been illegal retribution for exercising her rights). If the firm had had a clear-cut and well-enforced policy against sexual harassment, her decision would have been easier, but it did not.

It is clear from Anita's case that there is an atmosphere conducive to women employees that goes beyond hiring, pay, and benefits. The absence of sexual harassment is one such factor, but there are others as well. Factor 5 attempts to evaluate some of these additional aspects that influence how congenial a corporate climate is to its female employees.

We label these additional influences that make up a positive climate the *Femlin Factor*.[8] Both negative and positive aspects were included in our ratings. For example, we looked at whether the company had explicit policies with respect to sexual harassment and whether they were in compliance with the EEOC regarding discrimination laws. On the more positive side, we also evaluated whether the company had something truly special, namely, a woman as founder, president, CEO, or division president. We also looked at whether the company had received any recognition from such groups as the Department of Labor, or inclusion on the Domini Social Index (an organization that rates corporations on a variety of factors, including treatment of and opportunities for women), or in *Working Mother* magazine, which publishes a yearly list of the 100 best companies to work for.[9]

HOW FEM RATINGS FOR CORPORATIONS WERE DERIVED

Each of the five categories was scored, 10 being the average score for the sample of all companies. Almost all scores fell between 0 (the worst) and 20 (the best).[10] Scores on each of the five factors were then added together for a total FEM score, which could range from 0 to 100, with an average of 50.

To illustrate this system, we present a bar graph that contains ratings for two different hypothetical companies, ABC Widgets and XYZ Gidgets.

Both companies are below average (average is 10) in the area of charitable giving to women. ABC Widgets is clearly above average, however, on the other four factors—and thus scores much higher than XYZ Gidgets. Based on the scores, XYZ Gidgets would receive a FEM rating of only 40, whereas ABC Widgets would receive a much more respectable score of 65. Clearly, if their products are comparable, ABC Widgets is the better company to patronize. Note that not only does this reward ABC Widgets, it also should encourage XYZ Gidgets to change its policies.

In order to keep these ratings relatively simple, we have assumed that all five factors are of equal importance. We have tried, in the chapter text,

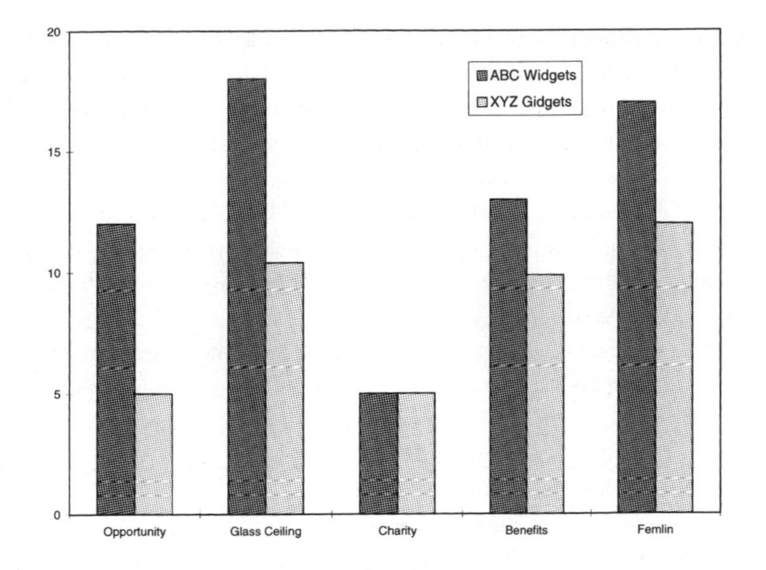

to elaborate highs and lows within a company's score, so if you do not agree with this underlying assumption, you can base your decisions on whatever factor you consider most important. Further, in Appendix Two, we show the individual ratings for each company alphabetically: if one factor is of particular interest or importance to you, you can rank the companies on your own.

FEM RATINGS OF STATES

Most tourist and business travel dollars are spent within the United States. Although women are a minority of business travelers, their numbers are increasing. Women also often have primary responsibility for setting up conferences and business trips for co-workers and managers. In these various capacities, they often have choices about locations, hotels, and airlines. Why not use this spending power to support travel organizations and states most congenial to women?

States themselves differ widely in their laws affecting women. They also differ quite a bit in the degree of female involvement in governance. Finally, state governments are also significant employers of women, and thus are faced with the same issues that corporations have to deal with concerning affirmative action, pay equity, and promotion. Although federal laws and judicial decisions clearly impact all state employment practices (often much more than private employers), there is still considerable variation in terms of how these are enacted and enforced.

Large hotel chains, airline companies, and rental car agencies are rated in Chapter Six, using the corporate FEMME system described above. Chapter Seven, however, required a somewhat different set of factors to rate states, one that represented a mixture of the various roles that states play vis-à-vis their female citizens.

The first factor, *state as employer*, deals with the state government's role as an employer of women. In this regard, it is comparable to the corporate ratings. The ranking considered such things as percentage of women in high-level positions and the availability of family-oriented benefits such as child care options or family leave.

The second factor rated is *economic climate for women in the private sector*. What we looked at here were disparities in salaries between male and female employees[11] within the private sector, as well as how many women-owned businesses the state had.

The third factor rates *women in governance*. Examined in this factor were such things as the percentages of women occupying elected positions, such as congressional representatives or state legislators, and appointed government positions, such as advisory committee members.[12]

The fourth factor, *legal rights for women*, deals with two different aspects of states' laws affecting women: *civil rights*, such as whether the states augment or attempt to retract rights concerning reproductive choice and family, and *family rights*, such as state laws regarding divorce, property distribution, child support, and domestic violence. We looked at these laws in terms of how positive or negative their effects were on women.

A fifth factor, *general climate*, contains unusual features, such as whether the state has a female governor or U.S. Senator (or had one in the last 5 years). This is comparable to whether a company had a female CEO, president, or founder.

Scores based on each of these five categories are presented separately and in combined form for all 50 states. As for corporations, scores for each of the five factors yielded an overall FEM rating between 0 and 100, with an average of 50.

FEM RATINGS FOR COUNTRIES

The factors involved in evaluating how countries treat their female citizens overlap quite a bit with states. Countries show considerably more variation, however. There are certain rights that all women in the United States have, like voting, unfettered travel, equality in marriage, divorce, and education that do not yet exist in some other countries. For this reason, the factors chosen to rate countries differ somewhat from those used to rate states or corporations.

In order to keep our procedures as similar (and comparable) as possible, five factors were chosen to evaluate the favorability of a country's treatment of its female citizens.

The first factor, *legal rights for women*, is comparable to the one used to rate states. It includes laws governing voting rights, participation in paid employment, property ownership, and the presence or absence of gender discrimination laws.

Factor two, *family laws affecting women*, is also analogous to the state ratings. It evaluates laws governing marriage, divorce, reproductive choice, property distribution, child support, and domestic violence. It also scores the presence or absence of laws protecting girls against genital mutilation (where such practices are culturally relevant).

The third factor, *reproductive health status*, considers such things as maternal mortality rates, as well as fertility rate and household size. The last two indices have been shown to relate to the amount of stress women experience with regard to their family roles.

The fourth factor, *economic and educational climate for women*, evaluates the extent of gender differentials in wages, positions of authority, and participation in higher education.

General climate for women is the last factor rated. Much as we did with regard to corporations, with this factor we tried to capture how female friendly the general atmosphere was. Included for rating in this broad and varied factor was whether the particular country ratified the United Nations' Statement on the Non-Discriminatory Treatment of Women, whether governmental policies promote progress for women and women's life expectancy, literacy rate, and access to basic education. Not all countries, for example, mandate universal education for both boys and girls. Also included was the degree of women's participation in the governance structure (in those countries where women have already obtained the vote). Although women constitute more than half of the world's population, their political influence is dishearteningly negligible. A recent compilation by the United Nations, for example, reports that *only 6 out of 159 of its member states (3.8 percent) were headed by a woman*, and that women are shut out of the highest levels of decision-making in 49 countries.[13] There is still enough variation, however, to pay attention to.

As was done in the earlier ratings, all five factors were rated for each country, and the sum of the ratings equals the overall FEM score. Thus, corporations, states, and countries all have FEM scores ranging from 0 to 100.

Space limitations obviously preclude the inclusion of all countries in our ratings. Instead, we focus on several from each major geographical area, with an emphasis on those most frequently visited by U. S. tourists and those that export a substantial number of products to us.

Although our primary purpose in rating countries was for consumer and business travel decision-making, the ratings could also be used to influence political decision-making. The United States provides billions of dollars of foreign aid to many countries, with little or no regard for how they treat female citizens. We also assisted countries during the Gulf War, such as Kuwait and Saudi Arabia, with similar disregard for the lack of civil rights for their female citizens. If you think that your elected representatives should be paying attention to gender discrimination issues in the allocation of aid, these ratings should help bolster your correspondence. Just as corporations keep tabs on their female consumers, congressional representatives are also aware of their female constituents, even if they too often ignore their concerns. A threatened loss of female votes would most likely increase their attention substantially.

HOW INFORMATION WAS GATHERED FOR THIS BOOK

As women comprise more than half of the world's population, it is not surprising that many nonprofit and government groups spend much of their time collecting data about their status in the workplace, at home, and in the political arena. Our intent at the outset was to tap into these very spread-out data bases and combine them in ways that would be meaningful to consumers who wished to further women's advancement through their purchases. This task proved to be more formidable than initially thought. Furthermore, some of the information we had hoped to obtain, particularly concerning corporate practices, proved to be almost impossible to get.

We would like to share some of our experiences, because we believe that these data should be more readily available to the public than they presently are—particularly in a country that prides itself on freedom and equality.

Federal law stipulates that all corporations with 50 or more employees must file both an Affirmative Action Plan and a form describing the numbers of female and ethnic minority employees at various job levels within their organizations. The agency responsible for obtaining and analyzing this information is the Equal Employment Opportunity Commission of the Department of Labor. The very same law that mandates the annual filing of this information, however, also stipulates that federal employees may be subject to criminal penalties if they release any of it to the public. This is a specific exemption from the Freedom of Information Act (FOIA) under which interested citizens, researchers, and journalists may obtain an extraordinary amount of information from all government agencies. If companies do more than $50,000 of business a year with the federal government, however, some information about employment practices can be obtained from the Office of Federal Contract Compliance Programs (OFCCP), albeit with difficulty.

Despite the time constraints specifically outlined in the FOIA law (10 business days), it takes a great deal of patience and persistence to obtain the forms; it took us a minimum of 1 year, and in some cases more than 2 years, to obtain some of these forms. Agency policy, it appears, is more sensitive to business paranoia than to the public right to information.[14] Companies are notified when someone has requested their form and they have the right to protest—an action we saw frequently in our quest.

Information relevant to the average salaries of men and women within a given corporation, which is contained in the corporate Affirmative Action Plan, has never been released by the Department of Labor, because their legal department believes this information to be exempt under the designations of "trade secrets" or "confidential information."[15] As mentioned before, companies also have the opportunity to try to block any requests filed under the Freedom of Information Act. As a result, even basic data about differential pay scales for men and women were unavailable on individual companies unless a particular company itself agreed to supply the data to us.

Although we originally planned to use only data collected by others, it became clear that this would not suffice. Accordingly, we decided to try to obtain the information from corporations through the use of a short questionnaire. After repeated mailings, faxes, and follow-up calls, however, many corporations declined to answer our questions about female employees, charitable giving, and benefits. In many instances we were able to obtain some of this information from other sources (see Appendix One for specific listings). Our ratings for each company, therefore, take cooperation—or lack of cooperation—into account.

We hope you will use this volume frequently. We realize that it may take some time to do so and that many of you are already overburdened. Keep in mind, however, that whenever you buy something, you are exercising power. Why not do this in a way that will help make businesses and countries more socially responsible to women?

There are many success stories associated with socially responsible consumerism, but few have anything to do with women. This is somewhat ironic, given our huge purchasing power. Perhaps this is because women have not been socialized to view money as power, or to use their power to further their own self-interest. In order to help the "Charlenes" and "Anitas" of this country, it is time we started doing so.

There is nothing to be lost but inequality.

Part Two

Women as Domestic Spenders

Chapter Three

The Kitchen CEO

There are very few jobs that actually require a penis or vagina. All other jobs should be open to everybody.

Florynce Kennedy

The most popular labor-saving device is still money.

Phyllis George

It was over 20 years ago when I (first author Phyllis Katz) first ventured into the corporate world.[1] At the time, I was a professor of psychology and was asked to consult for a large corporation to help them develop a prototype child care center. I worked with a senior executive, who assigned several employees, all men, to work with me.

Based on their reactions, you might have thought I was delivering pink slips rather than designing a child care center. The people assigned to work with me never did what I asked. Moreover, they snubbed me: they were always busy when I approached. No one would even eat lunch with me.

I had never experienced this kind of reaction before—my co-workers at the university seemed to like me—and I began to wonder if this was happening simply because I was the only woman in the corporation at a management level. When I discussed this theory with my husband, he advised me to try to understand how difficult it must be for the men at this office to be supervised by a woman. Perhaps it was, but it was at least equally difficult for me to be ostracized and constantly left in the lurch.

The behavior of these mid-level managers was quite inimical to the aims of the corporation, for a goodly sum of money was being invested in this project and their failure to produce was clearly dooming its chances of success.

Why were they behaving in this unproductive and defensive way?

THE CURRENT ISSUES REGARDING WOMEN IN BUSINESS

It would be nice to think that this kind of thing, this shunning of women as managers of men, is a relic of the 1970s. Alas, many vestiges are still with us. Certainly, more women have entered the ranks of jobs classified as managerial, but current research shows that there remains a large gap in both earnings and authority. Moreover, it is still rare for women, even those in managerial positions, to supervise men directly.[2]

Since the 1970s, many more women have entered the ranks of management. Because half of the business school graduates are women, they are contributing more to the pool of management candidates than was true in the past. Furthermore, all companies with more than 50 employees are obliged by law to file reports about the numbers of women and people of color in various job categories. Some researchers have suggested that this requirement has resulted in an artificial inflation of job titles without the corresponding rewards. But the fact that more women have found their way into lower- and mid-level corporate managerial ranks is indisputable—as is the fact that their salaries are still lagging well behind those of their male counterparts.

In 1994, female managers earned 72 percent of what their male counterparts did.[3] This is certainly an improvement over 1980, when their salary differential was 56 percent, and an even more dramatic improvement over the findings of a Department of Labor survey in 1940, which showed that most companies simply wouldn't admit women to their managerial ranks. It is progress of sorts, but women in the work force would probably like to know what it will take to make their paychecks equal to men's.

Part of the discrepancy reflects the so-called glass ceiling effect, which has meant that, despite appearances, there remain "invisible" and subtle barriers to the promotion of women. The result is that women are kept off the highest rungs of executive and corporate ladders, and thus out of the most important decision-making loops. The phrase "glass ceiling" may be misleading, as it implies that it can be easily broken; it cannot, however, as the data clearly indicate. According to one widely cited source,[4] women occupy *fewer than 5 percent* of senior executive positions. It has been estimated that at the current rate of progress, it will take women 475 years to achieve equality.[5] Many of us do not wish to wait that long.

This still does not explain the earnings gap, which exists even when men and women occupy the *same* level.

Technically, groups should differ in earnings as a function of different levels of skill or training. But economists have never found meaningful differences in skill levels between men and women. For example, years of schooling are usually related to wages, and women as a group have *more* years of schooling than men. Still, they earn less. To demonstrate this, consider the median annual earnings of female and male full-time workers in 1994[6]:

EDUCATIONAL LEVEL	MEN	WOMEN
High school graduate	$25,038	$14,955
College graduate	46,278	26,483
Master's degree	55,296	35,770

These figures show that more education is associated with higher earnings, but it makes a much more significant difference for men. The wage gap exists at all educational levels. Women with a college degree earn just somewhat more than men *who have not gone to college*. Women with postgraduate degrees earn considerably *less* than men with only a college degree.

Other technical explanations for the gaps in both management and salaries have not successfully accounted for the results, leading many researchers to conclude that women face discrimination because stereotypes still exist that managers should be men.

ANOTHER APPROACH TO THE PROBLEM

Academics will continue to debate the reasons why gender is such an important factor in the workplace, but many women are living the reality: they are underpaid and undervalued, and they are tired of it. Since legislation has not yet succeeded in solving this problem, we suggest an alternative course of action. Let us use our economic power in a concerted way *as consumers* to reward those companies that treat women well and boycott those that do not. The purpose of this and the next three chapters is to give you the information you will need to pursue this course.

This information is not easy to come by. Although companies are required to provide some of it to the government, they are not required to make it public and they usually do not. Often they even fight to keep it private. There are a few other groups that rate companies in terms of their women-friendly policies, but they do not go far enough. The Council of Economic Priorities in its excellent book, *Shopping for a Better World*, rates companies on a number of dimensions, including their treatment of women. This fine organization, however, only uses one index to rate women's progress: how many women are on the board of directors or among the top officers of the organization. Also, it ranks more than eight categories, such as environmental performance and community outreach, and treatment of women thus does not figure as prominently as we think it should. In contrast, we focus our ratings *only on women-relevant issues*.

Another widely followed report that appears every year in *Working Mother* magazine is entitled "The 100 Best Companies for Working Mothers." This report is valuable, but it is written for potential female employees, not for female consumers. It also avoids the issue of naming the *worst* companies. From the consumer's point of view, most companies are average (although, as we found out, average for one industry is not for another), some companies are better, and some are worse. It is important to see the full range before making up your mind.

In the next two chapters, we rate companies associated with products that women typically buy and that are necessary in their daily lives. In this chapter, we include food and beverages, as well as the grocery store chains themselves. In the following chapter, we rate clothing stores and companies that make clothing, drugs, cosmetics, and children's products such as clothing and toys.

HOW WE DERIVED OUR RATINGS

We use five categories in our FEM ratings for each company. The first we call *management opportunities*. There are three measures included in this category: the proportion of female officials and managers (as provided either to the government or to us in our survey), the proportion of female professionals, and the type and extent of training, recruitment, and advancement programs available for female employees at the company.

The second category includes measures that describe the *glass ceiling* effect. How many women have made it to the highest rungs of the corporate ladder? We include here the proportion of women occupying positions on the board of directors. Although it is becoming more popular for companies to have at least one female director,[7] women still constitute a very small proportion of most boards—the national average is about 5 percent. We also include as measures for this category the percentage of women among the highest paid and most senior managerial positions. This last measure, senior management, is different from the management rating used in category one; instead of managers at all levels, it relates to the group the company designates as its top officers. Many companies that sell products primarily aimed at women still have all-male boards and/or all-male senior management. We identify these companies for you. They are our "zero percenters." Because it is so rare for large companies to have a female president or CEO, we gave those that do special recognition in our fifth category.

The third category was initially designed to deal with pay disparities between men and women at the company. Unfortunately, despite repeated and varied attempts, we were simply not able to obtain this information for most companies. The government will not provide it, except in an aggregated form by industry, maintaining that such information on individual companies constitutes confidential information. Although our questionnaires, sent to every one of the approximately 450 companies in our original sample, included requests for this information, only 16 supplied it. This was not enough to base a category on, so our third category became instead *support of women*. We include in this category the proportion of the company's charitable contributions (if they give) that target women's issues or organizations. The Foundation Center, a group that analyzes the giving patterns of corporations and foundations for a variety of factors, was kind enough to share some of their extensive research with us. It may interest readers to know that the average proportion of giving aimed specifically at women and girls is only 5 percent. Another measure that we include in Category 3 is a subjective one: how cooperative companies were in dealing with our survey. Some were very forthcoming and helpful, others ignored our request, and some were downright hostile. This is reflected in the score they received for Category 3.

The fourth category dealt with the types of *benefit programs* the company offers that are particularly helpful to women. The first measure concerns leave policies. Because all companies with more than 50 employees are required by the 1993 Family and Medical Leave Act (FMLA) to provide 12 weeks of unpaid, job-guaranteed leave to care for children or other family members, we give "credit" only for programs that go beyond that. Companies that offer more than 12 weeks receive additional points, as do companies that offer paid leave for new parents other than short-term disability leave. Another factor considered is how far companies go in helping employees find day care. The possibilities range from offering on-site centers to reimbursing employees for child care expenses to helping pay for sick-child care. We also consider what, if any, scheduling flexibility a company offers; the options can include flex time, job sharing, part-time, and work-at-home scheduling. The last measure included in this category is how many "family benefits" each company offers. These range from pretax spending accounts to elder care resource and referral information to education programs for pregnant employees.

The final category we use is labeled the *femlin factor*. This category includes things that are generally indicative of the company's female-friendly attitude, but do not fit readily into any of the other categories. Examples are whether the company has been given recognition by others for positive policies toward women in the form of awards by the Department of Labor or inclusion in reports like the "100 Best Companies for Working Mothers," whether the company has been included in socially responsible investment funds such as the Domini Index, whether the company had a female founder, or has a female CEO, president, or division president, and, on the other side of the ledger, whether the company has been cited for lack of compliance with the laws relevant to equal opportunity.

Scores were derived for each of these five categories and were subjected to mathematical transformations so that we could compare the companies objectively across a wide variety of measures. It is hard to compare apples and oranges, so the transformations normalized the distributions for each category. The average score for each of the categories was 10 with a maximum score of 20 in most instances[8]; these five category scores were then added to yield the total FEM score. By definition, the average FEM score would be 50, but the actual range of scores obtained is between 34 and 87. The averages were obtained on the entire sample of 386 companies. Thus, for any particular industry, the average could be lower or higher—and this happened in a number of instances.

We tried very diligently to obtain information needed from a variety of sources, including the company itself. Each company had at least three opportunities to provide us with the information. On our last try, we sent the information we had to all companies in the survey, and gave them a chance to correct it. Many did, and some didn't. Thus, we were not always successful. For a number of companies, we have only partial information, often obtained from other sources. For example, if the company is publicly traded (most of our sample), some information, such as the identity of directors, is readily available from government sources or annual reports. If the company was a government contractor (most in our sample), we were ultimately able to obtain information from the Office of Federal Contract Compliance Programs (OFCCP) about women in various job categories. The company itself, however, might have refused to cooperate with us—by not providing information in other categories. The question then became how to deal with missing data in assigning ratings.

Several options were available. The first one was to use the statistically determined average for the category (i.e., 10) that was missing. A second possibility was to use what information we did have available for the company to substitute for the missing categories. A third option was to use the industry average for the category. The last possibility was to give a zero for missing categories.

Our first inclination was to use zeros. If you are a student taking a final exam, and you choose not to answer particular questions, your grade for the test will reflect zeros given for those unanswered questions. Because there were a number of reasons for lack of cooperation, and companies were under no obligation to respond to our survey, however, we decided that this alternative might be too punitive.

Instead, we used a combination of alternatives, depending on the category that was missing. When data were missing for Category 4 (benefits), we assigned the lowest score possible, i.e., we assumed that they were in compliance with their legal obligations without offering anything extra to their employees. If there was evidence to the contrary from other sources, we used that information. In its absence, however, there seemed no good reason to give "extra credit" to companies that refused to give us information. With regard to missing data from other categories, we assigned scores that corresponded to the average score for the relevant industry, but with one qualification, namely, such assigned scores would not be higher than the overall sample average (i.e., 10). Our reasoning for this was similar to that in the previous discussion: we did not believe that noncooperating companies should receive any additional credits in the absence of specific information that confirmed it. Companies without information in all five categories are marked with an asterisk. Companies with either three or more missing category scores or those missing information in both categories one and two were dropped from the sample.

FOOD INDUSTRY COMPANIES

Because women do most of the food buying for their households, one might think that food producers would be particularly sensitive to the needs of their primary consumers and would reflect this by having adequate representation of women at the highest decision-making levels. Sadly, this seems to be the exception rather than the rule.

Translating Products into Companies: A Grocery Store Guide

Before you can use our ratings of food companies for your supermarket decisions, you must first have some idea of who produces the name brand products that you buy. In many cases, this is far from obvious. There are about a dozen large companies that are responsible for thousands of products. Some are not even primarily food companies. Philip Morris, for example, is the parent corporation of many of the brands you buy.

In the "Buying Guide" chapter, we have tried to focus on the most common products in 23 categories and have provided the reader with the parent companies for each.[9] We suggest that you look over this guide. You might also want to keep it with you for your shopping trips. The following sections deal with our ratings of these parent companies.

Food Industry Averages

We collected information on 49 national food and beverage companies. The overall average FEM score for these companies was about average (49), but the median is lower (47), which means that a few very good companies are pulling up the average. The worst category for the industry was Category 1, female advancement.

The following table lists the companies we rated, together with their overall FEM scores, from best to worst.

	FOOD COMPANIES	FEM	RANK	SOME BRAND NAMES
●	Monsanto	65.15	1	NutraSweet, Simplesse
●	Ben & Jerry's Homemade	64.96	2	Ben & Jerry's ice cream
●	Procter & Gamble	64.45	3	Crisco, Folgers, Jif, Pringles, Duncan Hines
●	Hershey Foods	63.70	4	Hershey's, Reese's, American Beauty
●	PepsiCo	63.45	5	Frito-Lay, Rold Gold, Ruffles
●	Alberto-Culver	59.99	6	Baker's Joy, Sugar Twin, Mrs. Dash
●	Newman's Own†	59.85	7	Newman's popcorn, salad dressing
●	Campbell Soup	58.93	8	Soup, Pepperidge Farm, Prego, Swanson
●	Wm. Wrigley Jr.	58.20	9	Wrigley gum
●	Warner-Lambert	57.81	10	Dentyne, Trident, Junior Mints
●	Sara Lee	55.67	11	Sara Lee, Ball Park, Jimmy Dean
◐	General Mills	54.04	12	Betty Crocker, Cheerios, Yoplait, Bisquick

	FOOD COMPANIES	FEM	RANK	SOME BRAND NAMES
◐	PET	53.99	13	Progresso, Hain, Van de Kamp's, Ole El Paso
◐	Grand Metropolitan	53.73	14	Häagen-Dazs, Pillsbury
◐	Quaker Oats	53.40	15	Rice-a-Roni, Aunt Jemima, Cap'n Crunch
◐	Health Valley	53.38	16	Health Valley
◐	McCormick	52.05	17	McCormick spices
◐	Anheuser-Busch	51.42	18	Eagle Snacks, El Charrito
◐	Philip Morris	50.72	19	Kraft, General Foods, Post, Oscar Mayer
◐	RJR Nabisco‡	50.43	20	Ritz, Planters, Ortega, Oreos, A-1, Snack Well's
◐	American Home Products	50.01	21	Chef Boyardee, PAM, Jiffy pop
◐	Clorox	49.23	22	Hidden Valley, KC Masterpiece
◐	Church & Dwight	48.68	23	Arm & Hammer baking soda
◐	H.J. Heinz	47.77	24	StarKist, Weight Watchers, Ore-Ida
◐	Chiquita Brands	46.86	25	Chiquita fruits, juices
◐	Dole Food	46.57	26	Dole
◐	Kellogg	46.43	27	Kellogg cereal, Mrs. Smith's, Eggo
◐	Cadbury Schweppes§	46.34	28	Cadbury Schweppes
◐	CPC	45.34	29	Skippy, Knorr, Maxola, Hellmann's
◐	Borden	45.29	30	Wise, Wylers, Creamette
◐	Ralston Purina	45.03	31	Hostess, Wonder
○	ConAgra	44.69	32	Healthy Choice, Peter Pan, Hunt's
○	Nestlé	44.44	33	Stouffer, Contadina, Carnation, Hills Bros.

	FOOD COMPANIES	FEM	RANK	SOME BRAND NAMES
○	Dial	43.96	34	Appian Way Pizza Mix
○	Mars	43.23	35	Kudos, Uncle Ben's
○	Hormel Foods	43.12	36	Hormel, Dinty Moore
○	IBP	42.99*	37	IBP meats
○	Tyson Foods	42.87	38	Tyson chickens, Holly Farms
○	Pilgrim's Pride	42.58*	39	Pilgrim's Pride chicken
○	Unilever	42.54	40	Lipton, Ragu, Lawry's, Wish-Bone, Breyers
○	Allied Domecq	42.04	41	Maryland Cookies
○	Welch Foods	40.64*	42	Welch Foods juice, jam
○	Universal Foods	39.89*	43	Red Star yeast
○	United Biscuit Holdings	39.61*	44	Keebler
○	Archer-Daniels-Midland	39.16	45	Gooch pasta
○	Dreyer's Grand Ice Cream	38.48	46	Dreyer's ice cream
○	Reckitt & Colman	38.46	47	French's, Colman's mustard
○	The Topps	36.39	48	Topp's chewing gum
○	Smithfield Foods	36.19	49	Smithfield hams, John Morrell

*Did not have scores for all five categories.
†Newman's Own is a special case: because it is so small, it does not offer the kinds of benefits associated with a larger company and thus was missing information. It does give all its profits to charity, however, and is half-owned by women, so it is likely a better company than is revealed by this rating.
‡RJR Nabisco is a majority owner and holding company for Nabisco.
§Dr. Pepper and 7Up are now owned by Cadbury Schweppes.

Rating Key:
● Female-friendly company; buy products when possible
◐ Average; buy with good conscience
○ Below average; avoid buying these products

From the point of view of treatment of women, two of the four best companies are Ben & Jerry's Homemade and Hershey Foods—good news for chocolate and ice cream lovers. Interestingly, when the actual scores of the categories are considered, they are not uniformly good but tend to be outstanding in one of the five categories.

For example, Hershey's high score occurs on benefits—it gives its employees terrific family-friendly benefits. It is, however, below average on charitable giving targeted to women. Ben & Jerry's Homemade high score is in Category 2—it has more women in top management and board positions than most other companies in this category. Yet it has no special programs for advancing women, and was not cooperative with our survey.

Still, based on their total FEM scores, they are the best companies to patronize when you have a sweet tooth. We would recommend that you purchase Ben & Jerry's ice cream rather than others, such as Breyers or Dreyer's, whose parent companies, Unilever and Dreyer's Grand Ice Cream, have much less impressive records with regard to their female employees and score way below average. Häagen-Dazs (Grand Metropolitan) is somewhat above average, but not in the same league as Ben & Jerry's. Although both Ben & Jerry's and Grand Metropolitan have only one woman on their boards of directors, which is about par for the course, Ben & Jerry's has 32 percent female officers, versus 9 percent for Grand Metropolitan. They are both about average in charitable giving patterns aimed at women's issues—6 percent for Ben & Jerry's versus 5 percent for Grand Metropolitan—but at Ben & Jerry's, women occupy *50 percent* of the top-salaried positions, whereas for Häagen-Dazs the percentage is zero. (It should also be noted that Ben & Jerry's, although it only targets an average percentage of its charitable giving toward women, gives an extraordinarily high amount of its overall earnings to charity.) Grand Metropolitan was cited for noncompliance with the equal opportunity laws in 1993.

If you like candy and chocolate, we recommend purchasing Hershey's brands over, say, Mars or Nestlé candies, as the latter two companies receive considerably below average scores for their treatment of women. (Hershey also makes several items in the spaghetti aisle, including Delmonico, Ronzoni, American Beauty, Light 'N Fluffy, Pastamania, and San Giorgio.) Hershey's board of directors is 25 percent female; Nestlé has no women on its board. Hershey, as previously mentioned, has good benefits, receiving the industry high in this category because it offers employees a full year of job-guaranteed parent leave (both mothers and fathers are eligible for this), easy access to day care, job sharing, flex time, and the opportunity to work at home if necessary. Nestlé, on the other hand, is about average in terms of family-friendly benefits, and we had no information in this category for Mars. Moreover, neither Nestlé nor Mars was cooperative with us, and we had to rely on other sources for our data.

Three other companies deserve mention for being among the top five best rated companies in the industry: Procter & Gamble, Monsanto, and PepsiCo. Procter & Gamble makes a broad range of products—including Duncan Hines, Folgers, Jif, Pringles, Crisco, and Hawaiian Punch, among many others—and we suggest that you seek them out the next time you are in a supermarket. The company's record in the glass ceiling and charitable giving categories leaves much to be desired, but Procter & Gamble offers excellent benefits to employees and has good programs for women's advancement. Although we spent much time puzzling over why a company like Procter & Gamble, with only two women on its 7-member board (they had only one when we began collecting data), no women among its top-salaried officers and a lower than average percentage (27 percent) of female managers, would receive an Eve award from the Department of Labor, we need to recall that we are dealing with *relative* rather than *absolute* numbers—and there are many food companies that have far worse records. We think, therefore, that you should buy their products rather than others, but it is also clear that some of the awards being handed out for treatment of women are premature.

Monsanto, our top-rated company, posts some real highs and lows. Its lows came in Categories 1 and 5. Its low score in category 5 indicated an absence of any special recognition by other raters for its treatment of women. In the management opportunities category, women account for only 18 percent (the national average is 41 percent!) of Monsanto's officers and managers. It does make some amends, however, by having two women on its 14-member board, 9 percent women among its top officials, and the highest industry score for excellent family-friendly benefits. Therefore, despite its low score for women in management, Monsanto, maker of Simplesse and NutraSweet, still scored higher than others in the food industry.

PepsiCo, with its overall FEM score of 63 and its 33 percent female manager rate (third highest in the industry), also deserves consumer support for the snacks it produces: Frito Lay, Chee-tos, Doritos, Rold Gold, Sun Chips, Ruffles, and Lay's. Unlike some of our other top scorers with uneven patterns, PepsiCo's profile is consistently good. It offers above-average benefits, contributes at least an average amount of its charitable giving to women's causes, has good programs for women's advancement and has a woman running one of its subsidiaries.

One of the companies surveyed represents a special case, namely, Newman's Own, seventh ranked overall. Because the company is so small, not all of the categories are applicable, and in those that are, it received very high scores. For example, 50 percent of Newman's board is female, a record in the food industry, where the next highest score is 28 percent. Because the company has only 11 employees, however, the fringe benefit category is not completely applicable, nor is the percentage of women in management. In order to arrive at a total score, we used the industry average for these categories. We were unable to ascertain exactly what percentage of its charitable giving is targeted to women, but we must mention that *all* of its profits go to charity. On the basis of this, we would advise buying Newman's Own products—salad dressing, popcorn, lemonade, salsa, and spaghetti sauce.

Most of the other companies that earned low FEM ratings did so either because their records were very poor or because they were very uncooperative in our survey and did not supply us with the data needed to rate them. In some cases, the data obtainable from other sources were insufficient to get a rating, and we could not include the company (such as Dannon). Where we had data for some categories, we used the industry average for the missing category. This was done for companies marked with an asterisk in the tables. It should be noted that, as a result of our procedures, some of these scores may look better than they actually are, since many of the uncooperative companies had below-average scores for the categories for which we obtained information from other sources. In only one instance did any of the latter non-cooperating companies have a score that brushed average.

Lest you think you might starve to death by avoiding the foods sold by companies that scored poorly, we assure you that there are excellent alternatives for all of their products, made by companies that are much more caring about women. Moreover, if these uncooperative and poor-treatment companies lose significant sales as a result of increased consumer awareness, they will probably improve. So your deprivation may only be temporary.

The companies that occupy the bottom of our grocery company barrel have deficient policies toward women. They may have been cooperative or uncooperative, but if the latter, we were able to obtain the information needed for rating from other sources.

The range of FEM scores obtained from our total sample was from 32 to 88. As expected, most of the scores aggregated around the average of 50. In evaluating the scores, we considered scores between 46 and 54 to be average. Those between 42 and 46 were regarded as somewhat below average, whereas those between 54 and 58 were considered to be somewhat above average. Any scores below 42 were very much below average; correspondingly, scores over 58 were considered to be highly above average.

Thus, we suggest that you pay particular attention to scores below 42—which are the bottom eight of the food group. The lowest score in this group was that of Smithfield Foods, which was considerably below average on all five categories rated. Women constitute only 12 percent of managers at the company, or about one-fourth the national average. Smithfield was also cited for noncompliance by the EEOC and did not cooperate in our survey. Additionally, there are no women on its board. We suggest, therefore, that you buy ham from other companies.

Although Smithfield obtained the lowest total scores, there were several others that were even worse in particular categories. Hormel took last place in the female advancement category with a meager 6 percent female managers (contrasted with a high of 75 percent for Alberto-Culver). Thus, we would recommend not buying from this company either.

Two other companies with complete information that are close to the bottom are Topps and Dreyer's Grand Ice Cream, with FEM scores of 36 and 38, respectively. Their brands of chewing gum and ice cream are relatively easy to identify, and therefore easy not to purchase.

There are several multiple-product companies that scored considerably below average and should be avoided whenever possible. ConAgra and Nestlé just missed receiving our average rating. However, Tyson and Unilever were considerably below average. In addition to those products that bear the name of the parent company, we recommend that you also avoid the following: Toll House Morsels, Kit Kat and Butterfinger candy, Carnation and Stouffer frozen foods, all made by Nestlé; Breyers ice cream,

Lipton, Wish-Bone, and I Can't Believe It's Not Butter, all made by Unilever; and ConAgra's Hunt's, Healthy Choice and Banquet frozen foods, Reddi-whip, and Armour meats. If you want processed meats, your best choice as a feminist consumer is Oscar Mayer (made by Philip Morris)—but don't serve it with French's or Colman's mustards. These condiments are made by Reckitt & Colman, the third lowest ranked company in our sample. Grey Poupon (made by RJR Nabisco) is a better choice. One note regarding Philip Morris: although a solidly average company, one way in which they strongly support women is through their productions of the *National Directory of Women Elected Officials* in cooperation with the National Women's Political Caucus. They have helped make this directory (which was invaluable in our research for Chapter Seven) possible every other year since 1980.

The important thing to remember is that in every category of product, there are companies that are associated with both favorable and unfavorable policies toward women. The companies themselves do not make the task of informed consumerism an easy one, but the results are worth the effort. So take your shopping guide with you to the grocery stores.

Of the companies we rated, 15 made what we call our *zero percent club*. They are companies that have *no women* either on their boards or in upper management positions or upper salary echelons.

Zero Percent Club—Food Companies

COMPANY	FEMALE DIRECTORS	TOP FEMALE OFFICERS OR TOP-SALARIED WOMEN
Allied Domecq	0	
Archer-Daniels-Midland		0
CPC		0
Dreyer's Grand Ice Cream	0	0
Health Valley	0	
Hormel Foods		0
Nestlé	0	0
PET		0
Reckitt & Colman	0	
RJR Nabisco		0

COMPANY	FEMALE DIRECTORS	TOP FEMALE OFFICERS OR TOP-SALARIED WOMEN
Smithfield Foods	0	
The Topps	0	
Unilever	0	
Universal Foods		0
Welch Foods	0	

As can be seen, the zero percent club has many members from the food industry, almost a third of all food companies rated. In assembling this list, we found two things quite puzzling. The first concerns why the equal employment laws are not enforced very adequately by the responsible government agencies (see Chapter Seven for a summary of laws in this area). The second is how some of our zero percenters manage to receive awards. RJR Nabisco, for example, was awarded the Department of Labor's Eve Award in 1993. The award is given to companies that have implemented effective programs to enhance workplace opportunities for women and minorities, yet RJR Nabisco has no female senior officers, one female board member out of 15, and 19 percent female managers. We are not particularly impressed by awards to companies with such mediocre records—especially when they come from the very agencies charged with enforcing the laws. Nestlé's inclusion in the zero percent club shouldn't be too much of a surprise when you consider it is a Swiss company. As discussed in Chapter Eight, women in Switzerland did not get the right to vote until 1971!

BEVERAGES

We rate a number of companies that manufacture both alcoholic and nonalcoholic beverages. Although there are fewer companies in this section than in the previous one, the consumer is still faced with the problem of deciphering the parent company from the often unrelated brand name. We provide this information in the Buying Guide chapter, and our summary is in the table on the next page.

	NONALCOHOLIC BEVERAGE COMPANIES	FEM	RANK	SOME BRAND NAMES
●	Procter & Gamble	64.45	1	Folgers, Sunny Delight, Hawaiian Punch
●	Hershey Foods	63.70	2	Hershey Hot Cocoa
●	PepsiCo	63.45	3	Pepsi, Mountain Dew, Slice
●	Celestial Seasonings	62.02	4	Red Zinger, other herbal teas
●	Newman's Own	59.85*	5	Newman's Own Lemonade
●	Campbell Soup	58.93	6	V-8
◐	Quaker Oats	53.40	7	Snapple, Gatorade
◐	Philip Morris	50.72	8	Maxwell House, Sanka, Brim
◐	Clorox	49.23	9	Deer Park mineral water
◐	Coca-Cola	47.71	10	Coke, Minute Maid, Sprite, Tab
◐	Chiquita Brands	46.86	11	Chiquita juices, Naked Juice
◐	Dole Food	46.57	12	Dole juices
◐	Cadbury Schweppes	46.34	13	Canada Dry, A&W, Crush, Hires
○	ConAgra	44.69	14	Swiss Miss
○	Nestlé	44.44	15	Nestea, Libby's, Taster's Choice, Carnation
○	Seagram	44.20*	16	Tropicana Orange Juice
○	Unilever	42.54	17	Lipton
○	Allied Domecq[†]	42.04	18	Tetley tea
○	Triarc	40.89	19	Royal Crown Cola
○	Dr Pepper/7Up[‡]	40.14	20	Dr Pepper, 7Up

	ALCOHOLIC BEVERAGE COMPANIES	FEM	RANK	SOME BRAND NAMES
◐	Grand Metropolitan	53.73	1	Smirnoff, Bombay, Baileys, Cuervo

	ALCOHOLIC BEVERAGE COMPANIES	FEM	RANK	SOME BRAND NAMES
◐	Anheuser-Busch	51.42	2	Budweiser, Michelob
◐	Philip Morris	50.72	3	Miller, Löwenbräu, Milwaukee's Best
◐	Adolph Coors	48.96	4	Coors
○	Seagram	44.20*	5	Crown Royal, VO, Chivas Regal
○	American Brands	43.10	6	Jim Beam, Gilbey's, Ronrico
○	Allied Domecq†	42.04	7	Carlsburg, Ballantines, Beefeater, Kahlúa
○	Guinness*	40.07*	8	Guinness Stout, Harp

*Did not have scores for all five categories.
†Formerly Allied Lyons.
‡Dr Pepper/7Up has been acquired by Cadbury Schweppes.

Rating Key:
● Female-friendly company; buy products when possible
◐ Average; buy with good conscience
○ Below average; avoid buying these products

As a group, the beverage manufacturers did not have high ratings. This group contains a few companies, such as PepsiCo, that were discussed in connection with their food products. The average FEM score for the nonalcoholic beverage manufacturers is 50.61, and for the alcoholic beverage producers, 46.78; not one liquor maker earned our highest rating.

As is true for all consumer products, however, there is always a range of scores, and the scores of beverage companies are presented above in the hope that the reader will seek out products manufactured by companies with relatively high scores, and avoid those with low ones.

For the nonalcoholic category, the four best companies are Procter & Gamble, Hershey Foods, PepsiCo, and Celestial Seasonings. Three of these four were also high scorers in the food group.

Procter & Gamble, with a FEM score of 64, is the best in this group and excels in three categories: women in management, fringe benefits, and overall female-friendly atmosphere. It is below average, however, in charitable giving to women and in women in upper management.

As noted in the food section, Hershey's, the second place entry, has an outstanding record in the benefits category, providing its employees with a varied menu of family-friendly benefits. It also has a higher percentage of women on its board than do most companies in this category. It is not overly generous to women's groups in its charitable giving, however, and is somewhat below average in terms of women in management.

The third-place beverage company, PepsiCo, is very close in overall score and has a more well-balanced profile. It is above average in all five categories. And PepsiCo is ahead of archrival Coca-Cola, which earns a lower than average rating on the FEM scale.

Celestial Seasonings, a Boulder, Colorado-based company, also deserves mention for its considerably above-average total FEM score. It is especially strong in the charitable area (first in this category), but is also above average in the glass ceiling category, with a rather uneven profile. It has no women on the board; nevertheless, it has the second highest percentage of female officers and high salary earners.

Thus, on the basis of our rankings, we would advise that you purchase Hershey's cocoa over competing brands Swiss Miss or Carnation; Pepsi and its other brands of soda (Mountain Dew, Slice, and Mug root beer) are better choices than Coca-Cola, Canada Dry, Sunkist, Dr Pepper, 7Up, Hires and A&W root beers, or Royal Crown Cola. Buy Procter & Gamble's Hawaiian Punch, Sunny Delight, and Folgers over Nestlé, Taster's Choice, or Hills Bros. For tea drinkers, Celestial Seasonings is the choice over Lipton or Tetley brands.

The four lowest-scoring companies for nonalcoholic beverages were Dr Pepper (FEM score of 40), Triarc (41), Allied Domecq (42), and Unilever (43). Their products to avoid include Dr Pepper, 7Up, Royal Crown Cola, and Tetley tea.

In the alcoholic beverage category we rated eight companies. All eight received FEM scores below 55, so none earned our female-friendly rating. The top-rated company was Grand Metropolitan, which had good ratings for female managers, but below average scores for senior-level women. None of the other companies had even an average score for the women in management category. Our second highest-ranked company, Anheuser-Busch, was below average in three of the five categories, but did have an excellent female-friendly benefits package. Of the group we rated, we recommend that you purchase products made by Grand Metropolitan and Anheuser-Busch. These include Smirnoff, Bombay, Baileys, Budweiser, and Michelob. (Who says feminist consumers have no fun?)

As a group, manufacturers of alcoholic beverages were not very female-friendly. They had below average scores in all categories. Many were uncooperative (we were not able to obtain information from Seagram and Guinness) and a few were zero percent club members. Beyond recommending products made by Grand Metropolitan and Anheuser-Busch, we suggest that you cut down your purchases of alcoholic beverages from the companies in our sample. This may not only help women, it may also be good for your health.

Zero Percent Club—Beverages

COMPANY	FEMALE DIRECTORS	TOP FEMALE OFFICERS OR TOP-SALARIED WOMEN
Adolph Coors	0	
Allied Domecq	0	0
Celestial Seasonings	0	
Dr. Pepper/7Up	0	0
Guinness	0	0
Nestlé	0	0
Triarc	0	
Unilever	0	0

CIGARETTES

We are hard put to find anyone except a tobacco executive who would encourage you to buy cigarettes. On the other hand, people who smoke do spend a fair amount, so we thought it would be worthwhile to present the information we have about cigarette companies.

We rated only three companies in this category. Two earned an average rating and the third scored below average, giving the industry a below-average FEM score of 47.39. All three companies have female board members. On the other hand, the number of women in management for the industry is quite a bit below the national average.

	TOBACCO COMPANIES	FEM	RANK	SOME BRAND NAMES
◐	Philip Morris	50.72	1	Marlboro, Benson & Hedges, Parliament, Virginia Slims, Merit
◐	RJR Nabisco*	50.43	2	Camel, Now, Doral, Salem, Winston
○	Imasco	41.03	3	duMaurier, Players

*RJR Nabisco is a majority owner and holding company for Nabisco.

Rating Key:
● Female-friendly company; buy products when possible
◐ Average; buy with good conscience
○ Below average; avoid buying these products

RJR Nabisco and Philip Morris are names that may be familiar from the food industry. Although Philip Morris was our highest scorer in the cigarette industry, its FEM rating was only average. Makers of Benson & Hedges, Marlboro, Parliament, and Virginia Slims cigarettes, Philip Morris has two femalc board members out of 20 and almost one-fourth of its top officers are female—both respectable levels. And it does give an average amount to women in its charitable donations, the only company in the industry to do so. Of all of its total managers, however, only 21 percent are women, about half the national average.

RJR Nabisco has one woman on its 15-member board of directors. Four of the company's top 25 officers but none of its highest-paid officers are women. In the women in management category, RJR Nabisco has about 19 percent female managers, which is average for the cigarette industry but well below the 41 percent national average. It compensated slightly for this subpar performance by offering a good number of programs for training and recruiting women. The company has the best benefits in the industry, with formal flexible scheduling options available, but no extended leave is available for new parents. Finally, RJR Nabisco was one of the few companies we rated that earns a gold star for cooperation: in response to our request for information, they sent us several pounds' worth, including everything from product descriptions to the company's diversity calendar. RJR Nabisco manufactures Camel, Now, Doral, Salem, and Winston cigarettes.

Coming in with a below-average FEM score is Imasco. Imasco was uncooperative and for that reason information is missing. What we do know is that there are two women on Imasco's 15-member board, or 13 percent, which is above average for U.S. companies, and 7 percent of its most senior officers are women, which is just about average. But it earns two zeros for female programs, such as recruitment or training for women, and charitable giving. It is our recommendation that female-conscious consumers who can't stop smoking choose between Philip Morris and RJR Nabisco's brands of cigarettes.

GROCERY STORES

The results for this section—in which we rate 13 grocery and convenience store chains—were terribly disappointing. Anyone who has been to the grocery store knows that women make up the vast majority of shoppers, so the fact that the industry was so far below average and so unresponsive to our research was a particular disappointment.

Half of the companies we tried to rate did not supply information—more than in almost any other industry—and only one company of the entire lot returned our survey. One chain, Red Apple, had to be dropped from the sample for this reason. Only one company scored above average on the FEM scale; the rest fell below the mean. That brought the industry's average FEM score to a very low 45.45.

Looking at the breakdowns of measures, it's not hard to see why the industry fares so badly on the FEM scale. Eight of the thirteen companies have no women on their boards; eight of the ten for which we have information have no women among their highest-paid officials. Only one of the 13 (Southland) has an average number of women serving as managers or officers (42 percent). None of the 13 has a woman running either the company or a subsidiary company.

Kroger (along with subsidiaries King Soopers, Kroger, and Dillon grocery stores) earns our highest rating. With that one exception, however, this industry needs *a lot* of work to bring it up to even the most minimal standards.

	GROCERY STORE COMPANIES	FEM	RANK	STORE CHAIN NAMES
●	Kroger	60.33	1	King Soopers, Kroger, Dillon
◐	Great Atlantic & Pacific Tea	49.55*	2	A&P, Waldbaums, Food Emporium, Dominion, Super Fresh, Farmer Jack, Kohl's, Food Bazaar, Miracle, Food Mart, Big Star
◐	Giant Food	49.11	3	Giant Foods
◐	Albertson's	47.85	4	Jewell/Osco Foods, Albertson's
◐	Safeway	46.91	5	Safeway
○	Southland	44.73	6	7-Eleven
○	Publix Super Markets	43.83*	7	Publix
○	Investcorp	42.67*	8	Circle K
○	American Stores	42.40	9	Jewel Food, Star Market, Acme Market
○	Winn-Dixie Stores	42.38*	10	Winn-Dixie, Marketplace, Buddies
○	Stop & Shop	41.79	11	Stop & Shop
○	Grand Union	39.67*	12.5	Grand Union
○	Food Lion	39.67*	12.5	Food Lion

*Did not have scores for all five categories.

Rating Key:
● Female-friendly company; buy products when possible
◐ Average; buy with good conscience
○ Below average; avoid buying these products

Kroger was by far the best grocery store company rated and received enough points to earn our highest female-friendly rating. Three of its 15-member board are women, as are 7 percent of the company's top officials. Although its 31 percent women in management is below the national corporate average of 41 percent, Kroger does have programs targeting women and minorities for career development. Kroger is the only company in the industry to offer strong family benefits to employees. New parents may take up to 1 year of job-guaranteed leave, and there are flexible scheduling options available at supervisors' discretion. Because Kroger has distinguished itself as the only company that has made any strides in its treatment of female employees, we recommend shopping at its stores—Kroger, Dillon, and King Soopers.

Two chains were neck and neck for second place, Great Atlantic & Pacific Tea and Giant Foods. Giant Foods is a smaller chain that is centered mostly around Washington, D.C., Maryland, and Virginia. The company's seven-member board has no women. Its management is 25 percent female, or below average, but at least 16 of the 158 stores (10 percent) are run by women. In terms of benefits, new mothers receive some paid and extended leave and may return to work part-time. Job sharing and flex time are also available at some locations. Although the picture for women at Giant Foods is mixed, it still performs better than most of the competition.

The other, Great Atlantic & Pacific Tea, gets a relatively high rating despite the fact that the company declined to participate in our study. Three of its 13 board members are women (23 percent, the industry high), and it was one of three companies in the industry to give an average amount of its charitable donations to women's groups (the rest were far below the mean). Great Atlantic & Pacific Tea's retail chains include Waldbaums, Food Emporium, A&P, Dominion, Super Fresh, Farmer Jack, Kohl's, Food Bazaar, Miracle, Food Mart, and Big Star. Perhaps they will be more cooperative in our next edition.

In the convenience store category, we rated two chains, 7-Eleven and Circle K. Both scored pretty low. Southland, the parent company of 7-Eleven, emerged as the somewhat better one. Although Southland still has an all-male board, 6 percent of its top officers are women. It is also the only company in the industry with a slightly above-average number of female managers and officials. Investcorp, parent company of Circle K, scored similarly in the glass ceiling category: it has an all-male board and 9 percent of its top officers are women.

It should be noted that almost all of the grocery chains were uncooperative in our survey, and thus there is information missing unless we were able to obtain it from other sources. In the case of the companies marked with an asterisk, we used industry averages, which were pretty low, to estimate numbers for some categories (as explained in Chapter Two and Appendix Two). One exception was Safeway, which did return the questionnaire, and should get some recognition for this in an industry that manages to ignore us as easily as it does its female employees.

Based on the information gleaned from other sources, we were especially appalled at the absence of female representation at the higher levels of management.

More than half of the grocery chains are members of the zero percent club. In addition, Food Lion, Grand Union, Safeway, Stop & Shop, Winn-Dixie Stores, Southland, and Investcorp have no other redeeming features. Thus, we recommend not shopping at any of those chains.

Zero Percent Club—Grocery Stores

COMPANY	FEMALE DIRECTORS
Food Lion	0
Giant Foods	0
Grand Union	0
Investcorp	0
Safeway	0
Southland	0
Stop & Shop	0
Winn-Dixie Stores	0

SUMMARY

During the writing of this book, and particularly this chapter, people freqently asked us whether we thought their decisions in the grocery store could make a difference. After all, how could buying Hershey's Cocoa over Nestlé's Carnation really have an impact on the way corporate America is run?

The analogy we offer is that of your vote in an election. In the millions of votes that are cast for political elections, does your vote make a difference? The answer is yes. Chances are, you as an individual will not sway the results of an election, nor will the $1.89 you spend on something at the grocery store alone make a difference. But if you make female-friendly decisions time after time, and other women do the same, perhaps we can send a message to the men in the highest offices.

The message is that companies that want our hard-earned dollars will have to take some steps to show that they care about and respect women—enough to trust them with all job descriptions. Our hope, in writing this book, is that women will use their consumer power, beginning in as simple a place as the grocery store, to make the workplace more fair.

Chapter Four

Beyond the Grocery Store

Product Ratings for Household Goods, Stores, Clothes, and Children's Products

I hate housework! You make the beds, you do the dishes—and six months later you have to start all over again.

Joan Rivers

If you want anything said, ask a man. If you want anything done, ask a woman.

Margaret Thatcher

In this chapter, we review products that are bought primarily by women for other than the refrigerator. These products encompass both women's needs and those of their families. In the latter case, particularly with regard to children, women are usually the primary consumer and, therefore, the major decision makers as to what to buy.

The products we rate in this chapter include cosmetics and personal hygiene, healthcare products, household goods such as cleaning supplies, retail stores, shoe and clothing manufacturers and distributors, and products for children and infants.

Many of the manufacturers of these products advertise heavily on television and in magazines. We suggest that you ignore the ads and pay attention to their policies regarding women in making your spending decisions. In this way, you can wield influence instead of simply being influenced.

COSMETICS AND PERSONAL HYGIENE

This category includes products such as makeup, soap, lotion, feminine hygiene products, shampoo, and deodorant. Because women represent the major buyers of most products in this category—and the only buyers of others—it is a relief to learn that the companies in this industry were consistently above average in the categories rated. On the other hand, the fact that many companies that manufacture *women's products*—for example, tampons or lipstick—have no women on their boards of directors or among the highest ranked company officials, give nothing to women's causes, or provide nothing in the way of family benefits to employees, is that much more disappointing.

We have tried to choose primarily companies that are the giants in this industry, the ones that produce dozens of popular brand name products or that have outlets in malls in every major city across the country. These brand name products compete for shelf space in the supermarket and drugstore; they fight for a place in your memory while you watch television and read magazines. For example, seven of the companies we rated market deodorant: Lady's Choice and Arrid are made by Carter-Wallace; Colgate-Palmolive makes Speedstick and Mennen; Bristol-Myers Squibb makes Ban; Gillette makes Dry Idea, Right Guard, and Soft & Dri; Procter & Gamble offers Old Spice, Secret, and Sure; Revlon markets Mitchum; Johnson & Johnson produces Shower to Shower; and Eastman Kodak makes Tussy. Are there differences between the companies that make these popular brands? Absolutely.

Johnson & Johnson provides excellent family-oriented benefits to employees, such as on-site day care and a full year of job-guaranteed leave to new mothers and fathers. Three of the company's 15 members of the board of directors, or 20 percent, are women, a real triumph in a country where the average is 5 percent. One of its subsidiary company presidents is a woman, and the subsidiary has been listed on *Working Mother* magazine's "100 Best" list nine times—every year since the list began. Procter & Gamble also receives a respectable score. Carter-Wallace, Revlon, and

Gillette, on the other hand, placed close to the bottom of the list, all with below-average FEM scores. Gillette has one woman on its board of directors; the other two companies have none. All three companies have a below-average number of women in professional and management positions.

Sometimes, companies market under names that make it obvious who the manufacturer is; other times, it is much more difficult to tell. The "Buying Guide" chapter provides a detailed listing of product names and their parent companies along with FEM scores and ratings.

Below is a list of all the cosmetic companies we rated. Only a few product names are given for each company.

	COSMETIC COMPANIES	FEM	RANK	SOME BRAND NAMES
●	Avon Products	86.73	1	Skin-So-Soft
	Johnson & Johnson	81.81	2	Reach, Act, Stayfree, o.b.
	Body Shop	71.60	3	Body Shop shampoos, lotions, soap
●	Aveda	70.42	4	Aveda hair and skin care products
	Estée Lauder	65.50	5	Clinique, Origins, Prescriptives
	Cosmair	64.86	6	L'Oréal, Lancôme, Biotherm
	Procter & Gamble	64.45	7	Ivory, Cover Girl, Oil of Olay, Old Spice
	Johnson Products	61.14	8	Gloria Roberts, Ultrasheen, Soft 'n Free
	SmithKline Beecham	60.94	9	Aquafresh, Oxy, Ban
●	Alberto-Culver	59.99	10	Alberto VO5
	Bristol-Myers Squibb	59.62	11	Keri, Miss Clairol, Nice 'n Easy
	Eastman Kodak	58.54	12	Stridex, Tussy, pHisoderm
	Schering-Plough	58.51	13	Complex 15

	COSMETIC COMPANIES	FEM	RANK	SOME BRAND NAMES
●	Warner-Lambert	57.81	14	Lubriderm, Schick, Listerine
●	Pfizer	57.70	15	Lady Stetson, Coty 24, Barbasol
●	Johnson Publishing	57.34*	16	Fashion Fair, Duke, Raveen
●	John Paul Mitchell Systems	56.78*	17	Paul Mitchell
●	Newell	56.76	18	Goody hair accessories
●	Dow Chemical	56.24	19	Apple Pectin, Cepacol
●	Colgate-Palmolive	55.80	20	Mennen, Irish Spring, Colgate
◐	S.C. Johnson & Son	53.74	21	Edge, Agree, Halsa
◐	Nu Skin	51.68	22	Nu Skin
◐	Scott Paper	50.44	23	Promise, Poise
◐	American Home Products	50.01	24	Chap Stick, Denorex
◐	Neutrogena†	49.75	25	Neutrogena shampoo, soap, moisturizer
◐	Church & Dwight	48.68	26	Arm & Hammer
◐	Redken Laboratories	47.91*	27	Redken
◐	Abbott Laboratories	47.76	28	Selsun blue
◐	Helene Curtis	47.66	29	Suave, Finesse, Salon Selectives
◐	Maybelline	47.05*	30	Maybelline, Yardley
◐	Tambrands	46.54	31	Tampax
○	Kimberly-Clark	44.79	32	Kotex, Freedom, Lightdays, Depend
○	Gillette	44.05	33	Dry Idea,White Rain, Oral-B, Right Guard
○	Dial	43.96	34	Dial, Tone, Breck
○	Unilever	42.54	35	Vaseline, Caress, Dove, Listerine, Q-Tips

	COSMETIC COMPANIES	FEM	RANK	SOME BRAND NAMES
○	Revlon	39.59*	36	Mitchum, Flex, Almay
○	Carter-Wallace	39.18	37	Arrid, Lady's Choice, Nair, Sea & Ski
○	Reckitt & Colman	38.46	38	Neet, Binaca

*Did not have scores for all five categories.
†Neutrogena has been acquired by Johnson & Johnson, a company that earns a much higher FEM score.

Rating Key:
● Female-friendly company; buy products when possible
◐ Average; buy with good conscience
○ Below average; avoid buying these products

Our stars in the cosmetic industry are Avon Products, Johnson & Johnson, Body Shop, Aveda, and Estée Lauder. Cosmair and Procter & Gamble earn honorable mentions.

Avon Products was the world's largest manufacturer of toiletries and cosmetics and the top scorer of our study with an outstanding FEM score of 87. Four of the 12-member board of directors are women (33 percent), as are 41 percent of the company's top officers. Women have excellent opportunities to become managers—82 percent of the company's managers are women—and considerable freedom and flexibility regarding combining child care and work responsibilities. Avon offers very flexible work scheduling options as well as on-site day care. Eighty-nine percent of the company's charitable giving goes to women's causes. The fact that Avon was the world's largest cosmetics company until the late 1980s, demonstrates that female-friendly policies and financial success can go hand in hand—a lesson that some other companies have yet to learn. In fact, the Department of Labor's Glass Ceiling Commission reports, "A 1993 study of Standard and Poor 500 companies showed that firms that succeed in shattering their own glass ceilings racked up stock-market records that were nearly two and a half times better than otherwise comparable companies." So the next time the Avon lady rings your bell—offering you an alternative to some of the companies that ranked much lower on our FEM charts, such as Maybelline and Revlon—open the door.

Another good choice for cosmetics is Estée Lauder, which has a very impressive record of women in high positions in the company. Fifty-six percent of its top officers are women, and Estée Lauder, the company's founder, serves as chairman of the board. Furthermore, 46 percent of the company's management are women, better than the averages for both the cosmetic industry and corporate America.

The Body Shop and Aveda, both of which sell lines of skin and hair care products under their own names, are small companies that scored high—72 and 70, respectively—on our FEM scale. Despite their size, both show a strong commitment to women. Aveda, for example, scored above average in the management opportunities category: 48 percent of the company's officials and managers, and 68 percent of the company's professionals, are women. Although only one of the eight-member board of directors is a woman, almost 50 percent of the company's top officers are women—an unusually high percentage. The Body Shop earned the industry's highest rating in the management opportunity area, with women making up 80 percent of the company's officials and managers. The Body Shop got extra points for having two women on its seven-member board of directors, and its founder, Anita Roddick, has been named "Business Woman of the Year" in her native England. Both companies scored below average in the benefits category, but it should be noted that smaller companies often score lower than the giant because they have fewer formal programs to boost their scores.

Procter & Gamble and Johnson & Johnson are the two industry giants that also show a good commitment to women. Above-average scores in all five categories earned Johnson & Johnson the second highest score in the industry, with a FEM score of 82 (it was one of the highest of any of the companies we rated). Some highlights include: three female directors on the 15-member board, one female subsidiary president, a year of leave available for new parents or to care for sick family members, on-site day care, strong recruitment and training programs for female managers, and purchases of over $35 million each year from women-owned businesses. One-half of the company's charitable giving goes to women's and girls' groups. Some of Johnson & Johnson's products are: Shower to Shower deodorant powder, Johnson's Swabs (the competing brand, Q-Tips, is made by Unilever, which scored very much lower on the FEM scale), Reach toothbrushes, and sanitary products including Stayfree, Carefree, Sure & Natural, o.b., and Serenity.

Procter & Gamble, with a FEM score of 64, has a more mixed record with regard to women. The company receives an above-average score in three categories: management opportunities, benefits, and the femlin factor. Its top score is in benefits, the company having created a particularly pro-family work atmosphere. Procter & Gamble gives a full year of job-guaranteed leave to new mothers and fathers, enables a part-time return for mothers after their leave, and offers day-care facilities for employees. Its high score in the femlin category is the result of several recognitions and awards it has received from *Money* magazine, *Working Mother*, the Domini Social Index, and the Department of Labor. Some of Procter & Gamble's brand names include Noxzema, Oil of Olay, and Clearasil skin care products; Coast, Ivory, Zest, Camay, and Safeguard soaps; Crest and Gleem toothpastes; Ivory Shampoo, Head & Shoulders, Pantene, Pert, and Vidal Sassoon shampoos; Secret, Sure, and Old Spice; Always sanitary pads; and Cover Girl and Max Factor cosmetic lines.

Several other companies also scored well and are worth mentioning. Cosmair—the parent company of L'Oréal, Biotherm, and Lancôme and the maker of such fragrances as Cacharel, Giorgio Armani, and Guy Laroche—has very respectable numbers of women at the higher rungs of the corporate ladder: 47 percent of thc company's officers and managers and 37 percent of the top officers are women. SmithKline Beecham Consumer Brands, a company that makes products such as Oxy medicated face cream and Aquafresh toothpaste, excels in its family-friendly benefits, offering a full year of leave to new parents and others caring for an ill relative. It also provides day care and flexible scheduling. Some mention must be given to John Paul Mitchell Systems, where 50 percent of top management are women, and Johnson Publishing, maker of a line of products for African-American women, where 38 percent of the top officials are women. Johnson Products, maker of Ultrasheen, is unique in having three women among its top four officers. Women comprise 20 percent of its board.

One other company deserves mention here. Nu Skin, one of the few companies that sent us salary equity information, reports that female officers and managers earn 80 percent of their male colleagues' salaries and female professionals earn 90 percent—placing the company's female employees well above the 72 percent wage gap norm.

We now move to the bottom of the list. The worst companies for which we had scores for all five categories were Gillette (FEM score 44), Dial (44), Unilever (43), Carter-Wallace (39), and Reckitt & Colman (38). Neither Carter-Wallace nor Unilever has a single woman on its board of directors;

Gillette has only one. Similarly, just 6 percent of Gillette's top officers are women; the other two companies have no women as top officers. All three have managerial staffs that are three-quarters male. None of the three was particularly generous in its donations to women's causes. None cooperated with our research. And all offer worse benefits than the average company in the cosmetic industry.

To flex your economic muscle against companies that demonstrate a lack of commitment to women, we recommend that you avoid the following companies and products: Carter-Wallace, which makes Arrid and Lady's Choice deodorants, Nair hair remover and Pearl Drops toothpaste; Gillette, which makes a multitude of shaving products, including Foamy Atra, Good News!, Daisy, Sensor, and Trac, as well as Jafra skin care products, Dry Idea, Right Guard, and Soft & Dri deodorants, and Epic Waves and Dippity-Do for hair; Unilever, which markets Caress, Dove, Lifebuoy, Lux, and Shield soaps, Ponds, Cutex, Vaseline, Fabergé, Brut 33, and Aim, Close-Up, Mentadent and Pepsodent toothpastes; and Reckitt & Colman, which makes Binaca and Neet.

Also on our list are several companies for which we are missing information in two or more categories. None of these companies cooperated with our survey or requests for information, which was not unusual, but we were also unable to find this information from any other source, which happened less frequently. We did have some information about them, however, which was disappointing. Neither Revlon nor Maybelline, both of which make almost exclusively women's products, had a single female board member when we began gathering information for this book; Maybelline now has one woman on its board. Although having all-male boards is not uncommon in corporate America, it's certainly disappointing to find them running companies that make exclusively women's products. Thus, we have a hard time recommending their products.

Here are the members of our cosmetic industry zero percent club.

Zero Percent Club—Cosmetics

COMPANY	FEMALE DIRECTORS	FEMALE TOP OFFICERS OR TOP-SALARIED WOMEN
Carter-Wallace	0	0
Church & Dwight		0
Newell	0	

COMPANY	FEMALE DIRECTORS	FEMALE TOP OFFICERS OR TOP-SALARIED WOMEN
Reckitt & Colman	0	
Revlon	0	
S.C. Johnson & Son	0	
Scott Paper		0
SmithKline Beecham	0	0
Unilever	0	

It must be mentioned that some of the companies listed here were recommended earlier in this section. Recall that all of our numbers are relative, not absolute, and that we recommend those companies that are the best relative to the others we have rated. In many cases, it is obvious that even recommended companies can have a long way to go in the area of women's equality—or even the barest standards of women's representation.

DRUGS

Our list of drug companies includes those that make prescription as well as over-the-counter remedies, such as painkillers, birth control, sleeping aids, pregnancy tests, and vitamins. It is fairly simple, if you choose, to incorporate the following information into your over-the-counter drug buying decisions at the grocery or drugstore, but it gets a little trickier with prescription drugs, as the brand names are usually far less familiar and usually it is a physician who prescribes the medicines. Nonetheless, we have included information about several popularly prescribed prescription drugs and the companies that make them. You can discuss the issue of prescription drug possibilities with your physician or pharmacist, and make informed decisions when choices are available.

Below is the list of drug companies in our sample and how they scored:

	DRUG COMPANIES	FEM	RANK	SOME BRAND NAMES
●	Johnson & Johnson	81.81	1	Band-Aid, Tylenol, Retin-A, Hismanal

	DRUG COMPANIES	FEM	RANK	SOME BRAND NAMES
○	Merck	72.61	2	Prescription drugs, vaccines
○	Monsanto[†]	65.15	3	Searle, Lomotil, Flagyl
○	Procter & Gamble	64.45	4	NyQuil, Vicks, Pepto-Bismol
○	Hershey Foods	63.70	5	Luden's
○	3M	61.45	6	Titralac Antacid
●	SmithKline Beecham	60.94	7	Hold, Sucrets, Vivarin, Tagamet, Thorazine
○	Bristol-Myers Squibb	59.62	8	Bufferin, Nuprin, Excedrin, Theragran-M
○	Schering-Plough	58.51	9	Durasoft, Gyne-Lotrimin, Coppertone
●	Warner-Lambert	57.81	10	Sinutab, Halls, Proxacol, Hydrogen Peroxide, ept
○	Pfizer	57.70	11	Visine, Bengay, Unisom, Desitin
○	Syntex	56.93	12	Prescription drugs
○	Glaxo Wellcome	56.56	13	Ventolin (asthma), Zantac
○	Dow Chemical	56.24	14	Camphophenique, Benedictin
●	Colgate-Palmolive	55.80	15	Curad bandages
◐	Genentech	50.23	16	Prescription drugs
◐	American Home Products	50.01	17	Advil, Anacin, Dristan, Preparation H
◐	American Cyanamid[‡]	49.57	18	Lederle Labs, Centrum, Stresstabs
◐	Abbott Laboratories	47.76	19	Clear Eyes, Murine
◐	Bayer[§]	46.45	20	Alka-Seltzer, Bactine, One-A-Day, Cipro, Bayer, Mycelex-7
◐	Rhône-Poulenc Rorer	46.21	21	Prescription drugs

	DRUG COMPANIES	FEM	RANK	SOME BRAND NAMES
○	Nestlé	44.44	22	Alcon Labs (Opti-Free)
○	Pharmacia & Upjohn‖	42.77	23	Cortaid, Motrin IB, Kaopectate, Rogaine Depo-Provera, Halcion, Xanax
○	Unilever	42.54	24	Q-Tips, Vaseline
○	Eli Lilly	39.85	25	Prozac
○	Carter-Wallace	39.18	26	Trojan, First Response

†Monsanto is a wholly owned subsidiary of Bayer.
‡American Cyanamid has been bought by American Home Products.
§Bayer recently acquired Miles Laboratories.
‖Pharmacia & Upjohn was just Upjohn prior to November 1995.

Rating Key:
● Female-friendly company; buy products when possible
◐ Average; buy with good conscience
○ Below average; avoid buying these products

The good news about our survey of drug companies is that well over half of the companies earned our highest rating. Not surprisingly, because many of these big companies make products in several different industry categories, some of the companies are the shining stars and the duds of both the drug and cosmetic industries.

Once again, Johnson & Johnson and Procter & Gamble outscored much of the competition, with Johnson & Johnson's FEM score of 82 placing it atop the drug industry. Procter & Gamble, with a FEM score of 64, ranked fourth.

The other top drug companies are Merck, Monsanto, Hershey Foods, 3M, and SmithKline Beecham Consumer Brands.

Merck is the country's largest manufacturer of prescription drugs. With 8 percent of the board and about 12 percent of the company's top officers women, it earned a FEM score of 73, the second best after Johnson & Johnson. Twenty-six percent of the company's officials and managers and 47 percent of its professionals are women—both just above average for the drug industry. Merck scored very well in the area of benefits, offering 18 months of partially paid leave to new mothers, flexible scheduling options, and on-site day care. The company also posted a high score for the femlin factor, having been recognized for its fair treatment of women and minorities by the Department of Labor, *Hispanic* magazine, the Domini Social Index, and *Working Mother* (for all 9 years of their surveying). Although Merck received credit for furnishing us with salary equity numbers—thus distinguishing itself as one of the few companies willing to divulge such information—the picture revealed is a grim one: female employees' salaries are less than 60 percent of their male counterparts at the company—even lower than the dismal U.S. average of 72 percent.

Monsanto, the parent company of Searle drugs, got an outstanding score in the benefits category for allowing 1 year of unpaid personal leave for the care of a newborn, newly adopted child, or newly placed foster child, as well as for the care of a relative's serious health condition. The company also has day care and flexible schedule options.

Although Hershey Foods does not have a huge presence in the drug industry—their primary product is Luden's cough drops—the company's strong commitment to women might recommend Luden's over some other brands. Hershey has three women on its 12-member board—25 percent, well over the 5 percent national average. Company employees may take up to 1 year off to care for newborns or ill family members. They may also use an on-site day-care center and choose among scheduling options, including flex time and some job sharing.

3M, maker of Titralac Antacid, ranked sixth overall, and scored particularly well in the areas of benefits, support of women, and prior recognition. SmithKline Beecham Consumer Brands, with a FEM score of 61, was seventh overall. Its pattern of scoring was varied. In the glass ceiling category, SmithKline got the lowest possible score and qualified for the zero percent club (no women on its 15-member board or among its top officers). Its scores in the benefits and management opportunities categories, however, compensated somewhat: The company scored an industry high in the benefits department, offering a full year of leave not only for new parents but also for care of a sick family member. It also features on-site day care, flexible scheduling, and a plethora of family education, counseling, and subsidy programs.

Schering-Plough, which also earned top ratings, makes all kinds of consumer products and over-the-counter medicines as well as prescription drugs. The company received high scores in the support of women, benefits, and femlin categories. Although the company's charitable giving goes primarily to medical education and not specifically to women's causes, the company does rack up $1.5 million a year in purchases from women-owned businesses and was cooperative with our research. Schering-Plough gives new mothers 8 weeks of leave beyond that required by the Family and Medical Leave Act, some of that time *paid*, and has on-site child care, flexible scheduling options, and many family programs and other benefits.

One other drug company deserves mention. Bristol-Myers Squibb was one of a few companies to receive an above-average score in each separate category, even though it didn't obtain an outstanding score in any one.

Five companies in the drug industry warrant mention for scoring well below average on our FEM scale. They are Nestlé, Pharmacia & Upjohn, Unilever, Eli Lilly, and Carter-Wallace. Nestlé, Unilever, and Carter-Wallace have no women on their boards or among their top officers. Pharmacia does have two women on its 16-member board but has no recruitment or training programs aimed at women.

Carter-Wallace was the industry's lowest scorer. Besides being a zero percent club member, it received a below-average score for charitable giving to women's causes, less than 1 percent. We recommend that you not buy any Carter-Wallace products, which include First Response pregnancy tests, Trojan, Class Act, and Mentor condoms, and Sea & Ski suntan lotion.

Many of the companies in this industry, even some that earned our most female-friendly rating, were plagued with a common problem, namely, slow promotion of women into management. Whereas the national average is 41 percent female officers and managers, Bayer has 16 percent, Hershey 18 percent, 3M 15 percent, Monsanto 18 percent, Pharmacia 15 percent, and Rhône-Poulenc Rorer, 10 percent.

Zero Percent Club—Drugs

COMPANY	FEMALE DIRECTORS	FEMALE TOP OFFICERS OR TOP SALARIED WOMEN
Bayer	0	
Carter-Wallace	0	0
Nestlé	0	0
Rhône-Poulenc Rorer	0	
SmithKline Beecham	0	0
Unilever	0	

HOUSEHOLD GOODS

The household goods category, like drugs and groceries, offers many choices to consumers. Within each product category, from dish soap to paper towels, many companies are vying for your dollars. And, as we discovered, significant differences exist among companies regarding their treatment of women, their primary consumers.

As an industry, household goods companies showed an average FEM score, despite low average scores in two categories and the fact that only two companies in our study, Sara Lee and Alberto-Culver, can even boast meeting the national average for women in management positions, 41 percent. However, the industry does get a boost from fairly high scores in the other three categories, namely, benefits, support of women, and the femlin factor.

Below are listed the companies we studied and their ranking.

HOUSEHOLD GOODS COMPANIES	FEM	RANK	SOME BRAND NAMES
● Monsanto†	65.15	1	Ortho, Roundup, Lasso lawn care
● Procter & Gamble	64.45	2	Cascade, Ivory, Comet, Bounty
● Polaroid	61.88	3	Polaroid film
● General Electric	61.78	4	Lightbulbs
● 3M	61.45	5	Scotch tape, Post-it Notes
● SmithKline Beecham	60.94	6	Cling Free, Delicare
● Rubbermaid	60.21	7	Rubbermaid containers
● Alberto-Culver	59.99	8	Kleen Guard, Static Guard
● Eastman Kodak	58.54	9	Kodak film, d-Con, Lysol
● Schering-Plough	58.51	10	Muskol Insect Repellent, Dr. Scholl's
● Newell	56.76	11	Airbake, Wearever, Levelor
● Corning	56.40	12	Pyrex, Revere Ware, Corelle

HOUSEHOLD GOODS COMPANIES	FEM	RANK	SOME BRAND NAMES
● Mobil	56.31	13	Hefty bags, Baggies
● Dow Chemical	56.24	14	Fantastik, Glass Plus, Saran Wrap, Ziploc
● Colgate-Palmolive	55.80	15	Ajax, Murphy Oil Soap, Fab, Fresh Start
● Sara Lee	55.67	16	Fuller Brush, Ty-D-Bol, Kiwi, Hanes
◐ S.C. Johnson & Son	53.74	17	Glade, Off!, Pledge, Shout, Windex
◐ Scott Paper	50.44	18	Viva, Scott Towels, Cottonelle
◐ Clorox	49.23	19	Formula 409, Liquid-Plumr, Litter Green
◐ James River of Virginia	48.77	20	Northern, Marina, Dixie Cups, Brawny
◐ Church & Dwight	48.68	21	Arm & Hammer, Scoop Away
◐ Mead	48.66	22	Mead notebooks, school supplies
◐ Bayer	46.45	23	Agfa film, Cutter
◐ Georgia-Pacific	45.58	24	Angel Soft, Coronet, Sparkle
◐ Borden	45.29	25	Elmer's Glue, Krazy Glue
◐ Ralston Purina	45.03	26	Energizer, Eveready batteries
○ Kimberly-Clark	44.79	27	Kleenex, Hi-Dri
○ Gillette	44.05	28	Braun, Eraser Mate, Liquid Paper
○ Dial	43.96	29	Renuzit, Boraxo, Brillo, Sweetheart
○ Unilever	42.54	30	All, Wisk, Snuggle, Sun Light, Dove

	HOUSEHOLD GOODS COMPANIES	FEM	RANK	SOME BRAND NAMES
○	Armstrong World	40.54	31	Armstrong floor cleaners
○	Reynolds Metals	40.42	32	Reynolds Wrap, Cut-Rite, Sure-Seal
○	Farberware	40.11*	33	Farberware cookware
○	Carter-Wallace	39.18	34	Ear Rite, Fresh 'N Clean, Shield pet products
○	First Brands	39.00	35	Glad bags, Simonize, STP
○	Reckitt & Colman	38.46	36	Airwick, Black Flag, Woolite, Easy-Off
○	Matsushita Electric	36.56	37	Panasonic batteries

*Did not have scores for all five categories.

†Monsanto is a wholly owned subsidiary of Bayer, but is managed (and thus rated) separately.

Rating Key:

● Female-friendly company; buy products when possible

◐ Average; buy with good conscience

○ Below average; avoid buying these products

For household goods, the top-scoring companies are Monsanto and Procter & Gamble. Although the two did not do very well in both categories 1 and 2 (relating to female officers and managers), they excel in benefits provided. At Monsanto, women make up only 18 percent of the managers, less than half the national average, but the company does have two women on its board and outstanding family-friendly benefits.

Procter & Gamble, one of the industry giants, is ranked second in the industry, giving mindful consumers plentiful options for cleaning and household supplies. The company has, for this industry, a fairly strong record of women in management—28 percent total, 3 percent of top officers—although these figures are below average nationally. However, as discussed earlier, Procter & Gamble has an above-average number of women on its board and strong recruitment, management training, and promotion programs for female employees. It also has excellent benefits, including on-site day care and extended paternity leave, and has won several awards for its female-friendly work environment.

Some of our other best companies do not have a huge presence in the household goods industry, as the majority of their income is from unrelated industries: Polaroid makes film, General Electric (GE) makes lightbulbs, 3M makes Scotch tape and Post-it Notes, and SmithKline Beecham markets some laundry products. Nonetheless, the performance of these companies is good, so our recommendation would be to purchase their products over competing ones.

Polaroid scored below average in the first two categories, management opportunities and glass ceiling, but made up for that with an excellent score on benefits. The company guarantees jobs for 1 year for new parents and employees caring for family members. Although there is no day care on site, Polaroid does help employees find and pay for outside child care. Employees may also take advantage of several flexible scheduling options, including job sharing, flex time, and work-at-home programs.

We were glad to see GE's high ratings as the dominant producer of light bulbs. With competition only from Philips and Westinghouse and some large store brands, GE is sometimes the only choice when buying bulbs. GE distinguished itself by having a female subsidiary president and an industry-high 27 percent charitable donations to women's groups.

3M, rated fifth overall, also dominates in the products it makes. It distinguishes itself with a board that is 15 percent female, a female division president, and strong support of women through charitable giving and purchases.

Honorable mention goes to Alberto-Culver, Sara Lee, Colgate-Palmolive, and Rubbermaid for their commitment to women in middle and top management. Alberto-Culver, which is best known for its Alberto VO5 hair products, has two women on its nine-member board (22 percent, tying Colgate-Palmolive for the second highest in the industry), and one-quarter of its top executives and three-quarters of its managers are women—both impressive proportions. Furthermore, two of the company's corporate officers, including the president of the company's U.S. division, are among the 20 highest-paid female executives in the United States (as compiled by *Working Woman* magazine). Colgate-Palmolive provides good benefits and has received many previous awards. Rubbermaid set two industry highs: women make up 25 percent of its board and 27 percent of the company's top officers, again including a division president. All of these companies show a great commitment to women and are highly recommended by us.

The most disappointing companies for which we have complete information include Reynolds Metals, Reckitt & Colman, Carter-Wallace, First Brands, and Matsushita. These companies are all members of the zero percent club. First Brands and Carter-Wallace received double zeros (see table below). Other companies with all-male boards and management are listed below. We recommend that you avoid products made by these companies.

Zero Percent Club—Household Goods

COMPANY	FEMALE DIRECTORS	FEMALE TOP OFFICERS OR TOP-SALARIED WOMEN
Bayer	0	
Carter-Wallace	0	0
Church & Dwight		0
First Brands	0	0
James River of Virginia		0
Matsushita Electric	0	
Mobil		0
Newell	0	
Reckitt & Colman	0	
Reynolds Metals		0
S.C. Johnson & Son	0	
Scott Paper		0
Unilever	0	

RETAIL STORES

This section addresses the companies that own and run many of the large retail store chains where you shop. It includes several department stores, drugstores, health and beauty aid stores, discount stores, and hardware stores. As an industry, the companies come out slightly above average on overall score, with a FEM rating of 51.48. The worst categories for the industry are management opportunities and benefits. However, part of the reason for the low score in management opportunities was that many companies did not cooperate with our research and we were not able to find the information elsewhere; in fact, of the companies for which we do have information, three-quarters have an average or above-average number of women among their managers and officers, which may mean that the industry is better for women than its FEM average suggests.

As an industry, the retail store chains were among the least cooperative with our research: only 3 of 19 companies returned our survey. One company, industry high-scorer Dayton Hudson, which owns Target, was extremely cooperative and provided us with even more information than we asked for.

Here's how the companies compared:

	RETAIL STORES	FEM	RANK	SOME STORE NAMES
●	Dayton Hudson	71.71	1	Target, Mervyn's, Dayton's
●	Nordstrom	68.08	2	Nordstrom
●	Sears, Roebuck	63.55	3	Sears
●	Melville	61.27	4	Marshalls, Linens 'n Things
●	Broadway Stores†	59.41	5	The Broadway, Weinstock's
◐	J.C. Penney	53.85	6	J.C. Penney, Thrift Drug
◐	Grand Metropolitan	53.73	7	Pearle Vision
◐	Federated Department Stores	52.38	8	Bloomingdale's, Burdine's, Lazarus
◐	K-Mart‡	51.89	9	K-Mart, Builder's Square
◐	May Department Stores	50.51	10	Filene's, Famous-Barr, Lord & Taylor
◐	Neiman Marcus Group	49.73*	11	Neiman Marcus, Bergdorf Goodman
◐	Woolworth	48.40	12	Woolworth
◐	Walgreen	47.61	13	Walgreen
○	ConAgra	44.69	14	Country General stores
○	R.H. Macy§	43.94*	15	Macy's, I. Magnin
○	Service Merchandise	43.46*	16	Service Merchandise Catalog

	RETAIL STORES	FEM	RANK	SOME STORE NAMES
○	Montgomery Ward Holdings	42.58*	17	Montgomery Ward
○	American Stores	42.40	18	Sav-on, Osco Drug
○	Wal-Mart Stores	41.67	19	Wal-Mart, Sam's Club

*Did not have scores for all five categories.
†Broadway Stores was formerly known as Carter Hawley Hale.
‡K-Mart had plans to sell 51 percent of its Borders-Walden bookstores subsidiary.
§R.H. Macy, as part of its bankruptcy settlement, sold its debt to Federated Department Stores. Macy's and I. Magnin stores are now part of Federated.

Rating Key:
● Female-friendly company; buy products when possible
◐ Average; buy with good conscience
○ Below average; avoid buying these products

People who love to shop at Target will be pleased with the results of our survey: Dayton Hudson, which is the parent company to Target, Mervyn's, Dayton's, Hudson's, and Marshall Field, was the high scorer in the industry. One-fourth of the board that runs this 168,000-employee company are women, as are 54 percent of the company's managers and officials. Dayton Hudson scores very well on benefits, in part because of a strong commitment to child care. *Working Mother* praised the company in 1992 for earmarking millions of dollars to recruit and train child care providers for employees—a sure sign of a female-friendly company. Target is an excellent choice for shopping, much more female friendly than rival Wal-Mart.

Also scoring well on the FEM scale are Nordstrom; Sears, Roebuck; Melville; and the Broadway Stores. Nordstrom, the one high-end department store that rated well, has very impressive female representation among its management: 33 percent of the company's top 21 officers and 72 percent of the company's managers and officials are women (this includes store managers). Nordstrom has also been recognized for its pay, benefits, and opportunities for advancement for women in the book *100 Best Companies to Work for in America*. Nordstrom shows its commitment to women by supporting a college scholarship fund for black women and showing $80 million dollars in purchases of goods and services from female-owned businesses. If you have the choice in your area, Nordstrom is more female friendly than Neiman Marcus and Bergdorf Goodman, which earn average ratings, or Macy's and I. Magnin, which rate below average.

Sears, Roebuck, which ranked third, excels in its board composition, benefits, and purchases from women-owned businesses. The company has three women on its 12-member board and a woman did run its Consumer Financial subsidiary. (She now heads the Home Services division.) Female employees have excellent benefits: new mothers (and new fathers) may take up to a year of job-guaranteed leave, may return part-time after childbirth, and may take advantage of on-site day care and flex time to accommodate their families. Sears spends over $150 million in purchases from women-owned businesses.

Melville, the fourth of our top five companies owns the only drugstore chains—CVS and People's Drug Stores—that earn our highest female-friendly rating. Melville, which employs nearly 100,000 people, also owns Marshalls, Linens 'n Things, and This End Up. The company scores above average for percentage of women on its board and in its top management. Women make up 47 percent of its total management staff. Like Sears, Melville offers strong family benefits: 26 weeks for new parents and employees caring for sick family members and on-site day care and several options for flexible scheduling. For these reasons, CVS and People's Drug Stores are better choices than Osco Drug, which got our lowest rating.

The Broadway Stores, parent to The Broadway, Broadway Southwest, and Weinstock's department stores in the western United States, scored almost as well as Nordstrom in the women in management category (68 percent of its managers and officials are women). One woman serves on the company's eight-member board. The Broadway Stores received its lowest score on benefits, an area where it mostly adheres to the Family Leave Act for parental leave—although new mothers do receive some *paid* time off—and flexible scheduling options are not formal but rather are at the discretion of supervisors.

Although K-Mart didn't rank in the top five, it did earn our top female-friendly rating. Twenty percent of K-Mart's board and 50 percent of management are women. The company also gives an average amount to women's groups and makes substantial purchases, about $570 million, from women-owned businesses. K-Mart and its subsidiaries are solid choices for women-conscious consumers.

The companies at the bottom of our list were Macy's, Service Merchandise, Montgomery Ward, American Stores, and Wal-Mart. Some fared poorly because of a dearth of information; R.H. Macy, for example, responded to our request for information by explaining that the parent company had been in bankruptcy and was not able to furnish any accurate information (it was scheduled to emerge from bankruptcy in 1995 and has since been acquired by Federated Department Stores, which received our average rating)—a valid excuse.

For companies for which we had information, Wal-Mart was the lowest-scoring retail store chain. Although Wal-Mart's 13-member board has two women on it, only 2 percent of the company's top officials are women. And none of the over $10 million a year in charitable donations goes directly to women's groups (although it does go to social service and education groups, which usually benefit women). Furthermore, the company offers no advancement, recruitment or training programs targeted directly at women. Lastly, although not factored into our FEM ratings, the company made news 2 years ago by pulling a T-shirt line in response to customer complaints; the T-shirt proclaimed that some day, a woman might be president. We believe this action speaks for itself.

Zero Percent Club—Retail Stores

COMPANY	FEMALE DIRECTORS	TOP FEMALE OFFICERS OR TOP SALARIED WOMEN
Montgomery Ward	0	
Service Merchandise	0	0

CLOTHING

The companies in the clothing category vary as wildly from one another in their FEM rating as they do in the areas of price and style. At one end is top-scoring Moving Comfort, which has a board and top management that are two-thirds female and gives 100 percent of its charitable donations to women—boosting its FEM score to an outstanding 87. On the other hand, we found four companies in this category with no women on their boards and others without women at the top echelons—discouraging signs when one considers that most of the companies surveyed in this section are primarily women's clothing makers.

Nevertheless, the industry boasts by far the highest number of companies that have female CEOs, presidents or founders: seven of the companies either have women running the whole show or did in the past, and four others have women running one or more subsidiaries or major divisions. As a result, this industry's top score is in the femlin factor category. The low score is on benefits; many companies did not furnish this information and others are not doing much to help their employees balance work and families.

Below is a list of the companies surveyed and their ranking.

	CLOTHING COMPANIES	FEM	RANK	SOME BRAND NAMES
●	Moving Comfort	86.77	1	Moving Comfort athletic wear
●	Liz Claiborne	70.91	2	Liz & Co., Liz Claiborne
●	Benetton Group	68.49	3	Benetton
●	Ann Taylor Stores	64.38	4	Ann Taylor
●	Esprit de Corp	63.88	5	Esprit clothes and accessories
●	Melville	61.27	6	Chess King, Wilson's, Bob's Stores
●	Oshkosh B'Gosh	61.15	7	Oshkosh B'Gosh
●	The Gap	61.03	8	The Gap, Banana Republic
●	Danskin	60.02	9	Dance France; Round the Clock, Givenchy, Christian Dior, and Anne Klein hosiery
●	Levi Strauss	58.83	10	Levis, Dockers
●	Warnaco Group	58.46	11	Warner's, Fruit of the Loom bras
●	Sara Lee	55.67	12	Bali, Hanes, L'eggs, Playtex
◐	Leslie Fay	53.95	13	Head Sportswear, Breckenridge, Knitivo, Shapely, Victoire, Castlebury Knits
◐	VF	53.00	14	Lee, Wrangler, Jantzen, Barbizon, Vanity Fair, JanSport, Timber Creek
◐	Phillips-Van Heusen	51.46	15	Van Heusen, Geoffrey Beene, Pickwick, Somerset

	CLOTHING COMPANIES	FEM	RANK	SOME BRAND NAMES
◐	Reebok	50.91	16	Reebok, Tinley
◐	The Limited	48.72*	17	The Limited, Lane Bryant, Victoria's Secret, Abercrombie & Fitch, Henri Bendel
◐	Timberland	48.23	18	Timberland
◐	Laura Ashley	47.30*	19	Laura Ashley
◐	Lanz	47.01*	20	Lanz clothes and sleepwear
◐	Lands' End	45.96*	21	Lands' End catalog
○	Fruit of the Loom	40.62	22	Fruit of the Loom, Gitano Jeans
○	Gitano Group	39.51	23	Gitano

*Did not have scores for all five categories.

Rating Key:
● Female-friendly company; buy products when possible
◐ Average; buy with good conscience
○ Below average; avoid buying these products

The top makers of women's clothes in our survey are Moving Comfort, Liz Claiborne, Benetton, Ann Taylor, and Esprit. Three—Moving Comfort, Liz Claiborne, and Ann Taylor—either are run or were founded by women (Liz Claiborne is retired), a rare but refreshing sight in corporate America.

Moving Comfort is a very small company (about 25 employees) that makes high-performance women's athletic gear. The owners of the company—two of three of whom are women—believe strongly in their customers, as reflected in the FEM score of 87. Sixty-seven percent of the board and officers are women, tying (with Birkenstock Footprint Sandals) for first place in this category. Company officials were also happy to share their pay equity statistics with us and the numbers were heartening: women at the company make 100 percent of what their male counterparts earn. Furthermore, all of the company's charitable donations go to women's causes, including breast cancer research, domestic violence prevention programs, and women's fitness events. Moving Comfort gets a glowing recommendation from us; just look for its labels, which read "We believe a fit woman is a powerful woman."

Liz Claiborne and Benetton Group are a close second and third on our list. Although Liz Claiborne, the founder of the company that bears her name, retired several years ago as CEO, there is still a strong commitment to women. Twenty-two percent of the board are women, as are 58 percent of the managers and officers—an outstanding measure. The company gets the second highest score in the industry for benefits, providing a 26-week leave for new mothers and even some paid time off for new fathers, some flexible work scheduling options, on-site prenatal care classes, and mammography education programs. The company receives an above-average score for charitable donations, giving 6 percent to such causes as a garment industry day-care center, breast cancer organizations, and organizations that promote nontraditional employment for women.

Benetton Group's picture for women was a bit fuzzier. Forty percent of the company's officers and managers and 91 percent of the professional staff are women. Benetton Group was one of the rare companies that divulged salary information; unfortunately, the numbers—managerial women earn 63 percent of their male counterparts' salaries with professional women faring better at 85 percent—were not unequivocally impressive. Benetton Group did earn some points on benefits, however, for paying new mothers for part of their leave.

Ann Taylor and Esprit round out the top five. Three of Ann Taylor's seven board members are women (43 percent), as are 60 percent of the company's top officers. Furthermore, the CEO is a woman (who is one of *Working Woman* magazine's top 20 female corporate earners). Although we could not obtain information regarding Esprit's board, two of its top seven officers are women (29 percent). That combined with having two female subsidiary presidents and a well above-average 72 percent female managers is enough to boost Esprit to a fifth-place finish.

Also coming in well above average in the FEM survey are Melville, Oshokosh B'Gosh, The Gap, Danskin, Levi Strauss, and Warnaco Group (maker of Warner's bras). Danskin is best known for its dance and aerobic wear. Formerly Esmark, Danskin spent its first year as a public company in 1994. In addition to Danskin, its product lines include Dance France, Shape Activewear, and Custom Collection, as well as Round the Clock, Givenchy, Christian Dior, and Anne Klein Collections hosiery. Danskin's highlights include three women on its seven-member board (43 percent) and 50 percent female representation among officials and managers. The CEO and president of Warnaco Group is one of the few women running a

Fortune 500 company. Levi Strauss, although not particularly cooperative with our study, still came in slightly above average, as did Sara Lee, maker of Playtex, Bali, Hanes, L'eggs, and Aris-Isotoner.

The bottom of the list is occupied by Fruit of the Loom and Gitano. These two, along with Lands' End and Timberland, have all-male boards (to its credit, however, 20 percent of Timberland's top officers are women, and the company gives most of its charitable donations to a Boston urban youth service organization which was later used as a model for President Clinton's plan for a national youth service).

There were some companies for which we had incomplete information, but which nonetheless showed some promise. Laura Ashley was founded by a woman, and The Limited has *six* female division presidents among the 13 store and catalog divisions.

Zero Percent Club—Clothing

COMPANY	FEMALE DIRECTORS
Fruit of the Loom	0
Gitano Group	0
Land's End	0
Timberland	0

SHOES

This category surveyed three groups of companies: athletic shoes, other shoes, and shoe stores. Only one athletic shoe company, Nike, ranked with our best companies. All of the companies are catching on to the fact that women have become significant consumers of fitness equipment, as is evident from their large-scale advertising campaigns featuring and aimed at women. It is our hope to see as many women in the boardrooms and among the executives of these shoe companies as are seen biking and stepping in the ads.

The average FEM score for the footwear industry as a group was a slightly above-average 51.70. Three of the companies have all-male boards of directors; the remaining companies for which we have that information, however, all have more women on their boards than the national average.

	SHOE COMPANIES	FEM	RANK	SOME BRAND NAMES
●	Birkenstock Footprint Sandals	75.18	1	Birkenstock
●	Stride Rite	69.34	2	Sperry Top-Sider, Keds
●	Nike	64.83	3	Nike
◐	Phillips-Van Heusen	51.46	4	Van Heusen, Bass
◐	Reebok	50.91	5	Reebok, AVIA, Rockport
◐	Adidas	48.41	6	Adidas, Le Coq Sportif, Pony USA
◐	Timberland	48.23	7	Timberland
◐	L.A. Gear	47.40	8	L.A. Gear
○	Brown Group	44.09*	9	Air Step, Naturalizer, Connie
○	Nine West Group†	42.46	10	Easy Spirit, Capezio, Pappagallo
○	New Balance Athletic Shoes	41.32	11	New Balance
○	Genesco	39.26	12	Johnston & Murphy, Code West

	SHOE STORE COMPANIES	FEM	RANK	SOME STORE NAMES
●	Melville	61.27	1	Thom McAn, Meldisco, FootAction
◐	May Department Stores	50.51	2	Payless Shoesource
◐	Woolworth	48.40	3	Foot-Locker, Kinney
○	Brown Group	44.09*	4	Famous Footwear

*Did not have scores for all five categories.
†Nine West recently acquired the shoe divisions of U.S. Shoe.

Rating Key:
● Female-friendly company; buy products when possible
◐ Average; buy with good conscience
○ Below average; avoid buying these products

The top-ranked shoe company is Birkenstock Footprint Sandals, maker of those comfy-looking leather sandals frequently worn with wool socks. Two of the three company executives are women (67 percent), as are 69 percent of the company's officials and managers. For a small company—Birkenstock has just under 150 employees—it offers good family benefits: 16 weeks of maternity leave, part-time return after childbirth, job sharing, and flex time. It also gives an average amount of money to women's causes, including Planned Parenthood, scholarships aimed at women, and a school for single working women. These shoes may not be for everyone, but on the basis of the company's treatment of women, we give them our highest recommendation.

Ranked second is Stride Rite, maker of Sperry Top-Sider, Keds, Pro-Keds, and Grasshoppers. Although the composition of Stride Rite's management did not compare with Birkenstock's, it is still respectable. Thirteen percent of Stride Rite's board of directors, 24 percent of its top management and 43 percent of its total management are women—all at or above national averages. In the area of family benefits, the company offers 14 weeks of additional leave to women (and 6 to men) beyond the requirements of the Family Leave Act, a variety of flexible scheduling options, on-site day care, and other programs including adoption leave, allowable time off for school participation, and resource and referral services. Furthermore, the company has been recognized as above average in its treatment of women by the Domini Social Index and *Working Mother*.

Nike, which ranked third on our list, was the highest rated of the athletic shoe companies. Although the percentage of women among the company's managers and officials is, at 34 percent, below the national average of 41 percent, the company does have multiple programs for recruiting, training, and promoting women and minorities. Reebok also has a 34 percent female management staff, New Balance 25 percent, and Adidas 30 percent. Furthermore, Nike has more women among its top executives, at 24 percent, than its sneaker competition. Nike is a family-friendly employer, making available an assortment of flexible scheduling arrangements, on-site day care, some paid maternity leave, a fitness center, and an employee store. Nike received an average rating on charitable donations but earned extra points for being very cooperative and returning much additional information about the company. Based on all of these measures, Nike shoes would be a better choice than Adidas, Reebok, or New Balance.

The lowest-scoring companies are New Balance and Genesco. Genesco Inc. has no women on its board, and only 23 percent of the company's managers and officials are women, which is significantly below the national average. New Balance also posts a below-average number of female officers and managers. Other companies with no women in the boardroom are Timberland and Nine West, maker of Easy Spirit, 9&Co., Pappagallo, and Enzo Angiolini (although the company does have good female representation among its top officers).

To sum up, if you desire athletic shoes, your best bets are Nike and shoes sold at FootAction. The most female-friendly brand names in the rest of the shoe world are Birkenstock, Thom McAn, Meldisco, Sperry, Keds, and Grasshoppers. Other good choices are Van Heusen and Bass. We recommend not considering Air Step, Connie, Life Stride, and Naturalizer by the Brown Group; Johnston & Murphy, Jarmin, Laredo, Code West, and Mitre by Genesco; and New Balance sneakers.

Zero Percent Club—Footwear

COMPANY	FEMALE DIRECTORS
Genesco	0
Nine West	0
Timberland	0

BABY PRODUCTS

The companies that manufacture baby paraphernalia—items such as strollers, diapers, and formula—are a mixed bag. All of the companies for which we had complete information have above-average numbers of women on their boards—a pretty good record compared with some other industries. Four of the 19 companies rated—Johnson & Johnson, The Gap, Kimberly-Clark, and Huffy—have a woman running a subsidiary or a division. On the negative side, only two of the companies (for which we had that information) have *any* women among their top salaried officials.

Below are the companies surveyed and their ranking.

	BABY PRODUCTS COMPANIES	FEM	RANK	SOME PRODUCT NAMES
●	Johnson & Johnson	81.81	1	Baby Shampoo, Band-Aids

	BABY PRODUCTS COMPANIES	FEM	RANK	SOME PRODUCT NAMES
	Procter & Gamble	64.45	2	Pampers, Luvs
	Oshkosh B'Gosh	61.15	3	Oshkosh B'Gosh
	The Gap	61.03	4	babyGap
	Bristol-Myers Squibb	59.62	5	Enfamil, ProSobee
●	Eastman Kodak	58.54	6	Chubs baby wipes
	Schering-Plough	58.51	7	A&D Diaper Rash Ointment
	Pfizer	57.70	8	Desitin
	Colgate-Palmolive	55.80	9	Baby Magic
◐	William Carter	51.93	10	Carter's
◐	Scott Paper	50.44	11	Baby Fresh
◐	American Home Products	50.01	12	Nursoy, Promil, SMA
◐	Huffy	49.50	13	Gerry, Soft Gate
◐	Hasbro	48.62	14	Hugger, Pur
◐	Gerber Products†	47.85	15	Gerber baby food, Onesies, NUK
◐	H.J. Heinz	47.77	16	Heinz baby food
◐	Abbott Laboratories	47.76	17	Similac formula
○	Kimberly-Clark	44.79	18	Huggies diapers, Huggies Pull-Ups
○	Nestlé	44.44	19	Carnation formula

†Gerber has been acquired by a company that is not in our sample.

Rating Key:
● Female-friendly company; buy products when possible
◐ Average; buy with good conscience
○ Below average; avoid buying these products

The top scorer in the baby industry is Johnson & Johnson, its FEM score of 82 being more than 15 points higher than its closest competitor. Also rated as excellent companies are Procter & Gamble, Oshkosh B'Gosh, The Gap, Bristol-Myers Squibb, Eastman Kodak, and Schering-Plough. Pfizer and Colgate-Palmolive also earn our highest rating.

If your baby has diaper rash, we recommend the following three products: Johnson & Johnson's Baby Diaper Rash Relief, A&D Medicated Diaper Rash Ointment (Schering-Plough), and Desitin (Pfizer).

Consumers have two excellent choices for baby clothes: Oshkosh B'Gosh and babyGap, a miniaturized version of Gap clothes for adults, located as a store-within-a-store at GapKids. The Gap's top management is 38 percent female, and GapKids, the store's fastest growing division, is headed by a woman. Oshkosh B'Gosh has women in 62 percent of its general managerial positions and 11 percent of the board. A highlight for Oshkosh, a company best known for its striped kids' overalls, is in the charitable giving department.

Procter & Gamble, ranked second in the industry, gives parents two good choices for disposable diapers, Pampers and Luvs (the company is discussed in more detail earlier in this chapter, in the cosmetics and drugs sections). Eastman Kodak, which ranked sixth, sells Chubs, the winner in the baby wipes category. These products are more female-friendly choices than Kimberly-Clark's Huggies diapers or wipes, which earn our lowest rating. Having a female subsidiary president is a good start, but it could not overcome the company's short treatment of women on other measures.

Bristol-Myers Squibb, which makes ProSobee and Enfamil formulas, cannot boast too many women in its top management—one woman on the board (9 percent) and 6 percent female senior officers—but it does have good benefits. Although not included in the FEM rating system, we wanted to recognize Bristol-Myers as one of the few companies in our sample to have a special review committee to ensure that women and minorities are fairly represented in its ads. Bristol-Myers Squibb's baby formulas are more female-friendly choices than the competition, especially Nestlé's Carnation. Nestlé has no women on its board or among its senior officers.

Other highlights in the baby products industry are Scott Paper, Gerber Products, and Abbott Laboratories. Both Scott Paper and Gerber, like Johnson & Johnson, have 20 percent female membership on their boards. Abbott Laboratories, maker of Similac formula, has one woman on its 13-member board and two women among its top 32 officers. Abbott also scored above average in the support of women category for cooperating with our survey and giving 4 percent of its charitable donations (which total $5.7 million) to women's causes, including safehouses, Girl Scouts, and an anorexia nervosa research center.

Zero Percent Club—Baby Products

COMPANY	FEMALE DIRECTORS	FEMALE TOP OFFICERS TOP SALARIED WOMEN
James River of Virginia		0
Nestlé	0	0
Scott Paper		0

TOYS

For this category, we surveyed ten companies that manufacture kids' toys. As an industry, it was one of the least cooperative and the least responsive to this project. Only four of the companies returned our survey with information. On the positive side, the industry has one of the very few female CEOs or presidents: Mattel, which recently acquired Fisher Price, has a woman in its top office.

The companies in the toy industry range considerably in size, from 45 employees at Brio to nearly 100,000 for Melville, which owns Kay-Bee Toy stores.

Not surprisingly, few of the charitable donations made by toy companies are earmarked for women. The companies that do give, and would disclose the amounts or targets of their donations, give almost exclusively to children's and educational causes.

Below are the companies surveyed and their ranking.

TOY COMPANIES	FEM	RANK	SOME PRODUCT NAMES
● Children's Television Workshop	76.13	1	Sesame Street books
● Brio	69.74	2	Brio toys
● Mattel (Fisher Price)	67.78	3	Barbie, Hot Wheels, Disney toys
● Melville	61.27	4	Circus World, K & K, Kay-Bee toys and stores
● Rubbermaid	60.21	5	Little Tikes toys
◐ Huffy	49.50	6	Huffy bikes, basketball equipment

	TOY COMPANIES	FEM	RANK	SOME PRODUCT NAMES
◐	Hasbro	48.62	7	Barney, GI Joe, Playskool, Milton Bradley, Monopoly, Scrabble, Candy Land, Jurassic Park, Batman, Mr. Potato Head, Nerf, Parker Bros., Tonka
◐	Toys “R” Us	48.59	8	Toys “R” Us toys and stores
◐	Educational Insights	45.02*	9	Educational aids and kits for school/home
○	Tyco Toys	41.86	10	Tyco

*Did not have scores for all five categories.

Rating Key:
● Female-friendly company; buy products when possible
◐ Average; buy with good conscience
○ Below average; avoid buying these products

If your child is a fan of Big Bird or Oscar the Grouch, you will be pleased to know that the Children's Television Workshop, the company that brings those characters into your home via television and books, is our top scorer in this industry. Fifty percent of the company's board and 48 percent of its top officers are women. New mothers and fathers get 12 weeks of leave at full pay; mothers may take 8 additional weeks at 50 percent pay. Children's Television Workshop also provided salary information, and it was good news for women: female managers earn 90 percent, and female professionals 100 percent of what their male counterparts make.

Brio, which employs 45 employees, also has plenty of women running the show. Although Brio's four-member board is all male, one of the company's top two executives is a woman. The company also boasts a management staff that is 70 percent female and a professional staff that is 67 percent female. Brio was another company that shared salary equity information: female officers and managers make 86 percent of their male colleagues' salaries—well above the national average—the company reports that its female professionals earn 128 percent of male pay.

Ranked third was Mattel, whose President and COO, Jill Barad, is one of *Working Woman* magazine's 20 top-paid women in corporate America. She more than doubled the company's sales, from $485 million to over $1 billion per year, during her tenure. Although its numbers were not quite as impressive as Brio's, Mattel—which recently acquired Fisher Price—scores respectably in the female management categories: 20 percent of the company's top officers and almost 35 percent of the officers and managers are women. Mattel also received an excellent score for its benefits. Women get about 6 months of leave after childbirth, some of it paid. Mattel has on-site child care—unusual for a company of its size—and a variety of scheduling options and programs to accommodate families. Some Mattel products (the list of which may make hard core feminists cringe!) include Barbie, Hot Wheels, See 'N Say, Li'l Miss, Magic Nursery, and PJ Sparkles. Mattel also markets Disney toys and dolls, which include the characters from popular movies such as *The Little Mermaid* and *The Lion King*, as well as Fisher Price toys.

Also scoring high on the FEM scale were Melville, parent company of Kay-Bee Toy stores, and Rubbermaid, parent company of Little Tikes. Almost half of Melville's officials and managers are women and its benefits are the best in the industry. New mothers and fathers get 3 months of family leave beyond the FMLA, as do employees who take leave to care for a sick child or relative. The company has day care on site and several flexible scheduling options. Rubbermaid's performance with regard to women in management was not as strong as Melville's, but it does have three women on its 12-member board and 27 percent women among its top executives. It also has day care on site, but perhaps some of its general policies will be of greater interest to people considering buying Little Tikes products: the company does not make any war- or violence-related toys, does not advertise its toys on television, and makes a special effort in its sales materials and annual reports to show boys and girls playing together in non-stereotypical roles (in the world of Little Tikes's annual report, little girls are doctors and little boys make gourmet meals in the kitchen).

One other company will be mentioned, partly because its toys are so popular and plentiful. The company is Hasbro and some of its toys include: GI Joe, World Wrestling Federation action figures, Cabbage Patch kids, Milton Bradley games, Playskool, Lincoln Logs, Mr. Potato Head, Sesame Street characters, Barney, Tonka, Monopoly, Trivial Pursuit, Play-Doh, and Raggedy Ann, as well as Parker Brothers toys and games. Having a board of directors that is 24 percent female, a few high-ranking female executives, and charitable giving of 4 percent to women's groups (totaling about $4 million) was enough to earn Hasbro an average FEM score. You can buy its toys in good conscience.

Tyco Toys is our lowest scorer, and indeed, the only company in this industry to receive our lowest rating. It has an all-male board (as does Brio Corp.) and a low score in the support of women category.

Zero Percent Club—Toys

COMPANY	FEMALE DIRECTORS
Brio	0
Educational Insights	0
Tyco Toys	0

CHILDREN'S CLOTHES AND SHOES

Sixteen makers of kids' clothes and shoes were rated. The industry scored slightly above average in the glass ceiling and femlin factor categories. One company is run by a woman and five others have female division or subsidiary presidents—an encouraging sign. On the negative side, the industry as a whole earned a terrible score for benefits, partly because very few companies provided information in this category; with information missing, the best score a company could get was average.

Below are the companies surveyed and their ranking.

	KIDS' CLOTHES COMPANIES	FEM	RANK	SOME PRODUCT NAMES
●	Stride Rite	69.34	1	Keds
●	Esprit de Corp	63.88	2	Esprit Kids
●	Sears, Roebuck	63.55	3	Sears, Toughskins
●	Oshkosh B'Gosh	61.15	4	Oshkosh B'Gosh
●	The Gap	61.03	5	GapKids
●	Mousefeathers	57.84*	6	Mousefeathers dresses
◐	VF	53.00	7	Healthtex, Fisher-Price, KidsWear, Wrangler
◐	William Carter	51.93	8	Carter underwear, night and outerwear

	KIDS' CLOTHES COMPANIES	FEM	RANK	SOME PRODUCT NAMES
◐	Reebok	50.91	9	Weebok
◐	The Limited	48.72*	10	The Limited Too
◐	Toys "R" Us	48.59	11	Kids "R" Us, Babies "R" Us
◐	Woolworth	48.40	12	Kids Foot Locker, Kids Mart/Little Folks
◐	L.A. Gear	47.40	13	Bendable
◐	Lands' End	45.96*	14	Lands' End catalog
◐	Brown Group	44.09*	15	Buster Brown
○	Fruit of the Loom	40.62	16	Fruit of the Loom

*Did not have scores for all five categories.

Rating Key:
● Female-friendly company; buy products when possible
◐ Average; buy with good conscience
○ Below average; avoid buying these products

The five best companies on the FEM scale are Stride Rite, Esprit, Sears, Oshkosh B'Gosh, and The Gap. Stride Rite scored above average in each of the five categories. Although we could not obtain information regarding Esprit's board, two of the seven highest-paid officers are women (29 percent), as are two subsidiary presidents and a well above-average percentage of managers (72 percent), enough to boost Esprit to a FEM score of 64.

Oshkosh B'Gosh has 62 percent women among its general managerial staff. A highlight is its charitable giving: 27 percent of its donations go to women's groups. The Gap, parent company of GapKids, received a slightly below-average score for charitable giving, but compensated by having women in 38 percent of the company's top positions. GapKids, the Gap's fastest-growing division, is run by a woman. The Gap earned extra points by being most helpful with our research: in addition to returning our questionnaire, the company sent a great deal of information on its benefits, policies, and programs for employees. Sears has considerably above-average female representation across the board (and management too). Mousefeathers, a 30-odd-person dress and blouse company started in 1979, has a female president and one woman on its three-person board. Based on all of these factors, we highly recommend the products made by these companies.

Kids "R" Us, L.A. Gear, and VF all had average scores for the industry. However, these companies have an above-average number of women on their boards, and Kids "R" Us and VF also have more than 41 percent female managers (the national average is 41 percent). William Carter, another company that scored in the average range, has 78 percent female managers, far and away the industry record. The Limited, as mentioned earlier, has women running 6 of its 13 store and catalog divisions, and a board that is 15 percent female. You can buy their products with a clear conscience.

Fruit of the Loom, Lands' End, and the Brown Group occupy the bottom of the list. Fruit of the Loom has no women on its board or among its senior officers. We had incomplete information about the Brown Group and Lands' End, but we do know that Lands' End has an all-male board. We do not recommend products made by these companies.

Zero Percent Club—Kids' Clothes

COMPANY	FEMALE DIRECTORS
Fruit of the Loom	0
Lands' End	0

SUMMARY

By now, you may have become familiar with many of the names of the companies surveyed. There is quite a bit of crossover between industries, especially in areas such as drugs and cosmetics, or children's products and toys. Many of the companies appear in multiple industries, as do the bad ones. This makes your task as a consumer somewhat easier.

We hope that you take this information to heart, and then to the drugstore, the mall, and the hardware store. Your purchases could make a difference.

Chapter Five

Household-Based Consumption

Ratings of Other Products and Services Often Purchased by Women

The hardest thing about making money last is making it first.
Unknown

Women's discontent increases in exact proportion to her development.
Elizabeth Cady Stanton

In the previous two chapters, our argument has been that because women are the primary purchasers of food, household products, children's items, and their own clothing, the companies that produce these items should be particularly sensitive to the needs of female consumers. It is difficult to imagine how they could be, lacking significant (or in many cases, any) female managerial input. Because there are so many possible choices in the various categories rated in this book, we recommend that you select those made by companies that have demonstrated positive treatment of women. Your selections can make a real difference and produce changes in the business world such that women are no longer avoided or ignored.

In Chapter One, we remarked that the purchasing power of women extends far beyond that of food and clothing. In single-parent households run by mothers, they make *all* of the decisions regarding purchases. Additionally, women are increasingly becoming involved in—or in charge of—areas that have long been regarded as male domains. Automakers have begun to understand this and now direct some of their advertising at women. Unfortunately, as will soon become apparent, these companies have not, for the most part, made the required organizational changes to accommodate either women's changing consumer roles or the laws regarding equal opportunity for women.

Using the same FEM rating system discussed in previous chapters, this chapter assesses companies that supply your needs regarding transportation (automobiles, auto supplies, gasoline), saving work (appliances and electronic equipment), finance (banks, insurance, and financial services), and leisure time. Again, our focus is on national companies.

TRANSPORTATION

Automobile Companies

Women constituted 42 percent of car buyers in 1993 (according to a survey by D. Powers). There are many factors that determine one's choice of automobile, such as price, estimates of quality, gas mileage, size, and appearance. As most major automobile manufacturers try to appeal to a broad range of people with pricing and design, we suggest that you add one more factor to your decision: the company's FEM score.

We *tried* to rate eight major car companies, but five of the eight declined to cooperate with our survey. Most of the companies are publicly traded, and we were able to obtain *some* of the necessary information, but you might want to ask yourself why you should deal with a company that prefers not to reveal the record of its treatment of women. These uncooperative companies were Chrysler, Ford, Mitsubishi, Honda, and Nissan. It was particularly difficult to obtain information for the Japan-based companies, even though they are still subject to U.S. laws about disclosure if they have manufacturing plants here. As a group, the auto industry has one of the lower average FEM scores—43.22.

	AUTO COMPANIES	FEM	RANK	PRODUCTS
●	General Motors	63.90	1	Chevrolet, Buick, Pontiac, Cadillac, Saturn, Oldsmobile, GMC trucks
◐	Toyota	46.33	2	Toyota, Lexus, 4Runner, Avalon, Land Cruiser, Previa
○	Volkswagen	43.48	3	Volkswagen, Audi
○	Chrysler	41.26	4	Chrysler, Dodge, Plymouth, Jeep, Eagle
○	Mitsubishi Motors	39.57	5	Mitsubishi
○	Nissan	37.48*	6	Nissan, Infiniti, Pathfinder, Quest, Sentra
○	Ford Motor	37.22	7	Ford, Mercury, Lincoln, Jaguar
○	Honda of America	36.55*	8	Honda, Acura, Passport, Odyssey

*Did not have scores for all five categories.

Rating Key:
● Female-friendly company; buy products when possible
◐ Average; buy with good conscience
○ Below average; avoid buying these products

From the above, only one company can be recommended. General Motors is far superior to its competition with respect to treatment of women. Although it is not unlike its competitors on the glass ceiling factor, GM has two women on its 14-member board, compared with none for Toyota, Mitsubishi, Honda, Nissan, and Volkswagen (VW). Ford and Chrysler each had one woman on their boards when we began compiling data; Ford now has two. Although only 6 percent of General Motors's top officers are women, it is actually the best of the group. Chrysler has 3 percent female senior officers and Ford 4 percent. Similarly, the number of women in management positions at GM does not appear to be good at 11 percent (about one-fourth of the national average), but once again, it is better than VW's or Ford's 6 percent or Chrysler's 7 percent. Its charitable giving program is better than for either Ford or Chrysler, or the zeros for Toyota and Honda. GM also has special programs for recruiting and training women for management positions, which, among companies that provided us with information, is not true for anyone else except Toyota and Volkswagen. Another big plus for General Motors is the fact that, despite the considerable downsizing of their work force over the last 8 years, they have *increased* the percentages of both women and minority group employees.

Where GM really shines—in terms of our entire sample, not just the automobile industry—is its many family-friendly benefits. Its raw score on benefits was 401—the highest in our sample of almost 400 companies. By comparison, Ford's scores was 8, Toyota 40, and Mitsubishi 2. General Motors gives employees 2 years off for parental or family emergency leave (only 12 weeks are mandated by law), with a job guarantee at the end of that period. It also offers job sharing, flex time and some work-at-home options. A plus for consumers is that GM's car models cover the gamut, from luxurious Cadillacs to economical Geos, in style and price.

The second best company, just barely making it into the average range, is Toyota, with a FEM score of 46. Toyota is, unfortunately, a member of our zero percent club (no women on its board or among its top-salaried officers). Overall, however, 12 percent of Toyota's officers and managers are women, still well behind the national average of 41 percent, but the second best for the auto industry. Toyota has some recruitment and training programs for women, a 35 percent female professional staff, and reasonable benefits. It also cooperated with us by returning our questionnaire.

One company that gets a mention, if not quite positive one, is Mitsubishi. It made a strong first-place showing for the industry in the management opportunities category. Although its 27 percent female managers and 43 percent female professionals are below the national average, they are still well above any others in the automobile industry.

At the bottom of the barrel are two companies that refused to cooperate and for which we were unable to find complete information: Nissan and Honda. It should be noted, though, that in the categories where we were able to obtain information, such as their all-male boards of directors, their performance is poor, so it is possible they would occupy the same rankings, or worse since they received average scores for missing information, even if they had been more cooperative.

The worst company without missing information was Ford, with a FEM score of 37. At Ford, there are no women among the top-salaried officers and only a scant presence of women among the company's senior officers. Its percentage of female officials and managers—5 percent—is one of the lowest of *any* U.S. company, is only one-eighth the national average, and is clearly the lowest of the Big Three automakers. It doesn't redeem itself by having only 18 percent female professionals, well below the industry average. Furthermore, the company does not seem to be busily remedying the situation: it also lost points for its shortage of programs to train and recruit female managers.

For the most part, the automakers have terrible records, which is reflected in their FEM scores. One wonders why our government continues to purchase cars from some of these companies, when they—and now the reader—have a clear choice that will benefit women: *buy cars made by General Motors*. Avoid all others, with the possible exception of Toyota.

Zero Percent Club—Auto Makers

COMPANY	FEMALE DIRECTORS	FEMALE TOP OFFICERS OR TOP-SALARIED WOMEN
Honda	0	
Mitsubishi	0	
Nissan	0	
Toyota	0	0
Volkswagen	0	

Auto Suppliers

The major manufacturers of auto parts, such as tires and motor oil, are as bad as the automakers in their treatment of women, with a group average FEM score of 43.25.

AUTO SUPPLY COMPANY	FEM	RANK	PRODUCTS
◐ Cooper Tire & Rubber	53.70	1	Cooper tires
◐ Goodyear Tire & Rubber	46.69	2	Goodyear tires
○ Whitman	43.95	3	Midas mufflers, brakes, etc.
○ Cooper	43.63	4	Champion spark plugs, Anco
○ Quaker State	43.08	5	Motor oil, Champion wiper blades, brakes

AUTO SUPPLY COMPANY	FEM	RANK	PRODUCTS
○ Arvin	39.34	6	Exhaust pipes, shock absorbers
○ First Brands	39.00	7	Prestone antifreeze, STP, Simonize
○ Pep Boys—Manny, Moe & Jack	36.59*	8	Retail stores for parts and service

*Did not have scores for all five categories.

Rating Key:
● Female-friendly company; buy products when possible
◐ Average; buy with good conscience
○ Below average; avoid buying these products

Of the eight companies, Whitman, Arvin, First Brands, and Pep Boys—Manny, Moe & Jack did not cooperate with our survey.

Only two companies in this area are even in the average category: Cooper Tire & Rubber and Goodyear. Although Cooper Tire was a zero percent club member (no women on its board), it was still the top scorer for the industry. It offers employees good benefits and has many more programs to recruit and train female managers than any of its competition.

Goodyear Tire & Rubber, our other average finisher, has a small number of women (4 percent) represented in its top management. It has one woman serving on its 11-member board; interestingly, Cooper, Quaker State, and Whitman have boards of identical composition. Goodyear also offers workers above-average benefits.

The auto supply industry is obviously a male-dominated one, and these companies have a flat tire with regard to their many female consumers. The best female representation among middle management at any of these companies is First Brands' 11 percent, about one-fourth the national average. Half of the companies have no women on their boards. It is most disappointing when the only company we can recommend has, essentially, a dismal record with regard to women, and only looks good by comparison.

Zero Percent Club—Auto Suppliers

COMPANY	FEMALE DIRECTORS	FEMALE TOP OFFICERS OR TOP-SALARIED WOMEN
Arvin	0	

COMPANY	FEMALE DIRECTORS	FEMALE TOP OFFICERS OR TOP-SALARIED WOMEN
Cooper Tire & Rubber	0	
First Brands	0	0
Pep Boys—Manny, Moe & Jack	0	

Gasoline Companies

We rated 13 companies in this area. The average FEM score of the industry—49.38—was not far from average, and is better than that of the automakers and auto supply companies.

	GASOLINE COMPANIES	FEM	RANK	GAS STATIONS
●	E.I. DuPont	65.78	1	Conoco
●	Atlantic Richfield	62.06	2	ARCO
●	Mobil	56.31	3	Mobil
●	Chevron	55.86	4	Chevron
◐	Amoco	54.14	5	Amoco
◐	Exxon	52.49	6	Exxon
◐	Texaco	48.37	7	Texaco
◐	Phillips Petroleum	48.27	8	Phillips 66
○	British Petroleum	43.91	9	BP
○	Royal Dutch Petroleum/Shell Oil	41.59	10	Shell
○	Sun	41.12	11	Sunoco
○	Occidental Petroleum	37.72	12	Diamond Shamrock
○	Citgo Petroleum†	34.35	13	Citgo

†Citgo's parent company is Petróleos de Venezuela, S.A.

Rating Key:
● Female-friendly company; buy products when possible
◐ Average; buy with good conscience
○ Below average; avoid buying these products

From the above table, there are good choices: Conoco, ARCO, Mobil, and Chevron. Atlantic Richfield, parent company of ARCO, had the best management opportunities (number of female managers, officials, and professionals) for the industry and showed solid scores everywhere else. The company also purchased more than $43 million worth of goods and services from women-owned businesses. (Unfortunately, we were unable to obtain this information for most of the companies, although we do mention it when available and noteworthy.) DuPont, parent company of Conoco, has two women on its board and an impressive 41 percent female senior management. Unfortunately, DuPont, like all of its competitors except ARCO, has less than one-third the national average of female officials and managers. For DuPont, that percentage is 11; the national norm is 41 percent. Fortunately, DuPont has a number of recruitment and training programs to improve these statistics in the future.

Mobil is not exactly top heavy with women—all of the company's top executives are men—but it does have two women on its board. Where Mobil excels is in its family-friendly benefits, which are by far the best in the industry. It offers a year of leave for family emergencies and to new parents, as well as excellent day care, flex time, and the ability for some employees to work at home.

Chevron, ranked fourth, boasts a board that is 20 percent female—an industry high, and good by any industry's standards. It has no women serving as senior executives, but compensates somewhat by offering multiple programs aimed at promoting female managers. Chevron also targets women with an above-average percentage of its charitable giving, has good benefits, and cooperated with our survey.

The three lowest-scoring companies are Sun, Occidental Petroleum, and Citgo Petroleum. All three refused to participate in our survey. Royal Dutch Petroleum/Shell Oil and Citgo Petroleum are zero percent club members (their boards are all male). None of them target women's causes in their charitable giving, although Royal Dutch, the parent company of Shell Oil, does purchase from women-owned companies.

Mention should be made of Phillips Petroleum. Although its overall score is in the average range, its charitable giving to women's causes is 9 percent—the highest in the industry.

Regarding gasoline, pay less attention to ads and more to how these companies treat women. There are no tigers for your tank—only gasoline, and this is one product that doesn't differ very much from one company to another. The choices are clear: Conoco, ARCO, Mobil, or Chevron.

Zero Percent Club—Gasoline

COMPANY	FEMALE DIRECTORS	TOP FEMALE OFFICERS OR TOP-SALARIED WOMEN
Chevron		0
Citgo Petroleum/PDVSA	0	0
Exxon		0
Royal Dutch Petroleum/ Shell Oil	0	

HOUSEHOLD EQUIPMENT

Appliances

Women play a decisive role in the purchase of most appliances for the home. After all, many of these gadgets have been pitched as labor-saving devices, and the labor to be shortened has typically been associated with women—washing dishes, preparing meals, cleaning and doing laundry. Most households have many electric appliances in their kitchens, laundries, bathrooms, and elsewhere.

Given this huge market, it is not surprising that there is a broad range of products to choose from in all categories. What is surprising, however, is how few appliance companies employ women in decision-making positions. We rated 21 companies that make appliances of various types, and only 3 had FEM scores above our average of 50. The average FEM rating for the industry was 42.92, suggesting there is considerable room for improvement.

	APPLIANCE COMPANIES	FEM	RANK	SOME BRAND NAMES
●	Sears, Roebuck	63.55	1	Kenmore appliances
●	General Electric	61.78	2	Hotpoint and GE dishwashers, ranges
●	United Technologies	58.62	3	Carrier heating, air conditioning
◐	Hamilton Beach/ Proctor Silex†	48.37	4	Mixers, can openers, blenders, food processors, toasters, coffee makers, irons

	APPLIANCE COMPANIES	FEM	RANK	SOME BRAND NAMES
◐	Toro	46.73	5	Snow removers, Lawn-Boy mowers, weed trimmers, outdoor lighting
◐	Whirlpool	45.95	6	KitchenAid, Roper, Whirlpool
○	Gillette	44.05	7	Braun juicers, razors, coffee makers, blenders
○	Black & Decker	42.05	8	Spacemaker, Dustbuster, Workmate, can openers, coffee makers, toasters, irons, mixers, power tools
○	Maytag	41.84	9	Maytag washers and driers, Jenn-Air, Admiral, Hoover, Magic Chef
○	Toastmaster	40.56	10	Toasters and small ovens, clocks, griddles, breadmakers, heaters
○	Masco	39.94	11	Thermador ovens, Waste King disposals
○	Sunbeam	39.92	12	Osterizers, Mixmasters, clocks, timers
○	Raytheon	38.51	13	Amana, Caloric, Speed Queen
○	Dynamics of America	37.56*	14	Acme Juicer, Waring blenders
○	Electrolux	37.35*	15	Vacuum cleaners
○	Philips Electronics	37.00	16	Vacuum cleaners, irons, shavers
○	Fedders	36.72	17	Air conditioners
○	Matsushita Electric	36.56	18	Panasonic vacuums, razors, microwaves, dishwashers, washing machines
○	Emerson Electric	35.44	19	Fans, In-Sink-Erators, power tools

	APPLIANCE COMPANIES	FEM	RANK	SOME BRAND NAMES
○	York	34.52	20	Air conditioners
○	Sharp Electronics	34.40	21	Air conditioners, heaters, microwaves, toasters, dishwashers, vacuums, washer/dryers

*Did not have scores for all five categories.
†Hamilton Beach/Proctor-Silex is owned by NACCO.

Rating Key:
● Female-friendly company; buy products when possible
◐ Average; buy with good conscience
○ Below average; avoid buying these products

From the above list most of the companies rank in the lowest category. Three companies are above average: Sears, General Electric, and United Technologies (Carrier). All three make other products besides household appliances. Four reasons why we recommend Kenmore appliances: Sears has a board that is 25 percent female, a woman among its top nine officers, a management staff that is almost 50 percent female, and excellent employee benefits. Of the other top two, GE is somewhat better than Carrier for its record of women in both middle and senior management. Carrier has a slightly higher proportion of female board representation, 17 percent versus GE's 13 percent. General Electric's total score was elevated because it has the best benefits package in the industry, and also targets an extremely high percentage, 27 percent, of its corporate contributions to women. Carrier's high score is attributable to its willingness to share salary information with us, which shows that women earn 84 percent of what men earn—much better than the national average of 72 percent.

Hamilton Beach/Proctor-Silex and Toro were ranked average, though both of their scores, 48 and 47, respectively, are on the low side for that category. Hamilton Beach rates poorly in the senior management category, and its parent company, NACCO, is in the zero percent club for board members. However, the company was willing to share salary information with us. Toro's scores are far from great, but 13 percent of its board members and 12 percent of its top executives are women—not too bad for a company that markets many of its appliances primarily to men.

If Toro is far from stellar, the companies in our lowest category are positively earthbound; some are even under water. The three lowest-scoring companies, Emerson, York, and Sharp, had scores of 34 and 35 (recall that the average is 50). Sharp and York (along with Dynamics of America, Fedders, Hamilton Beach/Proctor-Silex, Matsushita Electric, Sunbeam, and Toastmaster) have no women on their boards. At York and Emerson, women comprise only 9 and 13 percent of their managerial staffs, and both have been cited for noncompliance with equal opportunity laws.

Many of these large corporations with terrible records of female advancement advertise heavily in magazines and on television. Don't let yourself be fooled by the image they try to portray in their ads. Consider what goes on behind the scenes and don't spend your money in ways that contribute to poor treatment of women. If you need an air conditioner, we recommend Carrier. Avoid York or Fedders. If you need a washing machine or a large kitchen appliance such as a refrigerator, stove, or oven, your best choices are clearly Sears and appliances by General Electric. For small appliances, our top companies are General Electric and Hamilton Beach/Proctor-Silex. Avoid purchasing appliances from companies with FEM scores way below average; although their products may be attractive, remember that most of the companies have a discriminatory pattern toward women that should not be reinforced by your patronage.

Zero Percent Club—Appliances

COMPANY	FEMALE DIRECTOR	TOP FEMALE OFFICERS OR TOP-SALARIED WOMEN
Dynamics of America	0	0
Electrolux		0
Fedders	0	
Hamilton Beach/Proctor-Silex	0	
Matsushita Electric	0	
Sharp Electronics	0	
Sunbeam	0	
Toastmaster	0	
Whirlpool		0
York	0	

Electronic Equipment

In this section, we rate 12 companies that manufacture an assortment of electronic equipment. Even though women are frequent purchasers in this category, the companies as a group are not very positive regarding treatment of women. The group FEM average was 44.30, one of the lowest in our survey (the reader should recall that these scores were designed so that the average would be 50 for the entire group of almost 400 companies that we rated). Despite a great deal of uncooperativeness—7 of the 12 companies did not return our questionnaire, even after repeated attempts—we still managed to find some companies that we can recommend. There was no middle ground for this industry: three companies score in the female-friendly range and all others are below average.

	ELECTRONICS COMPANIES	FEM	RANK	ELECTRONIC EQUIPMENT
●	AT&T†	66.18	1	Phones, answering machines
●	Hewlett-Packard	59.66	2	Calculators, computers
●	Motorola	57.34	3	Cellular phones, pagers
○	Pioneer Electronics	43.89	4	Stereos, car navigation systems, TVs, CD players, cellular phones, computers
○	Sony	42.90	5	TVs, phones, camcorders, stereos
○	Texas Instruments	42.06	6	Calculators
○	Canon USA	38.95	7	Cameras, camcorders
○	Zenith Electronics	37.16	8	VCRs, TVs, stereos
○	Philips Electronics	37.00	9	TVs, video equipment, car navigation systems
○	Matsushita Electric	36.56	10	Panasonic portable stereos, phones, CD players
○	Emerson Electric	35.44	11	VCRs, TVs, portable stereos

ELECTRONICS COMPANIES	FEM	RANK	ELECTRONIC EQUIPMENT
○ Sharp Electronics	34.40	12	TVs, VCRs, stereos, CD players, video, electronic personal organizers

†AT&T has divided into three separate companies, but was still one company when we collected this information. AT&T is still the communications company, NCR makes computers, and Lucent Technologies produces equipment.

Rating Key:
● Female-friendly company; buy products when possible
◐ Average; buy with good conscience
○ Below average; avoid buying these products

The three companies we can recommend are AT&T, Hewlett-Packard, and Motorola, as they are the only companies that stand out in a very poor field.

AT&T does not have a huge presence in the appliance industry, focusing only on telephone equipment. (AT&T has spun off its equipment division into a separate company, called Lucent Technologies; however, when we collected our data, AT&T was still one company. Thus, we give the information we had for AT&T.) AT&T, with two women on its board (13 percent), scores well above the national average, and its proportion of women in management, 45 percent, also places it above average. Furthermore, AT&T has female advancement programs and purchases generously from women-owned businesses. It also has excellent benefits for families: good leave policies and flexible scheduling options. The company shows a strong commitment to women, making their telephones a superior choice.

Hewlett-Packard, ranked second, gets off to a shaky start in the glass ceiling category with only one woman among its 28 most senior officers. The company does, however, have more women in entry-level and middle management positions than Motorola (25 percent versus 20 percent). Motorola has one woman on its 18-member board, and 3 percent of its top officers are women. Motorola also cooperated with our survey and has received considerable prior attention for its positive treatment of women. Both companies have good benefits for their employees and multiple affirmative action recruitment and training programs for women.

It should be noted that neither Hewlett-Packard nor Motorola was in strict compliance with the Department of Labor's policies governing affirmative action in 1993. This did not stop the same governmental group from giving its highest award concerning treatment of women to Motorola in that same year! Although we can recommend Motorola as one of the companies in the electronics industry that is generally better than its competitors, we have trouble understanding why the Secretary of Labor would have given a special award to the company.

Many companies scored far lower than Motorola. There were no women on the boards of Pioneer Electronics, Canon, Sony, Philips Electronics, Panasonic (Matsushita Electric), Sharp, or Emerson while we were collecting data; Emerson elected its first female board member in 1995. Two companies, Samsung—whose parent company is Korean but still must comply with U.S. affirmative action laws with regard to its U.S. subsidiary—and Kenwood are neither government contractors nor publicly traded companies. Because they refused to cooperate with us, we were unable to obtain data for some of our categories. Their scores reflect this.

The three worst electronics companies for which we have information are Matsushita Electric (37) (Panasonic's parent company), Emerson (35), and Sharp (34). Two of the three are members of our zero percent club, (no female board members). They all have very low percentages of women in management positions, ranging from 12 to 16 percent, and none of them gives money to women's causes. Additionally, Panasonic and Sharp have been cited for affirmative-action noncompliance. We can think of no good reason to buy their products when there are so many better alternatives.

Zero Percent Club—Electronic Equipment

COMPANY	FEMALE DIRECTORS	FEMALE TOP OFFICERS OR TOP-SALARIED WOMEN
Canon USA	0	
Matsushita Electric	0	
Philips Electronics	0	0
Pioneer Electronics	0	
Sharp Electronics	0	
Sony	0	
Texas Instruments		0

FINANCE

In this section we rate three types of institutions: banks; financial services, including stock brokerages and credit card providers; and major national insurance companies. As in previous sections, we reiterate two important facts: (1) there is a lot of variation in how these companies treat women and (2) your choices can reward companies that treat women well.

Banks

There are thousands of banks in the United States, and you may prefer your local one, although it may be owned by one of the biggies. You probably also deal with some of the large national banks because they often issue your credit cards. You may also live in or near a large city where you can do your banking at some of them. Many of you also are responsible for making decisions in business settings, and although we discuss most of our business-related companies in the following chapter, the information contained in this section may also be used in conjunction with your office decisions.

We rated 11 large banking organizations. They scored slightly above average as an industry. The good news is that not a single bank rated below average on the FEM scale. On the other hand, they were not cooperative with us. Only two, Chemical Bank and Citicorp, returned our questionnaire (Chemical Bank is now part of Chase Bank). The FEM scores are shown below.

BANKING COMPANIES	FEM	RANK
● American Express	64.10	1
● Citicorp	59.75	2
● Wells Fargo	58.95	3
● BankAmerica	57.38	4
◐ Bank of Boston	52.99	5
◐ Great Western Financial	52.40	6
◐ Travelers Group (Commercial Credit)	52.38	7
◐ Chase Manhattan†	51.08	8
◐ First Fidelity Bancorp	50.60	9

BANKING COMPANIES	FEM	RANK
◐ Mellon Bank	45.29	10
◐ Bank of New York	45.26	11

†Chemical Bank and Chase Manhattan Bank completed their merger in 1996. The new entity is called Chase Bank.

Rating Key:
● Female-friendly company; buy products when possible
◐ Average; buy with good conscience
○ Below average; avoid buying these products

As seen in the table, there are two groups of banks. The first are all somewhat above average. The second group is average (defined as the range of FEM scores between 45 and 54.99).

The best score went to American Express. The company earns kudos by having women in all the right places: two on the board, 6 percent in the executive suite, 51 percent as managers, and multiple division presidents. Furthermore, the company provides good benefits and has been listed by *Working Mother* 8 years running. American Express has multiple divisions relating to credit, financial services, and travel; all of them are good choices.

Our second-place finisher, Citicorp, has also been recognized by *Working Mother* magazine. The company got off to a good start by cooperating with our research and giving an average amount to women's charitable causes. Citicorp does have women represented at all of its top echelons, although not well represented. The company does offer multiple female advancement programs.

The third position goes to Wells Fargo. It has the best female representation on its board, at 13 percent, of any company in the industry. The percentage of female senior officers (22 percent) is also impressive. The company offers reasonably good benefits, and gives to women's causes. Wells Fargo also has been recognized as one of *Working Mother* magazine's 100 best companies for working women.

There is a highlight for this industry, which is that female managers are, for the most part, abundant. In only two banking companies does the percentage of female managers and officials dip below the national average of 41 percent. And American Express, Bank of Boston, BankAmerica, First Fidelity Bancorp, and Great Western Financial all have more women than men on their managerial staffs.

The three lowest-scoring banks are First Fidelity Bancorp, Mellon Bank, and Bank of New York. As the table above indicates, however, even our low scorers are still in the average category. Mellon did poorly in the glass ceiling category, with no female women among its senior officers or top-salaried employees. The Bank of New York does have women among its top officers but earmarks none of its charitable contributions to women. And First Fidelity, with a board that is only 4 percent female, is the lowest scorer in the industry. Our recommendation is that you choose from among our top four companies, or from the banks that earn our female-friendly rating, if either your personal or business need for financial services involves any of these large banks.

Zero Percent Club—Banks

COMPANY	TOP FEMALE OFFICERS OR TOP-SALARIED WOMEN
Mellon Bank	0

Financial Institutions

We rated 25 financial institutions, mostly investment houses. Some of them are the parent companies of stock brokerage firms and others are nationally based mutual fund managers that you may already be using for investment purposes. Our results show that there are some very good ones and some very bad ones. Looking at Wall Street as a whole, the picture for women is grim: the average FEM score is below average for our total sample. Lack of cooperation was typical, and unless the firm was either publicly traded or a government contractor, it was difficult to obtain information about them. Where the company does not have the same name as the entity you would be dealing with, we have provided the name of the relevant subsidiary.

	FINANCIAL COMPANIES	FEM	RANK	SUBSIDIARY (IF APPLICABLE)
●	Calvert Group (Social Investment Fund)	67.09	1	
●	Working Assets Funding Service	64.68	2	
●	American Express	64.10	3	

	FINANCIAL COMPANIES	FEM	RANK	SUBSIDIARY (IF APPLICABLE)
●	General Electric	61.78	4	GE Capital
●	Bankers Trust New York	61.45	5	
●	McGraw-Hill	58.32	6	DRI/McGraw-Hill, Standard & Poor's
●	Transamerica	58.31	7	
◐	Travelers Group	52.38	8	Smith Barney Shearson, Primerica Financial Services
◐	ITT	51.53	9	ITT Financial Corp.
◐	Charles Schwab	48.60	10	
◐	IDS Financial	47.91	11	
◐	PaineWebber Group	47.01*	12	
◐	Merrill Lynch	46.02	13	
◐	United Asset Management	45.62*	14	
◐	Dreyfus Service	45.31*	15	
◐	DeanWitter, Discover	45.24	16	
○	Goldman Sachs Group	43.19*	17	
○	Investcorp	42.67*	18	
○	American International Group	42.59	19	
○	American General	41.88	20	
○	Lehman Brothers Holding	41.83*	21	
○	Salomon	41.15	22	
○	Morgan Stanley Group	40.97	23	
○	Fidelity Investments	40.66*	24	
○	Bear Stearns	39.34	25	

*Did not have scores for all five categories.

Rating Key:
● Female-friendly company; buy products when possible
◐ Average; buy with good conscience
○ Below average; avoid buying these products

Not surprisingly, our winners in the investment category are two companies with socially responsible agendas: Calvert Group and Working

Assets Funding Service. Working Assets is a very small group, but one that has the highest percentage of women in the top echelons. One-third of its board members, top officers, and top-salaried employees are women, as are 75 percent of its managers. Calvert Group's Social Investment Fund, which is a member of the Acacia Group, is bigger, and also has an excellent record of women in high places. Almost 40 percent of its board members are women, as are 22 percent of its top brass. It also offers good benefits to employees. Working Assets and Calvert Group not only have good policies regarding women in the workplace, they also limit their fund purchases to companies that exhibit social responsibility, such as not polluting or manufacturing weapons. Thus, if you invest with them because you are concerned about treatment of women, you get a bonus in other areas as well. Moreover, the yields for these companies' funds are generally average or somewhat above average, so you really won't have to sacrifice financial returns for your principles. These two are clearly good choices for mutual funds.

Our third-place finisher was American Express. For more specific information about the company, see the preceding section on banks.

Some of the more traditional stock brokerage houses seem to have relatively negative records toward their female employees. Their scores ranged from below average at Goldman Sachs to last-place finisher Bear Stearns. Other well-known brokerage houses with very low scores include Lehman Brothers Holding, Salomon, Morgan Stanley Group, and Fidelity Investments.

Zero Percent Club—Financial Institutions

COMPANY	FEMALE DIRECTORS
American General	0
Investcorp	0
Morgan Stanley Group	0
PaineWebber	0

Insurance Companies

Most of us buy insurance to protect our homes, cars, other possessions, and the lives of our families. We also hopefully have health insurance, usually through our employers, and, occasionally, disability and professional liability. Although you will frequently purchase such insurance from a local agency, your policy will probably be issued by one of the larger, national insurance companies. In the past, you have probably either gone along with your agent's recommendation as to the underwriter, based perhaps on the price of the premium or the strength of the agent's relationship to the company. We have rated 19 of these national companies on a different dimension, namely, how well they treat their female employees. We suggest that you review this list and ask your agent to comply with *your* recommendations the next time you need insurance. Although we recognize that price is a consideration for most of us, eight companies are above average in our ratings and we hope that you can find some good values among them. Companies are often competitive on price when they know you are shopping around for individual policies.

Our ratings are shown below. Because of the diversified nature of many large corporations, you will have seen some of these names in other categories.

	INSURANCE COMPANIES	FEM	RANK	SUBSIDIARY (IF APPLICABLE)
●	Cigna	69.98	1	Connecticut General Life, Insurance Company of North America
●	Prudential Insurance Company of America	63.78	2	
●	General Electric	61.78	3	GNA, Harcourt General Insurance
●	John Hancock	61.66	4	
●	Chubb	61.09	5	
●	Aetna Life†	59.19	6	
●	Continental	58.60	7	Continental Insurance Companies
●	Transamerica	58.31	8	
●	Sears/Allstate‡	55.18	9	Allstate
◐	Metropolitan Life Insurance	54.41	10	
◐	Travelers Group	52.38	11	Traveler's Insurance
◐	ITT	51.53	12	ITT Hartford

	INSURANCE COMPANIES	FEM	RANK	SUBSIDIARY (IF APPLICABLE)
◐	Gerber Products[§]	47.85	13	Gerber Life, Grow-up, Young People's
◐	Equitable	45.32*	14	
○	GEICO	43.46*	15	
○	Paul Revere	43.39	16	
○	American International Group	42.59	17	American Home Assurance, Lexington, American Life Insurance
○	American General	41.88	18	Franklin Life, VALIC, American General Life & Accident
○	US Life	38.16*	19	

*Did not have scores for all five categories.
†Aetna has merged with Traveler's Group.
‡Allstate was formerly a division of Sears, Roebuck.
§Gerber has been acquired by a company that is not part of our sample.

Rating Key:
● Female-friendly company; buy products when possible
◐ Average; buy with good conscience
○ Below average; avoid buying these products

As a group, insurance companies were just above average in their FEM score of 53.19. Given that a majority of the workers at insurance underwriters are women, we might have expected—or hoped—that they would have scored even higher than that.

Our top-scoring companies are Cigna, Prudential, and GE. Cigna's score gives it a clear first-place finish. It offers its workers great family-friendly benefits and has been cited three times on *Working Mother* magazine's list of the 100 best companies for women. It also has the highest percentage of female managers, 53 percent, in the industry, and has good female representation on its board and executive staff.

Prudential's percentage of women in management is below average, but the company has the highest proportion of female professionals, 72 percent, in the industry. It also has great benefits, including on-site child care, good charitable giving, and lots of previous recognition for treating women well. Prudential scored a few extra points by returning our questionnaire.

GE, although its management opportunities for women are below average, makes up some points by having many female advancement programs and a female division president. Another highlight is its charitable giving, where it targets about four times the average toward women's causes.

Mention also goes to Chubb, one of the few companies to reveal its salary information. Female managers at Chubb earn 90 percent of their male counterparts' salaries and female professionals, 97 percent. Although 100 percent seems a reasonable goal, few companies are as close to offering salary parity as Chubb. We recommend this company highly.

Some mention should also be made of another good record. Aetna, while not in the top three because its record was inconsistent, still earned the highest score for the industry in the management opportunities category because of high percentages of women in both middle management and professional categories.

At the lower end of the scale, we would urge you not to buy insurance from American International or American General. American General has an all-male board. And American International has no female advancement programs and no women among its top-salaried officials. There are much better choices than these companies.

Zero Percent Club—Insurance

COMPANY	FEMALE DIRECTORS	TOP FEMALE OFFICERS OR TOP-SALARIED WOMEN
Allstate		0
American General	0	
Equitable	0	
USLife	0	

ENTERTAINMENT

The category of entertainment is a broad one, and includes restaurants, movie theaters, magazines, newspapers, books, television channels, sports equipment, and even long-distance telephone service. If you plan to spend upcoming leisure time on vacation, see the section on travel in Chapter Six, which rates airlines, hotel chains, and rental car agencies, and our analyses of states and countries in Chapters Seven and Eight.

The sorts of goods and services found in this section are not, in many cases, as easy to control as the more plentiful product choices that present themselves in the grocery store. Most cities and towns have only one newspaper. Most people would not decide to avoid a book because it was published by a low-rated company. On the other hand, if you feel strongly about the issues discussed in this book, there are always options: borrow the book from a friend or check it out from the library. Use this information to increase your awareness of the issues involved. Remember that most companies care a great deal about their public images, and you are always free to write letters and tell your friends about how the companies measure up.

Restaurants

We rated 12 companies that own restaurants. Primarily, the companies we have chosen own chains or fast-food restaurants that are found all over the country. The food they sell ranges from hamburgers to chop suey.

	RESTAURANT COMPANIES	FEM	RANK	RESTAURANTS
●	PepsiCo	63.45	1	Pizza Hut, Taco Bell, Kentucky Fried Chicken (KFC)
◐	Wendy's	54.10	2	Wendy's
◐	General Mills	54.04	3	Olive Garden, Red Lobster, China Coast
◐	Grand Metropolitan	53.73	4	Burger King
◐	McDonald's	53.62	5	McDonald's
◐	IHOP	48.73*	6	International House of Pancakes (IHOP)
◐	Flagstar	45.40	7	Denny's, Quincy's, El Pollo Loco
○	Domino's Pizza	44.51	8	Domino's
○	Allied Domecq†	42.04	9	Dunkin' Donuts, Baskin Robbins
○	Imasco USA	41.03	10	Hardee's
○	Int'l Dairy Queen	40.90	11	Dairy Queen, Orange Julius, Karmelkorn, Golden Skillet

RESTAURANT COMPANIES	FEM	RANK	RESTAURANTS
○ Triarc Companies	40.89	12	Arby's

*Did not have scores for all five categories.
†Formerly Allied Lyons.

Rating Key:
● Female-friendly company; buy products when possible
◐ Average; buy with good conscience
○ Below average; avoid buying these products

Only one company, PepsiCo, earns our female-friendly rating. PepsiCo, which owns Pizza Hut, Taco Bell, and Kentucky Fried Chicken restaurants in addition to its soft drink operations and Frito Lay snack foods, is one of the few companies in our entire sample that posts above-average scores in all five categories. About 13 percent of the company's senior and highest-salaried officers are women, which is well above the national average. The company targets an average amount of its charitable donations to women, purchases goods and services from women-owned businesses, and cooperated with our research. PepsiCo also had an industry-high score in the benefits category for offering extended leave for new parents (some of it paid), supporting day-care options for employees, and endorsing flex time and part-time return schedules.

If it's a burger you're craving, the three biggies—McDonald's, Burger King, and Wendy's—all score in the average category. Wendy's beat the other two in the rankings, but just barely. The patterns of scores differ somewhat. Wendy's has the most impressive record of women in high places, with two female board members out of 12, and was the only one of the three to cooperate with our research. McDonald's scores the best in the middle management category, with 46 percent female representation. It has also received quite a bit of prior recognition as a female-friendly company. Grand Metropolitan scores the highest in the benefits category (after PepsiCo), but McDonald's is a close third.

We had incomplete information for International House of Pancakes (IHOP) but some of the information we do have suggests the restaurant has some promise. Thirteen percent of its board is female, and it has an industry-high representation of women, 25 percent, among the company's top management. Had IHOP cooperated with our study, they might have scored much better, as both of those scores are well above average.

Our below-average companies include Domino's Pizza, Allied Domecq (Dunkin' Donuts and Baskin Robbins), Imasco (Hardee's), Dairy Queen, and Triarc (Arby's). Domino's, although it did respond to our survey, declined to provide information about how many women are in management or in executive positions. Domino's founder and 98 percent owner has extensively supported anti-choice causes. The company's unchallenged growth during the 1980s brought a large personal fortune to the president, who has taken a strong anti-abortion stance. That stance led the National Organization for Women to urge a boycott of Domino's during the late 1980s, a position they still maintain.

Zero Percent Club—Restaurants

COMPANY	FEMALE DIRECTORS
Allied Domecq	0
Triarc Companies	0

Telephone Services

The three companies rated here—AT&T, MCI Communications, and Sprint—as an industry earn the highest average FEM score of any industry considered in this book (61.83). All three, as you know from the constant barrage of mail and phone calls most of us get, are hungry for your business. Obviously, rates and special programs—some of which require you and the people you call to subscribe to the same company—will play a role in your decision about which company to patronize.

	TELEPHONE COMPANIES	FEM	RANK
●	Sprint	70.35	1
●	AT&T†	66.18	2
◐	MCI Communications	48.97	3

†AT&T has divided into three separate companies, but was still one company when we collected this information. AT&T is still the communications company, NCR makes computers, and Lucent Technologies produces equipment.

Rating Key:
● Female-friendly company; buy products when possible
◐ Average; buy with good conscience
○ Below average; avoid buying these products

Sprint and AT&T receive comparable—and very female-friendly—scores. That dime a minute you spend on Sprint long distance service is money well spent in support of women. Sprint's board of directors has a slightly higher percentage of women, 15 percent versus 13 percent for AT&T, but each of the three companies has two women on their boards. AT&T has a higher percentage of female managers, 42 percent to Sprint's 38 percent. AT&T also has more female advancement programs than Sprint, but they tie for women among top-salaried officials, at 8 percent. Sprint also has two female division presidents. Both companies buy goods and services from women-owned businesses and have excellent benefits for families: good leave policies and flexible scheduling options for employees; neither company has on-site day care. All in all, both AT&T and Sprint show a strong commitment to their female employees and earn our highest rating. Sprint scores a little higher on the FEM scale, but either one is an excellent choice.

MCI Communications received an average rating on the FEM scale. MCI actually has a higher percentage of women on its board and among its top executives than either Sprint or AT&T. MCI lost some points for failing to cooperate with our research and for giving below-average amounts of its charitable donations to women's groups. It also has not received the same kudos from other rating sources as the other two providers.

Sports Equipment

For this category, we rated six national companies that make sports gear such as golf balls, tennis racquets, bikes, and bowling balls. Many sports equipment manufacturers do not distribute nationally and thus we did not include them. If you are interested in athletic shoes or sportswear, see the sections on shoes and clothes in Chapter Four.

This group of companies was remarkably unresponsive to our research. In fact, only one company responded to our survey. As a result, our efforts to rate these companies were plagued by a real lack of information. What we did find out about them, however, indicates that this industry is sorely lacking in female representation: the average FEM score for the industry is a very low 43.15.

SPORTING GOODS COMPANIES	FEM	RANK	EQUIPMENT MADE
◐ Huffy	49.50	1	Bicycles, basketball backboards and accessories
○ Brunswick	44.59	2	Boats, boat motors, and bowling, fishing, golf, and billiards equipment
○ American Brands	43.10	3	Titleist golf equipment (balls, shoes, gloves, and clubs)
○ Head Sports	42.19*	4	Head, Tyrolia ski equipment, tennis gear
○ Black & Decker	42.05	5	Dynamic Gold and Dynalite Gold golf clubs
○ Prince Manufacturing	37.47*	6	Tennis equipment

*Did not have scores for all five categories.

Rating Key:
● Female-friendly company; buy products when possible
◐ Average; buy with good conscience
○ Below average; avoid buying these products

It is hard to recommend any of the sporting goods companies rated here; all but one received our lowest rating. In any event, Huffy is at the top of the list. Huffy scores considerably below average in the management opportunities category, with only 21 percent female managers, or less than half the national average. It does, however, earn a respectable score in the glass ceiling category. Huffy's board has two women on it, and 25 percent of the company's highest-paid officers are women, a high for this industry and a good score by any industry standard. There is also a woman running a division, and in the area of benefits, employees may make arrangements to work at home. These factors make Huffy a relatively female-friendly choice.

Brunswick, rated second, is one of the world's largest manufacturers of recreational boating equipment with such brands as Mercury, Mariner, Force, Bayliner, and Sea Ray. It is also involved in bowling (Brunswick), billiards, fishing (Zebco), and golf. Its record regarding women is very different from others in this industry. In the first two categories, involving women in management, Brunswick is most unimpressive. Its proportion of female managers, at 12 percent, is an industry low and there were no women on the board until 1993. However, it makes up some points in the areas of benefits—employees may arrange flexible scheduling and locations for their work—and cooperation: Brunswick was the only company in the industry to cooperate with our research. It is also the only company in the industry that gives an average amount to women's causes.

American Brands, once primarily in the tobacco industry, also has a presence in sporting goods, namely, with Titleist golf equipment. American Brands' numbers are similar to Huffy's: one woman on the board and 21 percent female representation among its managers (although none of its most senior managers).

Prince Manufacturing and Head Sports, both best known for their tennis equipment (although Head, parent of Tyrolia, now has a large presence in the ski industry as well), are also below average, as is Black & Decker. Each company has a proportion of women managers—18 percent for Black & Decker, 19 percent for Prince, 20 percent for Head—that is comparable with other companies in the industry but well below the national average. None of the companies gives generously to women's causes, nor have they been recognized by any of the other agencies or publications we consulted.

Publishing

We present three different types of publishers here: magazines, books, and newspapers. Magazine publishers make for a good comparison; in most cases, magazines on the same subject are published by competing companies, giving the consumer good choices. Book and newspaper publishers are clearly a little trickier as you often don't have an obvious alternative.

The good news is that women are for the most part very well represented in both middle and top management in the publishing industry. Very few of these companies found their way into the zero percent club.

Following are the rankings for the publishers, classified according to the three categories. Some of the companies are diversified and appear on more than one list.

	BOOK PUBLISHING COMPANIES	FEM	RANK	SOME LABELS
●	Time Warner	67.40	1	Book-of-the-Month, Time-Life Books
●	Walt Disney	60.12	2	Hyperion Press
●	Reader's Digest	59.58	3	Reader's Digest Association
●	McGraw-Hill	58.32	4	McGraw-Hill, Osborn Books
●	Capital Cities/ABC†	57.61	5	Chilton Guides
●	Johnson Publishing	57.34*	6	Johnson Publishing Co. Book Division
◐	Harcourt General	53.86	7	Harcourt Brace & Co.
◐	Houghton Mifflin	49.95	8	Houghton Mifflin
◐	Times Mirror	49.94	9	CRC, Jeppesen Sanderson
◐	Advance Publications	49.13	10	Random House, Fawcett
◐	Hearst	47.85*	11	Avon, Hearst, William Morrow
◐	Viacom	44.86	12	Simon & Schuster, Pocket Books, Prentice Hall
○	Pearson	43.68*	13	Penguin, Viking, Longman, Addison-Wesley, Signet
○	Bertelsmann	43.40	14	Bantam Doubleday Dell
○	Western Publishing	42.89*	15	Golden, Golden Books (kids)

	MAGAZINE PUBLISHERS	FEM	RANK	SOME MAGAZINES
●	Time Warner	67.40	1	*Time*, *People*, *Sports Illustrated*
●	MacDonald Communications‡	62.44*	2	*Ms.* Magazine, *Working Mother*

	MAGAZINE PUBLISHERS	FEM	RANK	SOME MAGAZINES
●	Walt Disney	60.12	3	*Discover*
●	Reader's Digest	59.58	4	*Reader's Digest*
●	Condé Nast	58.50	5	*Vogue, Vanity Fair, Gourmet, Glamour*
●	McGraw-Hill	58.32	6	*Business Week, BYTE*
●	Johnson Publishing	57.34*	7	*Jet, Ebony*
◐	Times Mirror	49.94	8	*Outdoor Life, Popular Science, Ski*
◐	Dow Jones	49.71	9	*American Demographics, Barron's*
◐	Washington Post	48.68	10	*Newsweek*
◐	Hearst	47.85*	11	*Cosmopolitan, Town & Country*
◐	New York Times	47.05	12	*Snow Country, Sailing World*
◐	Wenner Media§	45.80*	13	*Rolling Stone, US*
○	Meredith	44.04	14	*Ladies' Home Journal, Better Homes & Gardens*
○	Bertelsmann	43.40	15	*Parents, YM, Family Circle, McCall's*

	NEWSPAPER PUBLISHERS	FEM	RANK	SOME PAPERS
●	Gannett	79.35	1	*USA Today*
●	Knight-Ridder	59.07	2	*Miami Herald, Philadelphia Inquirer*
●	Capital Cities/ABC†	57.61	3	*Kansas City Star, Fort Worth Star Telegram*
◐	Times Mirror	49.94	4	*Los Angeles Times, Newsday*
◐	Dow Jones	49.71	5	*Wall Street Journal*
◐	Advance Publications	49.13	6	*The Oregonian, The* (New Orleans) *Times-Picayune*
◐	Washington Post	48.68*	7	*Washington Post*

	NEWSPAPER PUBLISHERS	FEM	RANK	SOME PAPERS
◐	Hearst	47.85*	8	*Houston Chronicle, Seattle Post-Intelligencer*
◐	New York Times	47.05	9	*The New York Times*
○	E.W. Scripps	41.86*	10	*Cincinnati Post, Rocky Mountain News*
○	Media General	38.35	11	Garden State Newspapers

*Did not have scores for all five categories.
†Capital Cities/ABC has been acquired by Walt Disney.
‡Formerly Lang Communication.
§Formerly Straight Arrow Publications.

Rating Key:
● Female-friendly company; buy products when possible
◐ Average; buy with good conscience
○ Below average; avoid buying these products

In newspaper publishing, there is one star: Gannett. Women make up 33 percent of its board, 34 percent of its top officers, and 27 percent of its top-salaried officers. Gannett offers a plethora of family-friendly benefits, including some extended maternity leave, on-site day care, flexible scheduling options, and education programs on family, maternity, wellness and child care. The company has been recognized by such organizations as *Working Mother* magazine (where Gannett has been listed among the 100 best companies to work for every year the list has been published), the Domini Social Index, the National Council of Adoption, and *Black Enterprise* magazine. Placing fourth in our total sample of almost 400 companies, Gannett was extremely cooperative with our research and also has several female division presidents. Gannett's *USA Today*, as well as its regional papers, are top-notch choices for the female-minded newspaper reader.

At the other end of the scale in the newspaper business are the companies that make up our zero percent club. They are: Capital Cities/ABC, which places third on our list but has no female executive officers (it does excel in other areas, however, and has also merged with Walt Disney, a company with a high FEM score); E.W. Scripps, which has an all-male board; and Media General, which has no women in its highest management or on its board. Scripps and Media General both receive below-average scores. If you read any of their newspapers (which are listed in the "Buying Guide"), you might consider a letter to the editor or publisher letting them know what you think of their policies toward women.

In the book publishers category, Time Warner, Walt Disney, and Reader's Digest were the highest scores. Thirteen percent of Time Warner's board and 11 percent of its senior officers are women. It has *three* female division presidents (of *People*, Elektra Entertainment, and DC Comics), gives an average amount to women's groups, and has excellent benefits. Walt Disney, which was separate from Capital Cities/ABC when we collected our data, scored well with strong female management, good benefits, and ample prior recognition. Reader's Digest Association has 20 percent female representation on its board, an above-average number of women among its top officers (a strong 36 percent), and an above-average number of female managers (43 percent). Books published or sold by Reader's Digest, Book-of-the-Month Club, Hyperion Press, Little, Brown, Time-Life, and Warner Books as well as Chilton Guides are good, female-friendly choices.

There are six magazine publishers—three of which publish many titles—that make it into our female-friendly category and offer lots of good choices for buying magazines of every kind. They are Time Warner, Lang Communications, Walt Disney, Condé Nast, Reader's Digest, and McGraw-Hill. Time Warner and Disney were discussed earlier in this section.

Lang Communications, publisher of *Ms.* Magazine, *Working Mother*, and *Working Woman*, earned a second-place ranking, but it has recently been acquired by MacDonald Communications. We were unable to find information for the glass ceiling category, which includes board of directors and senior officers, but the company does have two outstanding points, namely, it has 75 percent female managers and gives 100 percent of its charitable donations to women. A telephone interview with MacDonald Communications indicated that its organization was very similar to Lang's, but we were unable to check out their charitable giving. As we rely on *Working Mother* to recognize female-friendly companies, we were glad its parent company started its quest for outstanding companies close to (or at) home.

Condé Nast (which is a subsidiary of Advance Publications but is rated separately because it has independent management) has 61 percent female managers (the national corporate average is 41 percent) and an 80 percent female professional staff. In the upper echelons, 19 percent of the company's most senior positions are held by women. Condé Nast offers good benefits, including some paid and extended parental leave, plenty of flexible scheduling possibilities, and many education and referral programs regarding child care, adoption, elder care, and families. It made *Working Mother* magazine's "100 Best" list for the first time in 1994.

McGraw-Hill is comparable to Condé Nast in the glass ceiling category: 14 percent of the company's top positions are held by women, as are 14 percent of the highest-paid positions. McGraw-Hill excels on benefits: employees may take up to 6 months off to care for newborns or ill family members, may take advantage of flexible scheduling and work-at-home options, and have access to all sorts of education programs through work.

Given the strong performance of these companies, their magazines are better choices than the competition. Time Warner's *Time*, for example, is published by a much more female-friendly company than *Newsweek*. Condé Nast's *Vogue*, *GQ* and *Glamour* are better choices than *Cosmopolitan*, *Harper's Bazaar*, or *Esquire*.

Other excellent magazine choices for consumers include: *Business Week*, *BYTE*, *Ms.*, *Life*, *Fortune*, *Money*, *People*, *Entertainment Weekly*, *Sports Illustrated*, *Working Mother*, *Bride's*, *Vanity Fair*, *Parade*, *Gourmet*, *Self*, *Woman*, *Mademoiselle*, *New Yorker*, and *Reader's Digest*.

Zero Percent Club—Publishing

COMPANY	FEMALE DIRECTORS	FEMALE TOP OFFICERS OR TOP-SALARIED WOMEN
Bertelsmann		0
Capital Cities/ABC		0
E.W. Scripps	0	
Harcourt General	0	
Media General	0	0
Walt Disney		0
Western Publishing		0

Broadcasting

This industry is one of the most difficult areas for consumers to express themselves through purchases. Generally, people watch the shows they like, regardless of what channel they appear on. When you decide to subscribe to cable, you usually have one choice, namely, the system provider in your area. And with the exception of some of the premium movie channels, most cable channels come as packages, meaning the buyer does not have much say about which channels she would like—or would not like—to subscribe to. Nonetheless, we include our data on both cable and network channel parent companies and cable systems operators.

BROADCASTING COMPANIES	FEM	RANK	CHANNELS OR SUBSIDIARIES†
● Time Warner	67.40	1	HBO, Cinemax
● General Electric	61.78	2	NBC
● Walt Disney	60.12	3	Disney Channel
● Capital Cities/ABC‡	57.61	4	ABC
◐ Comcast	54.34	5	Comcast systems
◐ National Public Radio	54.00*	6	National Public Radio
◐ Public Broadcasting System	53.35*	7	PBS
◐ Tele-Communications	51.04	8	TCI, Heritage, WestMarc
◐ CBS	51.00	9	CBS
◐ Turner Broadcasting§	50.35	10	CNN, TNT, TBS
◐ Times Mirror	49.94	11	Community, Dimension cable
◐ Hearst	47.85*	12	ESPN, A&E, Lifetime (all partial)
○ Viacom	44.86	13	MTV, VH-1, Showtime, Nickelodeon
○ Meredith	44.04	14	Meredith
○ E.W. Scripps	41.86*	15	Scripps cable systems
○ QVC Network	38.51	16	QVC

*Did not have scores for all five categories.
†Sometimes these companies own small pieces of cable channels; if it was a small percent, such as less than 10 percent, we have not mentioned it.
‡Capital Cities/ABC has been acquired by Walt Disney.
§At press time, Time Warner and Turner Broadcasting were in the process of merging.

Rating Key:
● Female-friendly company; buy products when possible
◐ Average; buy with good conscience
○ Below average; avoid buying these products

The good news for couch potatoes is that two of the three television networks, NBC and ABC, earn our highest rating. General Electric, parent company of NBC, has two female board members; CBS and Capital Cities/ABC each had one. Capital Cities/ABC, however, is now part of Disney, which scores very well on the FEM chart. General Electric, despite having a very low percentage of female managers (17 percent compared with the national average of 41 percent), excels in the area of charitable giving. GE gives 27 percent of its donations to women's causes, or five times the average. It also has a female subsidiary president. CBS has 39 percent female managers and a female division president, and has also been recognized by *Business Week* as a "pacesetter on women's issues." CBS falls short in the charitable giving category and earned only an average score for benefits.

Disney's glass ceiling numbers are not quite as strong as Time Warner's: 7 percent of Disney's board is female as are 43 percent of the company's managers and officials. Disney was very cooperative with our requests for information, and we were pleased to see that this company, whose products all target families and children, shows a strong commitment to family life in its benefit programs. Employees who are new parents get an additional month of leave beyond what is mandated by the Family Leave Act, and employees have access to on-site day care. When choosing premium cable channels, Disney, HBO and Cinemax are top choices.

In addition to GE (NBC and Disney), the top scorer in this industry is Time Warner, which owns HBO and Cinemax. These types of cable channels are termed premium, meaning you have to pay extra for them when you subscribe to cable; you decide whether or not to support these companies.

As mentioned earlier, two of Time Warner's board members, or 13 percent, are women, as are 11 percent of the company's most senior officers. It offers excellent benefits, including on-site day care, up to 1 year of parental leave, and many educational and other beneficial programs to help families. It has been listed five times on *Working Mother* magazine's "100 Best." At HBO, one-quarter of the senior vice presidents and almost one-third of all vice presidents are women.

We had incomplete information for National Public Radio, Hearst, Public Broadcasting System, and E.W. Scripps. The first three, however, show significant promise. Thirty-five percent of NPR's board and 33 percent of its highest-paid officials are women; 26 percent of PBS's board are women; and Hearst has three women on its 13-member board (or 23 percent) and gives an above-average amount to women's causes. Unfortunately, without more complete information, these companies find themselves near the bottom of the industry in terms of ratings. With PBS and NPR, it was particularly disappointing that neither completed our survey; after all, they have no problem asking us for money during their fund-raising drives. We have no qualms with the programming of either station, and in fact they both contribute quality entertainment and news to a sometimes lacking industry. On the other hand, any company that expects you to sit down and write a check should, we feel, be more forthcoming with details about itself.

Our bottom companies are Scripps and QVC Network. Both companies have all-male boards of directors. QVC's low score was especially disappointing because its content is targeted so strongly toward women. Although the company did not respond to our questionnaire, we do know that it gives nothing to women's causes. You have the option of not purchasing any of the plethora of goods sold on the network.

Zero Percent Club—Broadcast

COMPANY	FEMALE DIRECTORS	FEMALE TOP OFFICERS OR TOP-SALARIED WOMEN
Capital Cities/ABC		0
E.W. Scripps	0	
QVC	0	
Walt Disney		0

Miscellaneous Entertainment

This category is a broad one, including some leisure time pursuits that don't fit under any other headings. Below are the different fields we rated, along with parent companies and ratings.

	ENTERTAINMENT COMPANIES	FEM	RANK	WHAT IT OWNS
●	Time Warner	67.40	1	Six Flags, Atari, Atlantic, Elektra

	ENTERTAINMENT COMPANIES	FEM	RANK	WHAT IT OWNS
●	Walt Disney	60.12	2	Disney World, Disneyland, Hollywood Records
◐	Anheuser-Busch	51.42	3	Sea World, Busch Gardens
○	Viacom	44.86	4	Blockbuster Video, Paramount Theme Parks
○	Bertelsmann	43.40	5	RCA, Windham Hill
○	Sony	42.90	6	Columbia, Epic, Loews Theatres
○	Philips Electronics	37.00	7	Motown, A&M, Polygram
○	Matsushita Electric	36.56	8	MCA, Geffen, Uptown

Rating Key:
● Female-friendly company; buy products when possible
◐ Average; buy with good conscience
○ Below average; avoid buying these products

Of the eight companies rated, Time Warner was the undisputed number one. Because Time Warner is such a diversified company, its performance gives people good choices in many fields. As noted before, two of Time Warner's 15 board members (13 percent) are women, as are 11 percent of the company's most senior officers. Although women do not find their way into the highest-paid category, the company does have an average number of female managers and gives an average amount of its charitable donations to women's groups, with an emphasis on breast cancer and a project concerning women in jail. It has excellent benefits, including on-site day care, up to 1 year of parental leave, and many educational and other beneficial programs to help families.

Walt Disney comes in second in our ratings and is the only other company in this industry to earn a female-friendly rating. As mentioned earlier, 7 percent of Disney's board is female. Like Time Warner, Disney has no women among its highest-paid officers, but it does have a slightly above average proportion of female managers (43 percent). Disney also has a program to create business opportunities for minority- and women-owned companies and a program to increase women among its screenwriters. Because the company's theme parks in Florida and California are targeted toward families, it was comforting to see that the company has strong benefits: new parents get an additional month of leave beyond what is mandated by the Family Leave Act, and employees have access to on-site day care.

At the bottom of the list are found Sony, Paramount, Philips, and Matsushita Electric. Sony and Matsushita Electric own multiple record labels: Geffen, Uptown, and GRP are owned by Matsushita Electric's MCA, and Sony owns Columbia and Epic. Both are Japanese and have no women in upper management. Sony does have some women in its management ranks, 29 percent, or about twice as many as Matsushita Electric. The remaining record labels, Bertelsmann, parent company of RCA and Arista record labels, and Disney's Hollywood Records are better choices than either Sony or Matsushita Electric.

Zero Percent Club—Entertainment

COMPANY	FEMALE DIRECTORS	TOP FEMALE OFFICERS OR TOP-SALARIED WOMEN
Bertelsmann		0
Matsushita Electric	0	
Philips Electronics	0	0
Sony	0	
Walt Disney		0

SUMMARY

The grocery store is an easy place to exercise your power as a consumer. You always have choices. Competing brand names exist in every category, and usually you have clear choices between companies that treat women well and those that seem indifferent at best.

The products in the categories discussed in this chapter present a greater challenge. In some cases, they are big-ticket items, such as a car or a washing machine; in other cases, they aren't items you have direct choices about, like cable TV. Large purchases most often require a lot more thought and research than simply picking out a can of baked beans, and our hope is that the information in this chapter will create an additional factor for you to consider before you make a purchase. All of these companies want your business and your money; our hope is that they will learn they must work a little harder on behalf of their female employees if they want to earn those dollars.

Some of the companies listed may be your only choices for a purchase, and others may seem irrelevant to your consideration, such as the record label of a band whose album you want to buy. If any company conducts itself in a way that bothers you, write letters or tell your friends. In this way, these companies may realize that smart leadership—as in having women help run companies that sell a lot to women—means smart business.

Part Three

Women as Business Spenders

CHAPTER SIX

Making Female-Friendly Consumerism Work from Work

Rating Products and Services Related to the Office

The only job for which no woman is or can be qualified is sperm donor.

Wilma Scott Heide

If women can sleep their way to the top, how come they aren't there? There must be an epidemic of insomnia out there.

Ellen Goodman

This chapter concentrates on women at work. More specifically, it provides information that allows informed decisions to be made about purchases for the office. We include information about office equipment, supplies, and services, including computers, and about business travel plans (airlines, hotel chains, and rental car agencies).

Another relevant issue is company banking and insurance. And, since people at work sometimes have fun too, we encourage you to review our restaurant ratings before going out to lunch with your co-workers. For more information about these subjects, see the finance and restaurant sections of Chapter Five.

In helping you to make informed decisions about these big-ticket items and expenses, we hope to reward those companies that have made an effort to create a female-friendly environment at the office and to encourage those companies that haven't yet created better work environments for their female employees to get started.

TRAVEL

Women make up an increasing percentage of business travelers every year. In addition, even when it is not women who will be traveling, they may very well make the arrangements, which means they have the power to choose various airlines, hotels, and rental car companies. Obviously, schedules, destinations, price, mileage club memberships, and habit are the biggest factors in making travel plans. But as anyone who has traveled recently knows, these companies are hungry for your business, especially if you're a business traveler, paying higher fares. Your choices when making reservations can make a difference.

Airlines

Although things are beginning to change, there was a time when travelers might have thought that airlines were a woman's business. After all, until fairly recently, most flight attendants and ticket agents were women. In the event that you extrapolated and created an incorrect image of the more senior positions at the airlines' headquarters, let us set the record straight.

Of the 11 airlines surveyed, only American Airlines made it into our highest-rated category. American is the only company that has women on its board, among its most senior officers, *and* among its top-salaried officers. Furthermore, not one airline has a percentage of female managers that is even *average*. And only two are in perfect compliance with EEOC filings or complaints. Many also have a history of—or in some cases may still be involved in—complaints made by flight attendants over weight and appearance regulations. In some cases, these weight regulations required women returning from maternity leave to meet prepregnancy requirements before they could work again.

On the positive side, one airline was founded by a woman and three others have female division or subsidiary presidents.

Below are the companies surveyed and their ranking.

	AIRLINE COMPANIES	FEM	RANK
●	American Airlines	64.49	1
◐	America West Airlines	49.60	2
◐	Southwest Airlines	48.09	3
◐	USAir Group	47.48	4
○	United Airlines	45.67	5
○	Kimberly Clark (Midwest Express)	44.79	6
○	Alaska Airlines	42.92*	7
○	Continental Airlines	40.92	8
○	Trans World Airlines	40.26	9
○	Delta Air Lines	38.49	10
○	Northwest Airlines	35.04	11

*Did not have scores for all five categories.

Rating Key:
● Female-friendly company; buy products when possible
◐ Average; buy with good conscience
○ Below average; avoid buying these products

American was far and away our industry leader, with a FEM score of 64, nearly 15 points higher than the nearest competitor. In 1992, the airline won the Catalyst Award for programs and policies that foster women's upward mobility (Catalyst is a nonprofit group that researches women's work and family issues). Although American still has plenty of room for improvement in advancing women—along with most of the other corporations in our sample—it has made steps in all of the areas we rate. At the company's upper levels, the airline has two female subsidiary presidents and women among its top officers and highest-paid officials. Two women are on the board of directors. At the middle management level, 29 percent of American's managers are women—well below the national average but slightly above the industry average. American has the most family-friendly benefits in the industry: women may take extended, job-guaranteed maternity leave, and all employees may make use of on-site day care and flexible job scheduling, including job sharing and part-time returns from leave. The benefits program also includes many family education programs, such as prenatal education, breast health awareness, and wellness and family counseling. All of these factors define American as a company that has made an effort to promote women and to help them balance work and family. If American flies where you are going, we recommend giving this airline your business.

America West Airlines, Southwest Airlines, and USAir, ranked second through fourth, all earn average ratings on the FEM scale. America West excels on benefits: the company offers on-site day care, part-time return after childbirth, work-at-home options, pretax spending accounts, job sharing, flex time, and family counseling. Those benefits have earned the airline a spot on *Working Mother* magazine's "100 Best" list 2 years in a row. America West did not do quite as well in upper management, where only one woman is on the company's 15-member board. However, 18 percent of the company's most-senior officers and 33 percent of the company's managers are women, which is below average nationally but good for this industry. The airline's commitment seems to be stronger than its competitions', making it a solid choice for travel.

Southwest Airlines, whose routes are centered in California and the Southwest, deserves mention. A strong point is that Morris Air, acquired by Southwest, was founded by a woman, a distinction not shared by any of its competition. Morris founder and former CEO June Morris is now a member of Southwest's board of directors. Furthermore, it has the highest percentage of female managers, 39 percent, in the industry. Those points, combined with the inexpensive fares the company is known for, make Southwest a solid travel choice.

USAir isn't as strong category by category as American, but it does have some good points. There is one woman on USAir's board of directors and 14 percent of its top officers are women. USAir has about the same percentage of female managers as American and some important family benefits, although those benefits are not as extensive as some other airlines'. USAir received credit for being the first carrier to voluntarily settle an EEOC suit over weight standards for flight attendants. The women filing the suit claimed they were denied employment or had their jobs suspended because of discriminatory weight standards—a complaint that is common in the industry. Until recently, many airlines required flight attendants to meet certain weight standards, citing the need to maneuver through small spaces on the plane. Unfortunately, these standards accounted neither for agile women who weigh in heavier nor, often, for women returning from maternity leave. USAir settled the suit and agreed to substitute a performance standard for the weight regulations.

The one airline for which we had incomplete information is Alaska Airlines. We do know that *Condé Nast Traveler* gives the airline its top rating for comfort: the planes are configured with fewer seats for more leg room and the airline spends about twice the industry average on meals. We also know that the president of subsidiary Horizon Air is a woman. Given this information, we can say that the airline shows promise.

The airlines that made the bottom of our list are Trans World Airlines, Delta, and Northwest. Northwest's board has no women, a distinction shared only by United Airlines. None of the three cooperated with our study and none disclosed information about charitable giving. Although Northwest and Delta have an average (for the industry) number of female managers, TWA has only 12 percent women among its management. Delta and Northwest both have an exceedingly small percentage of women who qualify as professionals within the company—2 percent for Delta and 4 percent for Northwest. To its credit, TWA scores an industry high for this measure, with a professional staff that is 35 percent female. Clearly, some other airlines have made better strides toward women's equality.

Zero Percent Club—Airlines

COMPANY	FEMALE DIRECTORS
Northwest	0
United Airlines	0

Lodging

Having addressed the question of how to get you and your fellow travelers where you're going, we now turn to the question of where to stay once you get there.

Our top rating was earned by three hotel companies, two of which are large chains that have hotels all over the country, giving female-minded consumers plenty of choices. As an industry, lodging has some good points.

Although the industry's average FEM score is low, the hotel chains were, for the most part, extremely cooperative: seven of the ten surveyed returned our questionnaire.

Regarding hotels abroad, you should be aware that some hotel chains are owned by one company in the United States and by a different company internationally. We have tried to match the parent companies with the hotel chains they own domestically.

Below are how the companies rated and their ranking.

	HOTEL COMPANIES	FEM	RANK	HOTELS OWNED
●	Manor Care	66.86	1	Clarion, Quality Inn, Comfort Inn, Rodeway Inn, Econo Lodge, Sleep Inn, Friendship Inn
●	Host Marriott	61.13	2	Marriott, Courtyard, Residence Inn
●	Walt Disney	60.12	3	Walt Disney World, Disneyland
◐	Hyatt Hotels	54.47	4	Hyatt, Grand Hyatt, Hyatt Regency, Hawthorne Suites
◐	Renaissance Hotels	50.17	5	Renaissance Hotels, Ramada International
◐	Promus Hotel	49.85	6	Embassy Suites, Hampton Inn, Homewood Suites
◐	HFS†	49.19	7	Days Inn, Howard Johnson, Ramada‡, Super 8, Travelodge, Knights Inn
◐	Hilton Hotels	45.84	8	Hilton, Hotel Conrad, Waldorf-Astoria
○	Westin Hotel	43.97*	9	Westin
○	ITT Sheraton	40.31*	10	Sheraton

*Did not have scores for all five categories.
†Formerly Hospitality Franchise Systems.
‡Ramada hotels in the U.S. are leased to Hospitality Franchise from Renaissance.

Rating Key:
● Female-friendly company; buy products when possible
◐ Average; buy with good conscience
○ Below average; avoid buying these products

Two of the three highest-rated companies are large hotel chains, namely, Manor Care and Host Marriott.

Manor Care, rated first in the industry, gives travelers several good choices for accommodations, targeting a range of budgets, with its Choice Hotels, Clarion, Quality Inn, Comfort Inn, Sleep Inn, Rodeway Inn, Econo Lodge, and Friendship hotels. Manor Care's top score is in the management opportunities category: 72 percent of the company's managers and 87 percent of its professionals are women. Manor Care also distinguished itself by being one of the very few companies to send salary equity information. Female managers earn slightly below average compared with their male counterparts, at 65 percent, and female professionals fare somewhat better, at 85 percent. Although the salary information may not be impressive, it is just below the national average of 72 percent in one case and above average in the other. Manor Care, which also owns a large number of nursing homes around the country, has one female division president.

Host Marriott, which owns Courtyard, Residence Inn, Fairfield Inn, and, of course, Marriott Hotels, has good female representation on its board. One of the company's division presidents is a woman. Marriott received the highest benefits score in the industry: Parents have access to on-site day care, job sharing, part-time return after childbirth, and work-at-home options, in addition to family counseling, wellness programs, and pretax spending accounts. Marriott is the only hotel chain surveyed that has appeared on *Working Mother* magazine's "100 Best" list.

Finishing in third is Walt Disney. Although Walt Disney World may not be a frequent destination for business gatherings, Disney is a major player in the resort industry, and has shown an excellent commitment to women. Disney has an above-average number of female managers and excellent benefits, including on-site day care. The company received bonus points by sending us far more information than we requested as part of our survey. The company also has been involved in an enterprise program since 1981 that creates business opportunities for women and minorities.

A second company that provided salary information was Renaissance Hotels, parent of Ramada International Hotels (the U.S. Ramada hotels are leased by HFS, which scored close to Renaissance on the FEM scale). Renaissance placed in our average category. Its numbers—there are no women on the company's board and women are lacking at the company's senior levels—suggest that the picture for female employees is not all rosy. However, the salary information gives a slightly better picture. Female managers earn 85 percent of what their male counterparts earn and professional female employees earn 75 percent of men's salaries, both above the national average of 72 percent. There is no differential for employees on an hourly wage.

The worst-rated hotels are Hilton, Westin, and Sheraton. We had incomplete information for Westin and Sheraton, so they received industry averages where information was missing. We do know that Sheraton has no women among its executive roster. However, its parent company, ITT, scores very well. In many instances in this book, we rate parent companies instead of subsidiaries; in other cases we rate the subsidiaries or rate both separately, depending on how much information we could obtain. Thus, we note that parent company ITT has female representation on its board and among both its most senior and highest-paid executives. It gives about two times the corporate average of charitable donations to women's groups and also cooperated with our research. Whether to base your opinion of Sheraton Hotels on ITT Sheraton or parent company ITT is your decision.

Zero Percent Club—Lodging

COMPANY	FEMALE DIRECTORS	TOP FEMALE OFFICERS OR TOP-SALARIED WOMEN
Hospitality Franchise	0	
Renaissance Hotels	0	
Walt Disney		0

Rental Cars

The five rental car companies surveyed were a rather unimpressive lot in terms of their treatment of women.

RENTAL CAR COMPANIES	FEM	RANK	RENTAL CAR COMPANIES
◐ Alamo Rent A Car†	50.95*	1	Alamo
◐ Avis‡	45.64	2	Avis

	RENTAL CAR COMPANIES	FEM	RANK	RENTAL CAR COMPANIES
○	Budget Rent A Car§	42.33	3	Budget
○	Chrysler	41.26	4	Dollar, Thrifty, Snappy
○	Hertz§	39.29	5	Hertz

*Did not have scores for all five categories.
†Alamo was acquired in late 1996 by Republic, a company for which we have no information.
‡HFS, which earns a 49 on the FEM scale, has owned Avis since late 1996.
§Budget's and Hertz's parent company, Ford, has announced plans to sell Budget. Ford earns a lower FEM score than either Budget or Hertz.

Rating Key:
● Female-friendly company; buy products when possible
◐ Average; buy with good conscience
○ Below average; avoid buying these products

It is hard to find a rental car company to recommend: not one earns our highest rating. In fact, only Alamo and Avis scored in the average range.

Although Avis, Budget, and Hertz span six points on the FEM scale, their numbers in the first two categories are quite similar. Budget has a slight edge at 37 percent female officials and managers; the numbers for Avis and Hertz were 31 and 30 percent. Budget also leads the three with a 55 percent female professional staff, Avis and Hertz each with 49 percent. Avis and Hertz have all-male boards; Budget has an all-male executive staff. Where Avis takes the lead over the competition, and moves into the average category, is by having some female advancement programs, including recruiting in advertising and on college campuses, internships, and work co-op programs.

Alamo was our top finisher, in spite of some missing information. The company received average scores for the categories in which it had gaps, and the information that we did have, such as women representing 33 percent of the company's top officials, was good news.

We also rate Chrysler, which is the parent of Dollar, Thrifty, Snappy, and General car rentals. Chrysler, with female representation on its board, looked promising early on; however, its commitment seems to have stopped there. There is only one woman among the company's top 30 officers and none among the highest-paid officers. And only 7 percent of Chrysler's managers are women—about one-sixth the national average.

Zero Percent Club—Rental Cars

COMPANY	FEMALE DIRECTOR	TOP FEMALE OFFICERS OR TOP-SALARIED WOMEN
Avis	0	
Budget Rent A Car		0
Hertz	0	

OFFICE EQUIPMENT

Computers

Computers and software are a significant purchase for most people and most businesses, and a great deal of sampling, discussion, polling, and research often goes into the process of buying them. We hope the following information will add another factor to your decision.

We rated 27 companies that manufacture and sell personal computers, software, and networking systems. As a group, these companies scored pretty well on the FEM scale, perhaps because few of the companies (except IBM, which has an outstanding score) has been in existence long enough to have an ensconced men's network. The average FEM score for the industry is 59.39, with 10 companies earning our female-friendly rating, and 14 others falling below average. Thirteen companies cooperated with our research. It is in this industry that the best benefits are found.

	COMPUTER COMPANIES	FEM	RANK	PRODUCTS, SPECIALTIES
●	IBM	70.32	1	IBM computers, ThinkPad
●	Xerox	69.78	2	Laser printers
●	AT&T†	66.18	3	NCR PCs, software
●	Lotus‡	62.97	4	Notes, Lotus 1-2-3, AmiPro word processing, SmartSuite, Soft*Switch, cc:Mail
●	Apple Computer	60.24	5	PowerBook, Apple, Macintosh PCs

	COMPUTER COMPANIES	FEM	RANK	PRODUCTS, SPECIALTIES
●	Hewlett-Packard	59.66	6	Computers, laser printers
●	Tektronix	58.52	7	High-resolution LCD monitors, printers
●	Motorola	57.34	8	Computers, modems
●	Digital Equipment	56.22	9	PCs, LinkWorks software, server systems
●	Dell Computer	55.62	10	Dimension, Latitude, OmniPlex, PowerEdge, PCs, notebook computers
◐	Computer Associates	48.71	11	Software for mainframes, PCs
◐	Novell	47.29	12	NetWare (links PCs), Quattro Pro
◐	Wang Laboratories	45.37	13	VS series computers, software
○	Oracle	44.93	14	Oracle software, database management
○	Microsoft	43.96	15	Windows operating systems
○	Pioneer Electronics	43.89	16	PCs, CD-ROM players
○	Sun Microsystems	43.31	17	SPARC workstations, Solaris software
○	Sony	42.90	18	CD-ROM players, computers
○	Compaq Computer	42.58	19	Laptop, portable computers, and PCs
○	Texas Instruments	42.06	20	Notebook PCs, printers
○	Intel	41.67	21	Chips, software, add-in enhancements
○	Borland	40.73*	22	Paradox, dBase, C++, Delphi

COMPUTER COMPANIES	FEM	RANK	PRODUCTS, SPECIALTIES
○ Ricoh	40.72	23	Computers, software, peripherals
○ Canon USA	38.95	24	Printers, notebook computers
○ Unisys	38.69	25	CTOS, software
○ Matsushita Electric	36.56	26	Panasonic computers
○ Sharp Electronics	34.40	27	Computers, printers

*Did not have scores for all five categories.

†AT&T has divided into three separate companies, but was still one company when we collected this information. AT&T is still the communications company, NCR makes computers, and Lucent Technologies produces equipment.

‡Lotus Development has been bought by IBM to develop and sell its Notes technology.

Rating Key:
● Female-friendly company; buy products when possible
◐ Average; buy with good conscience
○ Below average; avoid buying these products

Recommending a company from which to purchase computers and computer products is easy: ten of the companies earn a female-friendly, above-average rating. They are: IBM, Xerox, AT&T, Lotus, Apple, Hewlett-Packard, Tektronix, Motorola, Digital Equipment, and Dell Computer.

The company with the strongest commitment to women is IBM. Although it does not have a very strong representation of women either in management or at its highest levels, IBM does have women in each of the categories rated by us. Where IBM excels is in benefits. New parents may take up to *3 years* of leave, and employees may work part time (just over 20 hours per week) and still receive full benefits for 3 years. The company offers on-site day care at some locations, other flexible scheduling options, and many family benefits. IBM has been recognized by Kinder, Lydenberg, Domini & Co. (KLD), an organization that rates companies on social performance, as having the best benefits in the computer industry, has been listed for 9 years running in *Working Mother* magazine, and has won awards from the Department of Labor. Furthermore, the company has a female subsidiary president. Everyone knows IBM is a major player in the computer industry; now you know it's a good feminist consumer choice as well.

Xerox ranks second, only one-half point behind IBM on the FEM scale. Although best known for its copy machines, Xerox has a presence in the computer industry, including production of laser printers. The company has female representation at its top levels—not common in the industry—with a board that had two women on it (today it has one) and women making up about 12 percent of the company's top officers and highest-paid positions. Xerox scores just above average in the management opportunities category, but does well in other categories. The company gives an average amount of its charitable donations to women, has two female subsidiary presidents, and has been listed every year on *Working Mother* magazine's "Top 100" list. Xerox also provides outstanding benefits, offering new parents up to 1 year of leave, flexible scheduling options, and special programs that include mentoring, child care subsidies, adoption assistance, and part-time return after childbirth.

Many of the other companies in our top category have strong points that are worth mentioning. AT&T has an industry high percentage of female managers—42 percent—and 35 percent of the management employees hired in 1994 were women. It also has an industry-high number of female advancement programs, which include recruiting, mentoring, and special training for women.

Hewlett-Packard, the dominant maker of printers and laser printers in the industry, and our sixth-ranked company, has two women on its board and one among its most senior officers. It also scores well on benefits and charitable giving, where it gives to groups supporting women in science, women in business programs, and women's shelters. It has appeared on the "100 Best" list seven times.

Two other high-ranked companies, Lotus and Motorola, have appeared multiple times on the "100 Best" list. Fourth-place Lotus, the maker of software including Lotus 1-2-3 and Notes, stands out both in the top-paid officials measure—29 percent of its highest-paid officers are women—and in benefits, where new parents receive a full month of *paid* leave and then have access to on-site day care. Employees at Motorola also have on-site day care, as well as job sharing, flex time, and work-at-home scheduling options.

Tektronix, Apple, Digital Equipment, and Dell are the remaining companies that earn our highest recommendation. Tektronix, which makes high-resolution computer monitors and LCDs, distinguished itself by being one of very few companies in our sample to divulge its salary equity information. Women receive just over 80 percent of their male colleagues' salaries, significantly more than the national average. The company also scores well on charitable giving, donating scholarship money to female engineers.

Apple Computer earns a reasonable score in the glass ceiling category, and offers on-site day care and other good benefits. It also scores well in charitable giving and supports women-owned businesses. Digital Equipment has several good benefit programs and female advancement programs, which have earned the company recognition or awards from *New Woman*, *Savvy*, the Department of Labor, the Domini Social Index, and the Council on Economic Priorities.

We were pleased that so many of the big names in the computer industry earned our top rating. There were also a few industry giants that scored below average, receiving our don't-buy rating. The biggest of those—and the hardest to avoid—is Microsoft. Microsoft makes the operating systems and programming languages—including MS-DOS, Windows, Windows 95, MS Excel, MS Word, Basic, Fortran, and Pascal—that most people use. It suffers from having no female advancement programs, an all-male board of directors, and being in noncompliance with the EEOC in 1992. Although it does have several good benefits, it is less generous with leave and day care than its competition. It has also failed to win any recognition or award given for advancing women within the company or being a great place to work. Apple Computer, which scores very well, may be a good alternative for operating systems and software—although Apple's systems only work in Apple computers.

Others warranting negative mention include Ricoh, Canon, Unisys, Matsushita Electric, and Sharp. Four of these companies are Japanese-owned, and perpetuate that country's tradition of no women in top echelons.

Zero Percent Club—Computers

COMPANY	FEMALE DIRECTORS	TOP FEMALE OFFICERS OR TOP-SALARIED WOMEN
Borland	0	0
Canon	0	
Compaq	0	0
Computer Associates		0

COMPANY	FEMALE DIRECTORS	TOP FEMALE OFFICERS OR TOP-SALARIED WOMEN
Matsushita Electric	0	
Microsoft	0	
Oracle	0	
Pioneer	0	
Ricoh	0	
Sharp	0	
Sony	0	0
Sun Microsystems	0	
Texas Instruments		0
Wang	0	

Electronic Office Equipment

We rated 11 companies in this category. These companies make big-ticket electronic office equipment, such as copiers, scanners, printers, fax machines, calculators, and other processing equipment. There was quite a bit of variation found in the 11 companies. In fact, there were no average scores in this industry: the companies all earn either our highest female-friendly rating or a below-average rating, which should make choosing a bit easier. If you are in a position to influence decisions involving purchasing of such equipment at your office, we urge you to refer to and act on the information in the following table.

This industry presents an interesting pattern. The office equipment companies have greater female representation on the boards than is the norm for *all* of corporate America, and companies in this industry offer strong benefits. On the other hand, not one company even brushes the national average in terms of women in overall management—perhaps because female engineers are still a relatively new phenomenon at most companies.

ELECTRONICS COMPANIES	FEM	RANK	PRODUCTS
● IBM	70.32	1	Copiers, printers
● Xerox	69.78	2	Copiers, printers

	ELECTRONICS COMPANIES	FEM	RANK	PRODUCTS
●	Pitney Bowes	64.12	3	Copiers, fax machines, mailing systems
●	Hewlett-Packard	59.66	4	Printers, fax machines, copiers
●	Eastman Kodak	58.54	5	Copiers, printers, batteries, film
○	Texas Instruments	42.06	6	Printers, calculators
○	Ricoh	40.72	7	Copiers, fax machines, printers
○	Standard Register	39.04	8	Financial, bar code, document processing
○	Canon USA	38.95	9	Copiers, fax machines, typewriters, printers
○	Matsushita Electric	36.56	10	Copiers, fax machines, phones
○	Sharp Electronics	34.40	11	Copiers, fax machines

Rating Key:
● Female-friendly company; buy products when possible
◐ Average; buy with good conscience
○ Below average; avoid buying these products

The top two companies in this industry are IBM and Xerox, both of which were discussed in the preceding section on computer companies. They are excellent female-friendly choices.

Other top-rated companies include Pitney Bowes, Hewlett-Packard, and Eastman Kodak. Highlights for Pitney Bowes, which makes mailing systems, copiers, and fax machines, include having good female representation on its board of directors (2 of 12 members are women), good benefits, and a female division president. *The 100 Best Places To Work in America* reports that Pitney Bowes has had a policy in place since 1985 mandating that women make up at least 35 percent of management (although the actual number, according to statistics filed with the EEOC, is 27 percent). Eastman Kodak's board has the same composition as Pitney Bowes's, but Kodak has fewer women represented in top echelons and among managers. Kodak also has very good benefits, including a well-used job sharing program and some paid family leave. It also has the strongest female advancement programs in the industry. Overall, products made by Pitney Bowes, Hewlett-Packard (discussed in the preceding section on computers), and Eastman Kodak are strong choices.

Because there are so many good companies, avoiding the below-average ones, which include Texas Instruments, Ricoh, Standard Register, Canon, Matsushita Electric, and Sharp Electronics, should be easier. All six companies are members of our zero percent club: Matsushita, Canon, Ricoh, Standard Register and Sharp Electronics have all-male boards, and Texas Instruments has an all-male executive staff. Four of the first five—i.e., excluding Ricoh—also have female representation in management that is lower than even the low industry average (none of the companies in the industry meet the national average of 41 percent).

Zero Percent Club—Electronic Office Equipment

COMPANY	FEMALE DIRECTOR	TOP FEMALE OFFICERS OR TOP-SALARIED WOMEN
Canon	0	
Matsushita Electric	0	
Ricoh	0	
Sharp Electronics	0	
Standard Register	0	
Texas Instruments		0

Office Supplies

This is another highly competitive industry whose consumers are predominantly women. The average FEM score of the companies we rated is 47.60, which is on the low side of average. We surveyed 22 companies that either manufacture or sell office supplies on the national level. Fortunately, there are a number of good choices. As seen in the table below, there are several bad choices as well. Unlike some of the other categories that we rated, where consumers may well select certain products for their uniqueness, such as entertainment or food, office supplies are more readily interchangeable.

	OFFICE SUPPLY COMPANIES	FEM	RANK	PRODUCTS, BRAND NAMES, STORES
●	3M	61.45	1	Scotch tape, Post-its, diskettes, desk organizers, mailing supplies, transparency film
●	Rubbermaid	60.21	2	Eldon desk supplies
●	Staples†	59.66	3	Major office supply retail store
●	Newell	56.76	4	Eberhard Faber markers and pens, Stuart Hall notebooks and stationery, desk organizers
◐	K-Mart	51.89	5	OfficeMax office supply retail stores
◐	Boise Cascade	50.02	6	Paper and office products
◐	James River of Virginia	48.77	7	Paper: Eureka!, Curtis, Word Pro, Delta Brite
◐	Mead	48.66	8	Paper, notebooks
◐	Consolidated Papers	47.29	9	Paper, glossy paper
◐	A.T. Cross	46.38	10	Multiple office products, pens
◐	Georgia-Pacific	45.58	11	Paper, envelopes
◐	Alco Standard	45.46	12	Paper products
◐	Avery Dennison	45.36	13	Tabs, labels, markers, adhesives
○	Kimberly-Clark	44.79	14	Neenah paper
○	Moore	44.51	15	Business and electronic forms, labels, direct mail services
○	Gillette	44.05	16	Flair and Papermate pens, Liquid Paper
○	International Paper	43.21	17	Office products

OFFICE SUPPLY COMPANIES	FEM	RANK	PRODUCTS, BRAND NAMES, STORES
○ American Brands	43.10	18	ACCO fasteners and paperclips, Swingline staplers, Wilson Jones steno books and binders, Day-Timers
○ Tandy	42.97*	19	Computer City, Incredible Universe, and Radio Shack stores
○ Office Depot†	41.51*	20	Retail outlets for office supplies
○ Standard Register	39.04	21	Business forms
○ Chesapeake	36.57	22	Packaging, tissue, kraft paper

*Did not have scores for all five categories.
†In mid-1996 Office Depot and Staples announced plans to merge.

Rating Key:
● Female-friendly company; buy products when possible
◐ Average; buy with good conscience
○ Below average; avoid buying these products

The top three scorers are 3M, Rubbermaid, and Staples, one retail store chain and two manufacturers. Their FEM scores are all comfortably above average. 3M, ranked first, makes Scotch-brand tape, computer diskettes, and Post-its, products that are ubiquitous in offices and homes. Fifteen percent of 3M's board are women, and the company receives high scores for benefits, cooperation, and prior recognition. Although there are few female senior officers (3 percent) and only 15 percent of 3M's managers are women, the company does have several programs to remedy this situation and has received accolades from a number of magazines, including *Working Mother* and *Fortune*, for its enlightened policies about family-friendly benefits and its tolerant attitudes toward gays and lesbians.

The second-place finisher, Rubbermaid, is a diverse company that has been discussed in other chapters of this book (household products and toys). Twenty-five percent of Rubbermaid's board is female, the highest in the industry, and the president of the company's Office Products division is a woman. Peculiarly, its record with regard to middle management positions is not as good. Hopefully, the top-level women will put pressure on the company to remedy this situation. Rubbermaid has a good benefits package for women, second in the industry only to 3M. In general, it is an excellent company.

Staples, ranked third, does not have a consistent record across all categories. None of its charitable giving is earmarked to women's causes, and its benefits package is mediocre. It gets high grades in the other three categories, however. Staples was also very cooperative with our survey, and provided salary information, something that only 5 percent of the respondents in our entire sample did. Female managers at Staples earn 89 percent of what male managers do, compared with the national average of 72 percent.

Newell also earned our highest rating. Points to mention are its willingness to share its better-than-average salary equity data and its high percentage of charitable giving to day-care centers and women's organizations. We would have hoped, however, that a company that had made such good strides toward women's equality in some areas might have found a seat on its board of directors for a woman.

Chesapeake, Standard Register, American Brands, and International Paper were the four lowest-scoring companies for which no information was missing. Chesapeake, a paper supplier, has no women on its board or among its senior officers, has no programs for women, and did not cooperate with us. Standard Register and Tandy also have all-male boards. International Paper was equally uncooperative, and only 10 percent of its managerial ranks are women, less than one-quarter the national average. There is one woman on its board, however, and the company does purchase goods and services from women-owned businesses. American Brands has one woman on its 11-member board and no senior officers, no charitable giving to women's groups, and did not cooperate with our study.

We would recommend that you shop, for example, at Staples or Office Max rather than Computer City (Office Depot does not earn our recommendation now; however, it has announced plans to merge with higher-rated Staples); use Scotch-brand tape and Post-its rather than competing brands; buy computer diskettes manufactured by 3M; and buy paper from companies other than International Paper or Chesapeake. Boise Cascade, although it was just average, was the best paper manufacturer in terms of opportunities for women, followed closely by Mead, Consolidated Papers, and James River of Virginia.

Zero Percent Club—Office Supplies

COMPANY	FEMALE DIRECTORS	TOP FEMALE OFFICERS OR TOP-SALARIED WOMEN
A.T. Cross	0	
Chesapeake	0	0
Consolidated Papers		0
James River of Virgina		0
Newell	0	
Standard Register	0	
Tandy	0	

Office Furniture

We surveyed five companies that make and sell office furniture. Each is particularly unimpressive in its treatment of women, all but one scoring below average. It was very hard to assess the performances of any of these companies because none returned our questionnaire. The average FEM score for the industry was 40.57, the lowest of any industry rated.

Even though we are unable to make any recommendations in this industry, we present the results below. Perhaps you know of other companies or local outlets that might be more deserving of your business.

	COMPANY	FEM	RANK	PRODUCT LINES
◐	Herman Miller	45.25	1	Herman Miller, Phoenix Designs, Meridian Inc.
○	Hon	41.78*	2	Gunlocke, Holga, BPI, Ring King Visibles, Remington
○	Globe Business Furniture	40.64*	3	Globe
○	Anderson Hickey	38.49*	4	Anderson Hickey
○	O'Sullivan	36.58*	5	O'Sullivan

*Did not have scores for all five categories.

Rating Key:
- ● Female-friendly company; buy products when possible
- ◐ Average; buy with good conscience
- ○ Below average; avoid buying these products

There is little in the way of good news to report about this industry. Of the five companies, Herman Miller, manufacturer of high-end ergonomically designed furniture, earns the highest score, just making it into the average category. There is one woman on the company's 11-member board, but none among its top corporate officers. The management is 21 percent female, about half the national average. The company does have pretty good benefits: no extended paternity leave, but availability of job sharing, flex time, some part-time return after childbirth, and some family-oriented programs and resources. Herman Miller has been recognized by KLD and the Domini Index and has appeared twice on *Working Mother* magazine's "Best 100" list. Even though it earns an average score, it is by far the best in the industry. If your only choices are the ones we rate, Herman Miller is the clear pick.

The only other information worth mentioning is about Hon, maker of Gunlocke, Holga, BPI, and Ring King Visibles. Although we do not have complete information about this company, we do know that its board is 9 percent female, the same as Herman Miller, and that it gives an average amount of its charitable donations to women's causes. Those two positives earn it a second-place rating, but given the quality of the competition, it is probably safe to say that the whole industry needs a lot of work.

About the remaining companies, Anderson Hickey, Globe Business Furniture, and the ubiquitous O'Sullivan, we can only say that they did not cooperate with our research and we were unable to find the missing information from any other sources, so they received average scores in categories with missing information.

Zero Percent Club—Office Furniture

COMPANY	FEMALE DIRECTORS	TOP FEMALE OFFICERS OR TOP-SALARIED WOMEN
Herman Miller		0
O'Sullivan	0	0

MAILING SERVICES

It is rare to find an office that is so organized and works so far ahead that the staff never has to use any kind of express mail. Once upon a time, the differences in price and service led some offices to choose one mailing service over another. These days, however, many offices have multiple accounts with the companies, which are frequently quite competitive.

In this category we rate four private companies that specialize in delivering mail and important documents. Also included is a fifth company, Pitney Bowes, that specializes in mailing services, such as postage machines and shipping and weighing systems.

	COMPANY	FEM	RANK
●	Pitney Bowes	64.12	1
◐	Federal Express	46.41	2
○	DHL Worldwide Express	39.45*	3
○	Airborne Express	38.10*	4
○	United Parcel Service of America	36.34	5

*Did not have scores for all five categories.

Rating Key:
● Female-friendly company; buy products when possible
◐ Average; buy with good conscience
○ Below average; avoid buying these products

Our top rating goes to Pitney Bowes, which is not a company that competes particularly with the other four. Pitney Bowes specializes in mailing systems, from weighing machines to bar code software. Because its competition is slim, it was a relief to see the company's good treatment of women. Two women sit on the 12-member board, and women are represented among both the most senior and the highest-paid officials. It also cooperated with our work, has strong benefits, has appeared on *Working Mother* magazine's "100 Best" list seven times, and has a female division president.

Of the four mailing companies, Federal Express was the high scorer. Although the company receives an average rating on the FEM scale, it certainly goes beyond what its competition is doing. FedEx has one woman on its 14-member board, as well as women among both the most senior and highest-salaried officers. About 26 percent of the company's managers are women, more than for any other mailing company. It is the only company to give an average amount of its charitable donations to women's groups. Its benefits are not the best, but there is an established flex-time band for employees. Federal Express is the clear choice in the industry.

Both DHL and Airborne suffer in the ratings because of missing information; each company earns average scores in two categories, where information was missing. We do know that Airborne has no women on its board, and that it has received a negative rating from the KLD rating service. As for DHL, none of the company's top-salaried officials are women.

United Parcel Service also rates a below-average mark on the FEM scale. UPS does give something to women's causes, although less than the national average, and does have some family benefits. That isn't enough, however, to compensate for its poor performance in the glass ceiling category, where it has an all-male board. UPS also has a very low number of female managers, 13 percent, compared with the national average of 41 percent.

Zero Percent Club—Mailing Services

COMPANY	FEMALE DIRECTORS
Airborne	0
UPS	0

SUMMARY

We hope women will take advantage of their changing roles as consumers. So far, we are not making the big decisions from the corporate boardrooms. On the other hand, as most women who work in offices can attest to, most of the everyday purchasing decisions are not made in the boardrooms. They are made by human resources and personnel departments, in which women are the majority, and by assistants and office managers, many of whom are also women. The consumer choices relating to office expenditures is one area where women's historical lack of seniority actually gives them more power.

If you are making travel decisions for yourself and/or other employees at your office, ordering supplies, or choosing computers or furniture, it is our hope that you will carefully consider the information in this chapter. Each time you decide to buy something from a female-friendly company and not from a company with a worse record of treatment of women, the message that it pays to treat women as equal employees will disseminate further. With any luck, that message will even trickle back to the place where you work.

Part Four

Women as Travel Consumers

CHAPTER SEVEN

The States of Women's Status

'We, the people of the United States,' which 'We, the people?' The women were not included.

Lucy Stone (1818–1893)

Americans travel extensively and spend a great deal of money in the process. Over $430 billion was spent on travel within the United States in 1995, according to the Travel Industry Association of America.[1] Of this figure, 23 percent was designated as business travel and 70 percent, pleasure. All other categories comprise the remaining 7 percent. Travel decisions are increasingly being made by women, who represent 40 percent of all travelers. Thus, women control the spending of a substantial percentage of vacation and business travel dollars. In addition to themselves and their families, women also play a significant role arranging business travel for others, and often make decisions as to where meetings and conferences are held.

These varied travel choices that women make translate directly into economic power. As noted in earlier parts of this book, however, this power has rarely been used to benefit women.[2] In order to exercise this power, consumers need to make decisions that increase spending of travel dollars in places that treat women well and decrease spending in those destinations that do not.

In this chapter, we consider travel possibilities within the United States; in the next chapter, we focus on international travel. States compete strenuously with each other for both tourist and business dollars, so spending decisions that are made with women in mind can have a significant impact.

One example of how mindful travel can work: In 1992, Colorado passed an anti-gay rights amendment, which received negative attention nationwide. Boycotts were organized, vacationers chose other destinations, and companies canceled conventions. The Denver Metro Convention & Visitor Bureau reported that 31 conventions that had been booked in the city were canceled. The business community estimated that these cancellations represented a potential loss of $31 million. Prior to the U. S. Supreme Court decision that ruled the legislation unconstitutional,[3] consumer trip cancellations scared many politicians and business owners, even those who might have originally supported the legislation. Dollars speak loudly.

HOW STATES CAN DIFFER

Because all U. S. citizens and residents are subject to many of the same laws, some readers may be wondering if the states vary enough to bother paying attention to differences. The answer is yes: states differ from each other in several areas that meaningfully impact women.

This variation occurs partially because the government of the United States was originally organized into a two-tiered system. Under this system, states have the legislative power to control many of their internal concerns, while the federal government regulates issues of common concern. For example, interstate commerce is seen as a common concern and is controlled on a federal level; an area such as family law, on the other hand, is controlled individually by each state. Where these concerns overlap, a state may not legally *take away* any of an individual's rights specifically granted by the Constitution and interpreted by the Supreme Court. The Supremacy Clause (Article VI of the Constitution) invalidates any state laws that contradict federal ones, and the 14th Amendment provides equal protection against state violations of federal law.

Still, each state has considerable discretion and power to affect areas that concern women. Here are several reasons why:

- State legislatures are free to pass laws that help women in areas where no federal law exists. For example, before Congress passed the Family Leave and Medical Act of 1993, granting 12 weeks of leave to care for a newborn, an adopted child, or an ill family member, some states already had enacted such laws.
- States, however, are free to pass broader laws. Consider the same Family and Medical Leave Act, which stipulates that employees who work for companies with more than 50 employees are entitled to 12 weeks of leave. Some states have increased that time period, targeted smaller companies, or expanded the conditions applying to the leave. This is important because women are more likely to work in small businesses that are exempt under federal law.[4]
- States are expected to enforce all federal laws and mandates concerning women. This enforcement can proceed, however, with greater or lesser vigilance. For example, laws that mandate equal pay for female and male employees have not been adequately enforced to date, but some states are better than others.
- States have primary responsibility for legislation in multiple areas that directly affect women's lives, such as divorce, property distribution, alimony, child support, and domestic abuse. There are important differences in state laws concerning these issues, and some are more favorable to women than others.
- States can adopt policies, at least temporarily, that make it more difficult for women to exercise their federal rights. For example, although women have a constitutionally based right to reproductive choice, some states have passed laws forcing women to wait 24 hours before they can have an abortion.

This governance is complex and continuously evolving, since the courts define both the substance and the limits of various civil rights as a result of the cases brought before them. Such decisions are based on both legal precedent and judicial philosophies. If it were clear-cut, much less attention would be devoted to Supreme Court appointments.

A BRIEF HISTORY OF WOMEN'S RIGHTS IN THE UNITED STATES

The history of women's rights in this country reveals both the diversity and the complexity of the federal system. The right of women to vote, for example, was not constitutionally granted until the hard-fought passage of the 19th Amendment in 1920.[5] Before the amendment passed, 20 states had already granted women suffrage. Women were permitted to vote in Massachusetts in 1691, for example, though this right was taken back by state legislators in 1780. Kentucky first granted women the right to vote in local elections in 1838, and Wyoming granted broad suffrage in 1869, before it was even a state.

The 19th Amendment was not universally ratified; 14 state legislatures did not ratify it. However, once it was adopted by the other 34 (there were only 48 states in 1920), no state could *legally* deny women the right to vote.

This does not mean that states always enforce such federally granted rights willingly or without legal battles. In the case of the 15th Amendment, which gave male ex-slaves the right to vote after the Civil War, many southern states put numerous obstacles in the path of those who tried to exercise their right to vote, such as literacy requirements and poll taxes. These limits were adjudicated for many decades and upheld by the courts until Congress passed the Voting Rights Bill of 1965, which specifically disallowed such discriminatory practices.

Women may recognize a similar pattern with regard to their own freedom of reproductive choice, a freedom implied in the Constitution as interpreted by the 1973 U. S. Supreme Court decision in *Roe v. Wade*. The end of that story still remains to be told.

A number of other rights relevant to women's progress have not yet been legislated. The Equal Rights Amendment, for example, failed to pass the required number of state legislatures[6] and so did not become part of the Constitution. (Eleanor Smeal, then President of the National Organization for Women, noted in 1989, "Guns have a Constitutional amendment protecting them and women don't.") A further problem is that existing laws, such as salary equity, are often not enforced with sufficient vigor. If merely enacting legislation in areas of women's rights was sufficient, there would not continue to be such large gender discrepancies in job positions and pay scales.

Before providing data about specific states, it is important for women to understand what their basic, legislated rights are. Most rights that female employees exercise today with regard to equal employment opportunity date back to Title VII of the Civil Rights Act of 1964. This legislation prohibits discrimination in employment on the basis of race, color, religion, national origin, and sex[7] in private employment settings of 15 or more employees, and in state government and other local entities with more than 100 employees. The law mandates an independent commission to enforce this prohibition, known as the Equal Employment Opportunity Commission. The commission specifically addresses the terms and conditions of employment, affirmative action, penalties for retaliation against employees claiming rights under the statute, employer filing responsibilities, and remedies available. Later modifications added sexual harassment as an unlawful employment practice. Another important modification was contained in the Pregnancy Discrimination Act of 1978, which prohibited discrimination in hiring or firing as a result of pregnancy, protected a woman's right to reinstatement, and required employers to treat pregnancy like any other disability with regard to insurance benefits. As a result, certain state laws that denied women employment on such bases as "fetal protection" were struck down.

Additional federal legislation important for initiating rights for women was the Equal Pay Act of 1963, which made it illegal to pay members of one sex less than the other when their jobs required equal skill, effort, and responsibility.[8] This act applies to all employment settings with two or more employees.

The executive branch further affirmed women's rights with Executive Order 11246, initially written by President Lyndon B. Johnson in 1965 and later amended in various ways by successive presidents. This order provides for affirmative action and protects minorities and women employed by contractors who do business with the federal government. A 1978 revision by President Jimmy Carter strengthened the agency charged with its enforcement, the Office of Federal Contract Compliance Programs, a branch of the Department of Labor. Unfortunately, the enforcement of Executive Order 11246 was significantly weakened by Presidents Ronald Reagan and George Bush through a reduction in funds allocated to the enforcement agency and by active opposition to affirmative action programs. To date, it has not been significantly strengthened by the current administration, and the political future of affirmative action is uncertain as of this writing because of legislative pressures and recent Supreme Court decisions.

The history of civil rights suggests the need for continued vigilance. Women should not assume that their hard-won rights will be maintained without a struggle. As of this writing, they seem more in jeopardy than ever. Long-held liberal state affirmative action plans have been voted down by California voters and similar scenarios are being planned for other states. The Republican-controlled Congress attempted to dismantle federal affirmative action plans but have not yet succeeded. Although the rhetoric has been toned down in order to attract female voters, the platform adopted by the Republican Party at their 1996 convention called for a constitutional amendment that would completely abolish reproductive choice for women under all circumstances, even if her life is in danger. Thus, there is clearly the need for political action. Vote for your rights before they are taken away. But you can vote with your pocketbook as well as with a ballot.

Assuming that these various federal laws, earlier court decisions, and Executive Orders are not dismantled, women today are legally entitled to: (1) receive equal pay for equal work, (2) not be discriminated against in hiring or promotion because of their sex, (3) not be discriminated against as a result of pregnancy, (4) complain to the relevant federal or state agency if they feel they have been discriminated against, (5) bring suit for remedies without employer retaliation, and (6) not be subjected to sexual harassment in the workplace. The reality lags far behind in most of these areas, however.

THE FEM RATING SYSTEM FOR STATES

Some of these various concerns have been incorporated into the system used for rating states. Our evaluations have considered, for example, whether states have bolstered federal rights for women or instead have attempted to put obstacles in their path. As one example of augmentation, workers in Connecticut can obtain 24 weeks of unpaid family leave instead of the 12-week minimum guaranteed by the federal legislation. (Despite recent federal legislation, the United States still remains way behind Europe, where some countries give paid maternity leave for up to 11 months and other countries offer a fixed allowance for parental leave for up to 3 years.)

On the negative side, several states have attempted to diminish federal women's rights. One of the most common strategies is to impose additional conditions under which a right to terminate a pregnancy may be exercised. States that have instituted requirements such as parental consent and waiting periods, if ultimately upheld by the courts, substantially increase the difficulty women experience as they exercise their rights. These kinds of factors were considered in the ratings.

Diversity in the enforcement and granting of rights by states was, however, only one of several factors on which we focused. Many facets contribute to the overall climate affecting a woman's advancement opportunities within a given state, and we sampled a broad range of these. Some related to the state's divorce and spousal abuse laws, and how favorable those laws are to women. Other ratings considered how successful women have been economically within each state in both the public and private sectors. One area of concern was how many women currently occupy management or professional positions in state government. For example, in Utah's government offices, only 20 percent of managerial positions were held by women, versus 41 percent for Delaware.[9] Other areas of concern included the adequacy of the state as an employer of women, such as how much they are paid relative to men. Women's political power was also investigated, such as the degree to which women receive important appointments relative to men.[10]

Based on these various considerations, we developed a FEM rating system that uses five categories, each containing several measures. The average ranking on each category was 10. Ratings for each of the five categories were then totaled to find the FEM score for that state. An average total score would be 50, but total FEM scores could theoretically vary between 0 and 100.[11]

The five broad categories rated were: (1) performance by the state as an employer of women, (2) the state's economic climate for women, (3) the proportion of women in various levels of government, (4) women's civil and family rights, and (5) the degree to which the overall climate of the state (including its history) is female friendly. Each of these categories could be (and has been) the central topic of many books; their treatment here, therefore, is far from exhaustive. Because of limited resources for primary data collection, we relied heavily on the most current data bases that other existing organizations have collected and kindly shared with us.[12]

The employer category score was based on several measures. The first was the percentage of women occupying state management jobs. The second was the average salary of female state employees relative to their male counterparts, where equality would be 100 percent. The third measure in this category focused on how many family-friendly benefits the state provided for its own employees.[13]

The second category, favorability of economic climate ("economics"), included two measures: the percentage of women holding management positions in private industry and the percentage of women-owned businesses within the state.[14]

The third category, women in governance, relied on a number of factors. These included the percentage of women holding seats in Congress. In the summer of 1996, 49 women held seats in the House of Representatives (out of 435), and 8 in the Senate (out of 100).[15] Also measured were the proportion of women elected to the state legislature and the percentage of women occupying high-level political appointments, such as agency heads. It is interesting to note that some states with otherwise good records regarding women still do not elect them to Congress. As of this writing, Massachusetts and Minnesota, for example, still have all-male congressional delegations as do 24 other states.

The fourth category, women's legal and civil rights ("rights"), includes how marital property is treated and distributed within the state following divorce. Some states use a "community property" standard, which assumes equal ownership of all property acquired during the marriage. This is particularly helpful to women, because it ensures that the major assets of the family, such as the family home, will be divided evenly, and not in accordance with the financially more powerful partner, which is usually the husband. This kind of law does not punish women who do not work outside of the home. A second index in this category was whether the state has encouraged or discouraged the exercise of reproductive choice options, especially abortion. A third measure gauged the adequacy of legal protection following spousal abuse.[16]

The last category, favorability of general climate ("general"), rated a number of diverse areas. Some included relatively rare occurrences, such as whether women occupied the highest political offices, namely, governor, lieutenant governor, or U. S. senator. As of this writing, only one governor (New Jersey) and eight U. S. senators are women (although one new female governor and a net gain of one woman in the Senate occurred in the November 1996 election). The history of voting rights for women was also examined. As previously noted, some states extended suffrage to women long before passage of the 19th Amendment (and they received some bonus points for this), whereas others failed even to pass it.[17] Because the availability of child care programs is important to women, both economically and psychologically, the last measure employed in this category was the amount spent by the state on child care and education, based on figures provided by the Children's Defense League.

All 50 states were rated. Discussion of these ratings is organized according to the four geographical regions used by the Census Bureau: the Northeast (9 states), the Midwest (12 states), the South (16 states), and the West (13 states). Further Census Bureau subdivisions are used, where relevant, such as New England versus Middle Atlantic, or South Atlantic versus West South Central.

If your own state or region does not meet your expectations, you may wish to register your sentiments.[18] A list of current governors is provided in Appendix Four, together with addresses and telephone numbers.

A reminder to the reader: each category has an average of 10, and an average total score would be 50. The actual range of scores was 27 to 85. We have given all categories equal weight, but you may consider some more important than others and so may make your judgments accordingly.

The measures and categories do have some statistical overlap with each other. For example, states with high numbers of female business owners also have a higher proportion of female congressional representatives.[19] States with higher percentages of female managers within the government also pay women more relative to men.

Here's how the states ranked overall, with more specifics about each state and region following. Remember that the range of "average" is different for states than it was for companies.

Rankings of States by Overall FEM Scores

STATE	FEM	RANKING	REGION
● Washington	85.27	1	West
● California	78.68	2	West

	STATE	FEM	RANKING	REGION
●	Nevada	75.61	3	West
●	Maryland	70.36	4	South
●	Hawaii	66.62	5	West
●	Kansas	66.43	6	Midwest
●	Connecticut	64.60	7	Northeast
◐	New York	61.10	8	Northeast
◐	Oregon	59.88	9	West
◐	North Carolina	59.43	10	South
◐	New Mexico	58.89	11	West
◐	Illinois	58.86	12	Midwest
◐	Maine	58.13	13	Northeast
◐	Minnesota	57.86	14	Midwest
◐	Colorado	57.86	15	West
◐	Massachusetts	57.40	16	Northeast
◐	Florida	55.57	17	South
◐	New Jersey	55.30	18	Northeast
◐	Arizona	54.21	19	West
◐	Ohio	53.08	20	Midwest
◐	Iowa	52.34	21	Midwest
◐	Vermont	49.97	22	Northeast
◐	Wyoming	49.08	23	West
◐	Rhode Island	48.45	24	Northeast
◐	Alaska	48.29	25	West
◐	Texas	47.97	26	South
◐	Oklahoma	47.76	27	South
◐	Georgia	47.74	28	South
◐	North Dakota	47.72	29	Midwest
◐	Michigan	46.94	30	Midwest
◐	Idaho	45.68	31	West
◐	New Hampshire	45.29	32	Northeast
◐	Missouri	44.79	33	Midwest
◐	Virginia	44.47	34	South
◐	Wisconsin	44.16	35	Midwest
◐	South Dakota	42.95	36	Midwest
◐	Pennsylvania	42.39	37	Northeast
◐	Louisiana	40.14	38	South
◐	Delaware	38.64	39	South

	STATE	FEM	RANKING	REGION
◐	Indiana	38.18	40	Midwest
○	Tennessee	37.51	41	South
○	South Carolina	34.95	42	South
○	Nebraska	34.46	43	Midwest
○	Montana	34.21	44	West
○	Kentucky	33.75	45	South
○	Utah	32.99	46	West
○	Arkansas	31.36	47	South
○	West Virginia	31.22	48	South
○	Mississippi	30.88	49	South
○	Alabama	27.61	50	South

Rating Key:
● Female-friendly state; visit when possible
◐ Average; visit with good conscience
○ Below average; avoid visiting

THE NORTHEAST

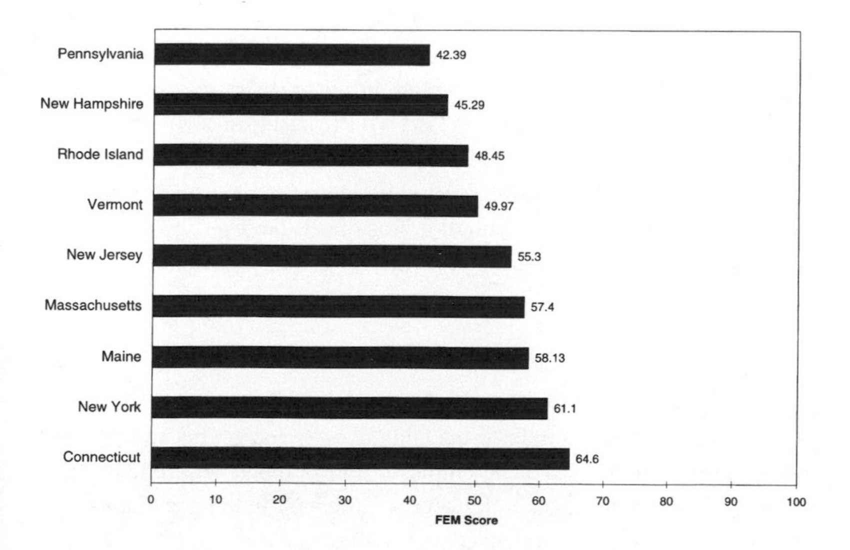

Nine states comprise this geographical area: Connecticut, Maine, Massachusetts, New Hampshire, New Jersey, New York, Rhode Island,

Pennsylvania, and Vermont. Their overall FEM ratings range from 42 to 65. A chart for each state is presented so that you can compare states regarding each of the categories. States are presented in alphabetical order within each geographical region.

CONNECTICUT

Rank: 7
Overall FEM Score: 64.60
- Employer: 15.68
- Economics: 13.61
- Governance: 17.33
- Rights: 8.21
- General: 9.8

Population: 3,287,116
Revenue: $9.8 billion

Connecticut's FEM score is above average for the country, and is the best of the northeastern states. Its profile is somewhat mixed, however. It gets high scores in the first three categories, but is slightly below average in the last two. As an employer, it has particularly female-friendly benefits. Connecticut offers its state employees 24 weeks of unpaid, job-guaranteed family leave—twice as much as that mandated by federal law. In the governance area, it has the highest percentage of female congressional representatives, 37 percent, compared with an average of 9 percent.

Its overall FEM score was lowered, however, by ratings in the rights and general climate categories. Its laws with regard to spousal abuse were less female-friendly than 39 other states (Category 4) and it and Vermont were the only northeastern states that failed to ratify the women's suffrage amendment. This historical limitation was offset, however, by a relatively high child care expenditure, thus yielding an average score in Category 5.

MAINE

Maine's overall score is respectably above average, but it, too, has a mixed picture. Its ratings as an employer and in the general climate area are slightly above average, and it scores quite well in the governance and rights categories. Maine received the highest score in the Northeast in rights for women, determined by having the most female-friendly domestic violence laws, and reasonably good access to reproductive choice. It falls

Rank: 13
Overall FEM Score: 58.13
Employer: 11.38
Economics: 5.49
Governance: 14.81
Rights: 14.65
General: 11.8
Population: 1,227,928
Revenue: $3.8 billion

down considerably, however, in the economic climate category. This is because it has the fifth lowest percentage of female business owners in the country. Maine seems to be doing many things well for women, but it needs to institute steps that will encourage more women's business activity. (In the 105th Congress, Maine is the second state to have two female Senators, an exciting election outcome.)

MASSACHUSETTS

Rank: 16
Overall FEM Score: 57.40
Employer: 6.3
Economics: 15.9
Governance: 14.38
Rights: 7.63
General: 13.2
Population: 6,026,425
Revenue: $20.5 billion

Like its Connecticut neighbor, Massachusetts has a respectably above-average FEM score and a very mixed picture. It rates very well in the economics and governance categories, with 32 percent female managers in private industry (compared with a national average of 29 percent), and a high percentage of women in appointed state cabinet-level positions (45 percent versus a national average of 23 percent). Peculiarly, however, it has practically no women at high elected levels of government: no female congressional representatives, senators, or governor. It is one of 26 states to still have an all-male congressional delegation in Washington, D.C., an incongruous phenomenon for such a traditionally liberal state.

Also contrary to expectation are its very low scores as a state employer and a provider of civil rights for women. As an employer, its score ranked 41st in the country, primarily because of low rankings in female managers (27 percent versus 31 percent nationally), and relatively low salary equity ratios. Female state employees earn 75 percent of what their male colleagues do, whereas the national average for government employees is 81 percent (the average in the private sector remains at 72 percent). Its below-average score in the rights category reflected its non-female-friendly practices with regard to reproductive rights. On the 8-point scale employed by NARAL to assess this category, Massachusetts received a 1 (for having a good policy regarding clinic violence).

Despite these obvious weaknesses, Massachusetts ranked fairly high in our general climate category, primarily because it leads the nation in child care expenditures. Like its Berkshire mountains, its profile with regard to women is one of peaks and valleys.

NEW HAMPSHIRE

Rank: 32
Overall FEM Score: 45.29
Employer: 4.29
Economics: 8.69
Governance: 11.53
Rights: 13.04
General: 7.75
Population: 1,109,252
Revenue: 2.73 billion

New Hampshire, one of the least populous of the northeastern states, has a below-average record with regard to women. It is particularly poor as a state employer with the worst scores of the region, and an overall ranking in this category of 46th of the 50 states. This reflects a very poor record of women in state managerial positions (25 percent versus 31 percent average) and a relatively poor showing in family-friendly benefits provided, relative to its neighbors (scoring 2 on a 6-point scale versus an average of 4.3 for the northeastern states).

Almost as dismal was its rating on general climate with no women holding high elective office and a low expenditure for preschool children's programs, the worst in the region. Its showing will be imiproved in 1997, however, because New Hampshire elected its first female govenor in November. Its best showing was in the rights area with very good legislation regarding domestic violence. Overall, however, the scenery in the state far exceeds its record toward women.

NEW JERSEY

New Jersey received an above-average FEM rating of 55.30. As can be seen from the chart, this state's rankings are extremely variable.

Rank: 18
Overall FEM Score: 55.30
Employer: 8.83
Economics: 12.39
Governance: 4.07
Rights: 13.04
General: 16.97
Population: 7,710,188
Revenue: $28.9 billion

New Jersey's poorest score is in the governance area, and it is the worst in the region. Only 6 percent of the state's congressional representatives are women, only 13 percent of state legislators are women (the national average is 21 percent), and only 11 percent of the state's political appointees are women, compared with a national average of 23 percent.

The state does, however, have the only female governor (who supports more progressive women's rights than her party's platform does), which is one of the factors involved in the relatively high score in the general climate category, together with New Jersey's relatively high child care expenditures. Govenor Whitman is now one of two (see New Hampshire). In view of the rest of its governance, the female governor may be more of a symbolic than a real victory for New Jersey's women. On a more optimistic note, however, perhaps she will improve things.

NEW YORK

Rank: 8
Overall FEM Score: 61.10
Employer: 5.83
Economics: 16.08
Governance: 11.42
Rights: 14.31
General: 13.47
Population: 17,990,455
Revenue: $74.9 billion

The most populous state in the Northeast, New York's FEM score is the second best in the region, and it ranks number 8 nationwide. Its only very weak category is as an employee, the second worst in the region. Compared with national averages, it has a very poor record of women in managerial positions (22 percent versus 31 percent nationally) and in pay equity (76 percent versus 82 percent nationally). It does, however, have very good fringe benefits, offering a wealth of options such as flex time, job sharing, and 7 months of job-protected leave.

Its strongest score came in the favorability of the economic climate for women (best in the region) with the fourth highest average of female managers in private industry in the country. The state is also quite strong in civil rights and general climate, with very female-friendly laws regarding abuse and reproductive choice, a good historical suffrage record, and high child care expenditures. In addition to all of the other excellent reasons for doing business in or visiting New York City, you can now add its good treatment of women.

PENNSYLVANIA

Rank: 37
Overall FEM Score: 42.39
Employer: 8.38
Economics: 6.9
Governance: 8.99
Rights: 9.59
General: 8.53
Population: 11,881,643
Revenue: $36.7 billion

Pennsylvania's FEM score of 42.39 is the worst in the Northeast, and is considerably below average for the nation.

It is not as variable as the other states: it is consistently below average in all categories. It is almost average in the rights area because of excellent policies on domestic abuse. Its relatively recent Protection from Abuse Act gives judges the power to issue an immediate order of protection based only on the testimony of a victim and enables judges to require defendants to pay legal, counseling, and temporary child support costs. The act is considered a model for other states, and replaces laws that were anything but female friendly.

In the reproductive rights area, however, Pennsylvania has been in the forefront of attempts to deprive women of their choices by passing legislation that impedes access to abortion services, both legally and practically. On the NARAL 8-point scale, it receives a score of 0. And this is in its best category.

The worst overall category was economic climate, where only 28 percent of private sector managers are women, and only 28 percent of businesses are owned by women (versus national averages of 33 and 30 percent, respectively). As its overall picture is not very female-friendly, we suggest that you spend your travel and business dollars elsewhere.

RHODE ISLAND

Rank: 24
Overall FEM Score: 48.45
  Employer: 9.23
  Economics: 8.8
  Governance: 12.84
  Rights: 7.29
  General: 10.3
Population: 1,001,464
Revenue: $3.6 billion

"Poor little Rhode Island," which is actually not so poor, receives a FEM score that is almost at the national average, but it ranks seventh of the nine states in the region.

Its worst category was the rights area, because of a poor rating on abortion rights legislation. Its best was governance. Although there are no women among its congressional representatives, Rhode Island does have a relatively high number of female political appointments, 35 percent versus 30 percent nationally.

VERMONT

Rank: 22
Overall FEM Score: 49.97
  Employer: 9.71
  Economics: 9.52
  Governance: 13.89
  Rights: 6.94
  General: 9.9
Population: 562,758
Revenue: $3.7 billion

Vermont is extremely average, both for the region and nationally. Its profile indicates approximately average scores as an employer, and in economic and general climate. It has the lowest score of the region in the rights area because of poor laws for women in both domestic abuse and reproductive choice. Its best category was governance because of a higher-than-average proportion of female state cabinet-level appointees. One other notable feature is that the state ranks third in the country in child care expenditures. This is the only indication, however, that it is by far the richest northeastern state on a per capita basis (with a per capita revenue of $6574 versus $4163 for runner-up New York).

Northeastern Region Summary

As a region, the northeastern states account for almost 51,000,000 people, or almost 25 percent of the nation's population. Its overall FEM rating is somewhat above average at 53.62. As a geographical region, the Northeast is the second best; its best category is governance, and its worst, employment. The Northeast has the lowest variability and its two regions, the New England and the Middle Atlantic states, do not differ appreciably.

The best state in the region is Connecticut, with New York the runner-up; the worst is Pennsylvania, with New Hampshire second from the bottom. Thus, if you wish to support good records and are planning a conference or trip to the Northeast, choose Hartford or New York City rather than Philadelphia. Take your vacation in the Connecticut Berkshires rather than the Poconos.

THE SOUTH

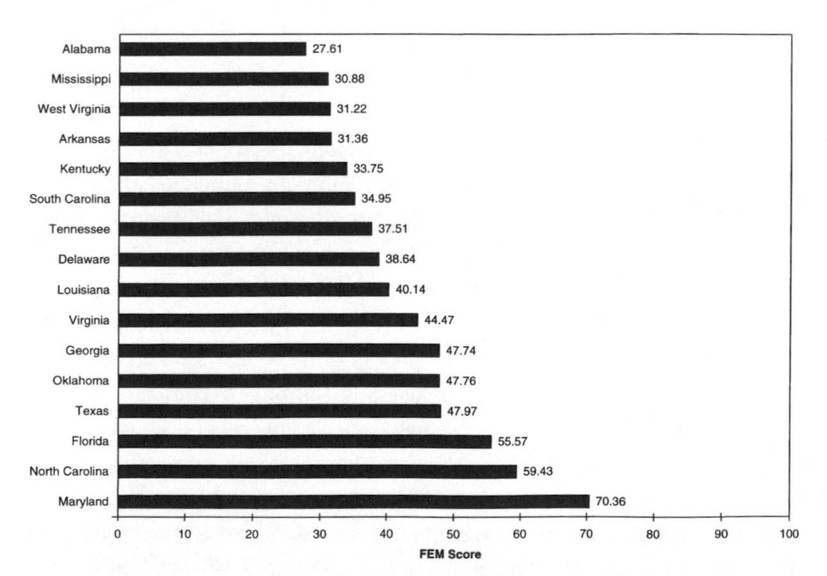

The South comprises Alabama, Arkansas, Delaware, Florida, Georgia, Kentucky, Louisiana, Maryland, Mississippi, North Carolina, Oklahoma, South Carolina, Tennessee, Texas, Virginia, and West Virginia. The presentation will mirror that of the previous section, with each state's results discussed in alphabetical order. Only 3 states were in the upper half of our rankings and the other 13 occupied many of the bottom slots. Seven of the ten worst states nationally were in this region.

ALABAMA

Rank: 50
Overall FEM Score: 27.61
  Employer: 13.85
  Economics: 2.13
  Governance: 0
  Rights: 7.63
  General: 4.53
Population: 4,040,587
Revenue: $10.5 billion

With its FEM score of 27.6, Alabama finished last in the nation. Four out of five categories reccived exceedingly low scores. The only one that did not was Category 1. State employees fared better in Alabama than did its other female citizens. There was a higher-than-average proportion of female managers (35 percent), and women earned 91 percent of what men did in the public sector. Once we get past this, however, the economic climate in the private sector is quite bad for women. Only 26 percent of private sector managers are women, and only 26 percent of the businesses in the state are owned by women (the third lowest in the country).

The rights category is the second best, although it is below average. Domestic abuse laws are reasonably protective of women, but reproductive rights are sorely lacking (Alabama scored a "1" out of a possible 8 on this variable). It is in the governance area that Alabama was truly awful, receiving a negative score in this category (a score this low would be expected less than 1 percent of the time). Alabama has no women in Congress, only 3.6 percent women in the state legislature (versus 21 percent nationally), and women receive only 9 percent of the state's political appointments, whereas the national average is more than 23 percent. In accord with this pattern, no women are high-level elected officials, child care expenditures are very low ($11 per child), and the state's historical record with regard to women's suffrage is negative, thereby resulting in the third lowest score for Category 5. We would recommend that you avoid doing business in Birmingham.

ARKANSAS

Rank: 47
Overall FEM Score: 31.36
Employer: 8.87
Economics: 1.46
Governance: 7.66
Rights: 7.45
General: 8.62
Population: 2,350,725
Revenue: $5.9 billion

Arkansas is a fairly small state that got little attention prior to President Clinton's election. Lest readers think that the state is home to lots of women like Hillary Clinton, they should take a look at the statistics. Its overall FEM score places it third from the bottom nationally, and it is consistently below average on all five categories rated. Its two worst were rights for women and economic climate. In the rights area, it is mediocre with regard to domestic abuse protection, and its score on reproductive rights was 0 on the 8-point scale used by the NARAL Foundation to rate states, a dubious distinction shared by only ten other states. Its state legislature, which is 87 percent male, has passed anti-choice legislation putting barriers in women's way as they attempt to exercise their rights. Moreover, the economic climate for women in Arkansas is even worse than it is in Alabama. Only 27 percent of private industry managers are women, and the state has the second lowest proportion of women-owned businesses in the country. Arkansas definitely needs to exert more effort toward its female citizens.

DELAWARE

Rank: 39
Overall FEM Score: 38.64
Employer: 11.14
Economics: 15.16
Governance: 4.29
Rights: 0
General: 8.13
Population: 666,168
Revenue: $4.8 billion

Delaware is a small but relatively affluent state (per capita revenue is $7200). Although this affluence may exert some influence in women's economic opportunities (a high score in Category 2), it does not seem to affect the other categories very much, leaving Delaware with a most undistinguished national ranking of 39th in overall FEM score. It is particularly weak in the governance and rights areas. Female citizens of Delaware receive no state protection in either the reproductive or domestic abuse area, giving the state the lowest score nationally for Category 4. Not surprisingly, women are very poorly represented in either state or national government. The congressional delegation is all male, and women receive only 6 percent of high-level political appointments. We recommend that you keep this in mind when planning your next business or vacation travel.

FLORIDA

Rank: 17
Overall FEM Score: 55.57
Employer: 10.13
Economics: 15.26
Governance: 13.88
Rights: 10.16
General: 6.13
Population: 12,937,926
Revenue: $28.3 billion

The record improves when we consider Florida. This state, with its above-average FEM score of 55.57, is the third best in this geographical region. It is above average on all categories but one, general climate. Although its average expenditures for child care are not as bad as most of its neighbors, it has no senior female elected officials and its historical record is negative. On the more positive side, Florida's economic climate is very favorable to women, as is its governance structure (24 percent female congressional delegation). On balance, we can find no reason not to support its tropical vacation spots or Disney World.

GEORGIA

Georgia is generally regarded as a more politically progressive state than its southern neighbor, Florida. Interestingly, however, its record toward women is not as good. Its overall score is somewhat below average and its pattern is quite different. It stands out as an employer, with 37 percent of state management jobs occupied by women, and shows an excellent gender wage disparity of 91 percent. Women do not fare as well in the private sector, where fewer than 29 percent of the managerial positions are held by women. They also do particularly poorly on our measures of

Rank: 28
Overall FEM Score: 47.74
Employer: 15.84
Economics: 8.1
Governance: 8.54
Rights: 10.51
General: 4.75
Population: 6,478,216
Revenue: $14.8 billion

general climate, with a ranking of 47th in this category. In addition to the absence of senior female officials and its historical stance against women's suffrage, Georgia only spends $35 per child on child care, which is in the lower third. Although millions of people visited Atlanta for the 1996 Olympics, our hope is for a future where these games are held only in places where women fare very well.

KENTUCKY

Rank: 45
Overall FEM Score: 33.75
Employer: 12.18
Economics: 4.79
Governance: 4.49
Rights: 7.45
General: 7.52
Population: 3,685,296
Revenue: $10.6 billion

Kentucky is 45th in the country on its record toward women. It does poorly on all categories except employer, where women are paid 88 percent of what men earn. It is extremely weak in the rights area, tying Arkansas and West Virginia for the second worst ranking in this geographical region, and receiving a zero rating on reproductive rights. Women in this state also fare badly in the private sector and in government, as there were no women in Congress (one was elected in November 1996) and only 8 percent in the state legislature. As tempting as the Kentucky Derby might be, we suggest that you sip your mint julep while watching the race on television.

LOUISIANA

Louisiana's FEM score places it in an unimpressive 38th position—sixth from the bottom for the region. Interestingly, it gets very good marks

Rank: 38
Overall FEM Score: 40.14
Employer: 17.94
Economics: 0
Governance: 3.44
Rights: 12.38
General: 6.84
Population: 4,219,975
Revenue: $11.8 billion

on its role as an employer of women, but rather bad ratings on most of the other categories.

Louisiana scores well as an employer for hiring more female employees than most other states, giving more women managerial positions, and paying women 91 percent of the average for men. Putting this score in context, unfortunately, diminishes how impressive it may at first seem. The average salary received by Louisiana state employees was the second lowest of all 50 states. It may be, therefore, that their seemingly general hiring and promoting of women occurs because men aren't attracted to these jobs. Women, of course, have always been overrepresented in low-paying jobs.

Louisiana is below average in three other categories, including economic opportunity, governance, and general climate. The state has few female managers in private industry and abused women receive scant legal protection. The state government has been consistently anti-abortion and has instituted many obstacles to women in this area, including passage of an unenforceable law that virtually bans all abortions within the state. In August 1994, it finally permitted the use of Medicaid money to finance abortions for rape and incest victims in order to avoid the cutoff of all Medicaid funds from the federal government.

Louisiana ranks 47th in the women in government category and 43rd in the general climate category. Only 18 percent of all political appointments are women (the national average is 27 percent). Only 9 percent of its congressional representatives are women, much lower than the 17 percent national average, and women comprise only 7 percent of state legislators, compared with the national average of 21 percent. (A November 1996 election update shows a loss of Louisiana's single female congressional Representative, but a gain of one female Senator.) Furthermore, very little is spent on child care.

Think carefully before you plan any vacation or business trips to New Orleans. There are other enticing cities with good food.

MARYLAND

Rank: 4
Overall FEM Score: 70.36
Employer: 11.34
Economics: 20
Governance: 18.46
Rights: 9.47
General: 10.97
Population: 4,781,468
Revenue: $13.7 billion

Maryland is the only state that mentions women in its motto: "Manly Deeds, Womanly Words." Its FEM score is better than others in the geographical region by at least ten points, and it places fourth nationally. Because of the state's unique history, Maryland arguably may not be a southern state. However, because most of the state is south of the Mason–Dixon Line, it has been categorized as such.

Maryland has the third best economic climate for women, is the best in the region for women in government with a 20 percent female representation in Congress, 29 percent in the legislature, and 39 percent in the governor's cabinet. The state scored additional points for having a female senator and lieutenant governor.

The state's weakness was in the area of women's rights. A low score in this category can be attributed to its neglectful policies toward domestic violence, its lack of community property laws, and its less-than-perfect regard for reproductive rights. Although more work is needed, on balance, Maryland is clearly a good place for women overall, and should be your first choice if you wish to travel to this geographical region.

MISSISSIPPI

There are 48 states with better records than this state. Mississippi scored way below average in every category except state as employer, where it maintains some respectability with its 40 percent of female managers and only a 9 percent difference between salaries for men and women.

Although this state does a good job combating domestic violence, it actively impedes a woman's right to choose, scoring zero in the subcategory of abortion rights. Like so many other states with aggressive stances against

Rank: 49
Overall FEM Score: 30.88
Employer: 16.64
Economics: 0
Governance: 2.77
Rights: 7.98
General: 4.48
Population: 2,673,216
Revenue: $6.2 billion

reproductive rights, Mississippi has one of the weakest records for financing child care, making the pro-child stance of its legislation questionable.

Nationally, this state surpasses only one (Alabama) in the governance category. Mississippi has no women in Congress and very few in the state legislature. It was 50th nationally in the economic category, with only 25 percent female managers and the smallest percentage of women-owned businesses overall. In short, there are few reasons to spend business or pleasure dollars in Mississippi.

NORTH CAROLINA

Rank: 10
Overall FEM Score: 59.43
Employer: 20
Economics: 5.68
Governance: 13.89
Rights: 14.31
General: 5.21
Population: 6,628,639
Revenue: $17.7 billion

North Carolina scored second best in the southern region, reflecting above-average scores nationally in three of the five categories. Its record as a state employer is most impressive, and is the best in the country. Female state employees make 93 percent much as men do, 36 percent of state managers are women, and North Carolina's family benefits are among the most generous in the country.

North Carolina's female record in government is also commendable when compared with the rest of the region: 15.4 percent of congressional representatives and 40 percent of cabinet-level appointees are women.

Most encouraging is this state's score in the women's rights category, which exceeds the national and regional averages with a strong record against domestic violence, and its score on reproductive choice issues is sixth best nationally.

Now for the bad news. North Carolina's economic climate for women is most unimpressive. Although it scored better than nine other states in the region, women make up only 27 percent of private industry managers, and only 28 percent of the businesses are women-owned. Given the publicity North Carolina gets for its "research triangle" and its economically progressive image, its performance with regard to women in the private sector is particularly disappointing. It needs to work harder promoting women in the private sector. Still, it's a good state on balance.

OKLAHOMA

Rank: 27
Overall FEM Score: 47.76
- Employer: 17.36
- Economics: 5.21
- Governance: 4.92
- Rights: 8.55
- General: 11.74

Population: 3,145,585
Revenue: $8.4 billion

Oklahoma's varied performance earns it a spot just below the middle of the pack, in 27th place. Its highlight came as an employer, where it ranked third nationally. The percentage of female managers within the state government is just slightly above average, but female employees are paid very well, earning almost 89 percent of what male employees are paid (well above the national average of 82 percent). Furthermore, the state is very generous with its family benefits, earning the highest possible score on this measure.

Oklahoma also earns an above-average score in the general climate category. The state has a female lieutenant governor and spends more than the average amount on child services.

The low points for Oklahoma are its economic opportunities and the number of political positions held by women. Women in the private sector are promoted into management positions less frequently than in other states and they also own relatively fewer businesses. The state has no women in Congress, and only 11 percent of state legislators are women, compared with a national average of 21 percent.

Although female state employees fare fairly well, women are not well represented in the business or political communities. The citizens of this state have some clear room for improvement before it becomes a female-friendly destination.

SOUTH CAROLINA

Rank: 42
Overall FEM Score: 34.95
Employer: 13.14
Economics: 0.88
Governance: 3.18
Rights: 12.12
General: 5.63
Population: 3,486,701
Revenue: $9.9 billion

South Carolina's treatment of women is not very impressive, landing it in 42nd place and second to last in the region (only West Virginia scores worse). With the exception of the state as employer category, where South Carolina performs better than average, and high scores in domestic violence and child care spending, South Carolina consistently scores near the bottom. Its scores in the political representation, economic opportunity, and general climate for women categories were particularly unimpressive.

The economic climate for women is a low point for the state. It ranks 49th (only Wyoming scores worse) in the percentage of women holding management jobs in the private sector. There are fewer women-owned businesses than in other states.

Women in political offices are even less visible than in the business arena. The state has no women in Congress, and the percentage of women in the state government is quite low.

South Carolina received another strike in the general climate category for its history of women's voting rights. In addition to not ratifying the 19th Amendment, South Carolina did not extend voting rights to women before the amendment was ratified, as many other states did.

South Carolina, despite some disappointing scores, did have a few good points. First, as previously mentioned, it is a good state employer. Even though women in the private sector may be lacking in the ranks of management, women who work for the state are promoted to those jobs at a rate that is better than average. Women working for the state also receive 88 percent of the average male employee's pay rate (keeping in mind that the average differential is 82 percent for government employees). Other good points include above-average spending for child services and excellent domestic violence laws. In fact, South Carolina earned the best possible score in the domestic violence category for having such protections as restraining orders that include mail and telephone contact, legal eviction, assumption of custody for the victim, continuation of child and other financial support, and provisions for reimbursement of attorney's fees. Hopefully, it will make progress in other areas as well.

TENNESSEE

Rank: 41
Overall FEM Score: 37.51
Employer: 13.63
Economics: 2.94
Governance: 3.61
Rights: 9.24
General: 8.09
Population: 4,877,185
Revenue: $11.1 billion

Tennessee, which ranks just one peg above South Carolina, has an almost identical pattern of scores and one that is common for many of the southern states. The state is a good employer of women but the number of women working as managers or business owners, in addition to the number of women in political office, is unimpressive.

Tennessee scores well below average for the number of female managers in the private sector and the number of women-owned businesses. It also scores well below average—in the bottom five nationally—for its lack of women in governance. No women from Tennessee serve in Congress, although this is equally true for most of the southern states. The number of women serving in the state legislature and as political appointees is far below average. Other blemishes include very restrictive abortion laws and per-child spending of only $16 a year, or about half the national average.

Women working for the state of Tennessee fare better, earning more relative to men than in other states and also receiving excellent benefits. Domestic violence victims may also fare better in Tennessee than in other states, thanks to very comprehensive protection laws. The state also earned some extra points for its history of voting rights. But other than these few strengths, Tennessee, like many of its southern neighbors, has a lot of room for improvement.

TEXAS

Relative to other states in the southern region, Texas put in a strong performance for women's progress. Interestingly, Texas had a terrible score

Rank: 26
Overall FEM Score: 47.97
Employer: 6.34
Economics: 6.87
Governance: 9.17
Rights: 13.3
General: 12.29
Population: 16,986,510
Revenue: $36.8 billion

as an employer—an area in which most southern states do very well—and excelled in women's rights and general climate.

Although Texas no longer has a female governor, it distinguishes itself by having a female senator. That combined with a per-child spending amount that is well above average pushed Texas to an impressive 11th place finish in the general climate category. Texas also performed fairly well in the women's rights area: having community property standards for divorce—rare in any state—was enough to overshadow slightly subpar records with regard to domestic violence and reproductive rights.

Although Texas outscored most of its southern neighbors in terms of business opportunities for women, it still did not score well nationally. The same was the case for the women in governance category, where Texas is in the southern minority by having women in the House of Representatives, but still falls below average for this category nationally.

Texas outperformed most of the other states in the South, and thus is a more desirable southern destination, but it could still improve relative to many other states.

VIRGINIA

Rank: 34
Overall FEM Score: 44.47
Employer: 7.7
Economics: 14.73
Governance: 7.55
Rights: 10.16
General: 4.32
Population: 6,187,358
Revenue: $15.3 billion

Virginia's best area for women is its economic climate, and its high number of women-owned businesses. It scored reasonably well in the women's rights category, with strong reproductive rights and domestic violence laws. Low scores as an employer, for women in the government, and for general climate, however, made for a relatively low score overall but a middle-of-the-pack ranking for the South.

Virginia has a poor record of women in governance. There are no women serving in Congress in nearby Washington, D.C., and the number of women in the state government is also well below average. On a positive note, Virginia has a large number of women serving as political appointees. However, the state ranks 50th in the general climate category, reflecting the absence of women as elected high officials, the historical opposition to women's suffrage, and an extremely low expenditure toward child care.

Offsetting its poor performance in the last category, Virginia has an extremely hospitable economic climate for women. The number of female managers in the private sector is above average and the number of women-owned businesses ranks fifth nationally.

WEST VIRGINIA

Rank: 48
Overall FEM Score: 31.22
- Employer: 11.81
- Economics: 2.37
- Governance: 4.73
- Rights: 7.45
- General: 7.56

Population: 1,793,477
Revenue: $5.5 billion

It is difficult to find many positive things to say about West Virginia. With the exception of a slightly above average score as an employer, the other indicators measuring women's progress are all wanting. West Virginia has a low number of female managers in the private sector, no women in Congress, few women in the state legislature and as political appointees, and no community property laws. Furthermore, the state has done everything it could to restrict a woman's access to an abortion, has done less than it could to protect women who are victims of domestic violence, and spends less than it could (and less than average) on children's services.

With the two exceptions of pay and benefits for female state employees, West Virginia earns feeble scores on issues relating to women's progress. Based on its performance, we recommend looking elsewhere to spend your business and vacation dollars.

Southern Region Summary

The South gets low marks on most categories except as an employer, where its performance exceeds that of other regions. Interestingly, however, a closer look reveals that its state employee salaries are the lowest of any region. Thus, men may be less attracted to such positions, and the pool of female candidates may be larger.

The South is divided into three smaller areas by the census: South Atlantic (eight states), West South Central (four states), and East South Central (four states). Although the South is at the bottom of the rankings and the West is at the top, the two areas share one interesting feature: the states bordering the ocean treat women better. In the South, the South Atlantic states have an overall FEM score of 47.8, just slightly below average, whereas the West and East South Central states have means of 41.8 and 32.4, respectively.

The three best southern states for women are Maryland (a leader by 11 points), North Carolina, and Florida. The worst are Alabama, Mississippi, and West Virginia, with Arkansas only a whisker away. For business travel plans, we suggest you consider Baltimore or Durham rather than Birmingham, New Orleans, or Wilmington. In most of the southern states, southern hospitality still needs to be combined with greater fairness toward women.

THE MIDWEST

The Midwest has been traditionally regarded as the nation's heartland. It includes Illinois, Indiana, Iowa, Kansas, Michigan, Minnesota, Missouri, Ohio, Nebraska, North Dakota, South Dakota, and Wisconsin. Historically, these states have very different traditions than the South. They have colder climates, more heavy industry, more agriculture, and many of them have had long liberal political traditions. So it was a bit surprising to learn that the midwestern states have FEM scores that are the second worst of the four geographical areas.

As a region, the Midwest scored the lowest in the category of women's rights, receiving an average category score of 8.28 out of a possible 20 points. Translation: These states, more often than not, do not support a

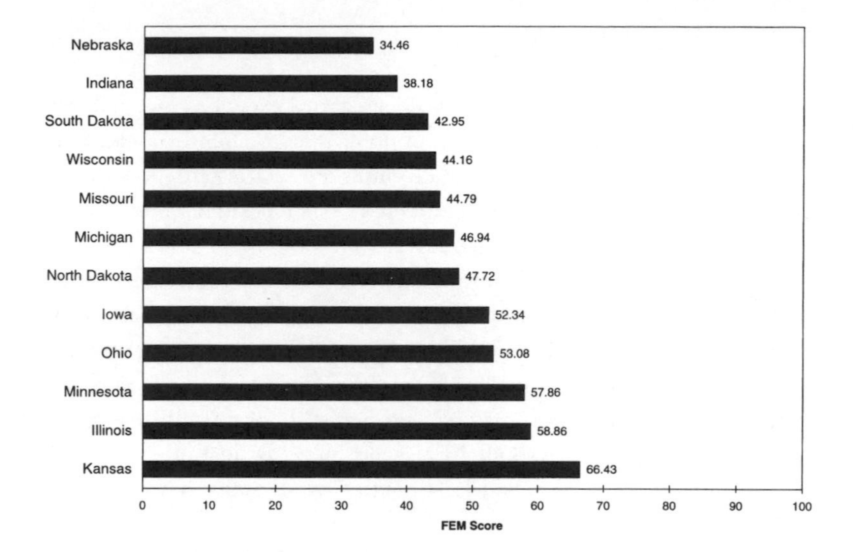

woman's right to choose, offer little protection in the case of domestic-violence, and most do not have community property laws. The Midwest also has a poor showing of women in political office (seven states have no female representation in Congress) and weak government support of child care.

In contrast, these states score relatively well in the economic category. The percentages of female managers are average, and many women own their own businesses. The state governments have positive records of paying women relatively higher salaries, and offering reasonable family leave benefits.

ILLINOIS

The FEM score for Illinois is above average, and the second best in the region. Although Illinois is not exactly a feminist's mecca, it has above-average scores in economic climate, governance, rights, and general climate. In general, none of the measures shows anything remarkable, but there is consistency. The highest score was in the last category, the result of having a female senator and a good historical record regarding suffrage.

Illinois ranked first in the Midwest on women's rights, but its picture here is mixed—excellent on domestic violence, and very mediocre on

Rank: 12
Overall FEM Score: 58.86
Employer: 7.35
Economics: 13.37
Governance: 12.51
Rights: 12.12
General: 13.51
Population: 11,430,602
Revenue: $27.9 billion

reproductive rights. It is a state that continues to challenge a woman's right to choose.

Illinois ranked above average in every category except one, state as employer. The gender pay disparity for state employees is particularly low at 74 percent, bringing down the total. On balance, however, Illinois is better than its neighbors.

INDIANA

Rank: 40
Overall FEM Score: 38.18
Employer: 11.22
Economics: 6.99
Governance: 7.76
Rights: 4.41
General: 7.8
Population: 5,544,159
Revenue: $13.5 billion

Indiana's showing in this study was quite poor. Its only above-average score was as a state employer. Specifically, it has a relatively high percentage of women in management positions (38 percent). Otherwise, its scores fall consistently in the bottom quarter nationally.

Indiana's private industry is dominated by men, who occupy 75 percent of management positions. There are only five other states whose records are worse in this regard.

Most disturbing is the fact that Indiana had no women in government—not in Congress or the governor's office (one female Representative was elected in November 1996). Its low score in the women's rights category reflects its anti-choice environment and its negligence in fully protecting battered women.

The general climate for women is poor and, not surprisingly, the state's funding of child care is embarrassingly low. Indiana is one of the worst of the midwestern states for women. Indianapolis tries very hard to appeal to convention and business travelers. We suggest that you steer clear.

IOWA

Rank: 21
Overall FEM Score: 52.34
Employer: 14.24
Economics: 9.22
Governance: 6.9
Rights: 11.43
General: 10.56
Population: 2,776,755
Revenue: $7.5 billion

Iowa is difficult to classify because of the wide disparity of scores across categories. It is the strongest in the region as a state employer, with a high percentage of women in management positions, but there are very few female managers in private industry. Although the lieutenant governor is a woman, the state has no female representation in Congress.

Like most other states, Iowa has no community property laws. In the midwestern region, however, this state has the best record on reproductive issues, and is above average with regard to domestic violence.

Iowa is straightforwardly average in all other measures rated, including female business owners and state legislators, and family leave policies. It seems to be on a better path than many other midwestern states, but still has some growing to do.

KANSAS

Rank: 6
Overall FEM Score: 66.43
Employer: 12.52
Economics: 11.32
Governance: 18.65
Rights: 8.9
General: 15.05
Population: 2,477,574
Revenue: $5.8 billion

Kansas scored number one overall in the Midwest. This state far exceeds regional and national averages in every category except women's rights, where its policies on abortion are less than exemplary. However, based on all of the other categories, the status of women is generally positive.

Take, for instance, its 29 percent showing of women in Congress and 28 percent in the legislature. The governor's cabinet is 33 percent women—outreaching many states nationally. Kansas also has a female senator and lieutenant governor. Unfortunately,

Senator Nancy Kassebaum will not be part of the 105th Congress, since she recently retired.

The picture is not all rosy. Low numbers of women in private industry management and the state's record on reproductive rights make it impossible to fully applaud this midwestern state. These factors are probably its biggest impediment toward progressiveness.

As a state employer, Kansas is keeping pace with the rest of the country and outpacing much of the region. Unlike many other states, Kansas shows some consistency in its treatment of women. For the most part, women should feel comfortable spending their money here, although it does not boast any major tourist attraction.

MICHIGAN

Rank: 30
Overall FEM Score: 46.94
Employer: 11.64
Economics: 9.94
Governance: 11.09
Rights: 3.14
General: 11.13
Population: 9,291,297
Revenue: $26.3 billion

Michigan makes a lot of cars, but it has not produced a particularly good climate for women. It ranks 7th in the region, 30th overall. This state's female representation in government is about average: 10 percent of the state's congressional representatives and 25 percent of the governor's cabinet. The lieutenant governor is a woman. Kudos also go to this state for giving women the right to vote 3 years prior to the 19th amendment.

All of these political positives, however, contrast sharply with the present political climate in Michigan. The governor and both houses in the state legislature favor criminalizing abortion. There are no community property laws and little legal intervention in domestic violence cases.

Michigan's rating as a state employer is average, but employment in the private sector (female managers and business owners) fell below the norm, as women occupy only 26 percent of private industry management positions. This may be attributed partially to the very low number of female managers in the automobile industry, which is male dominated.

MINNESOTA

Rank: 14
Overall FEM Score: 57.86
Employer: 9.84
Economics: 14.63
Governance: 13.11
Rights: 8.9
General: 11.38
Population: 4,375,099
Revenue: 15.1 billion

Minnesota, with a FEM score of 57.86, ranks 3rd in the Midwest and 14th overall. Regionally, the state is above average in the economic, governance, and general climate categories. It is average as an employer and somewhat below average in the women's rights area.

Although Minnesota is home to the Domestic Abuse Intervention Program, generally considered a model of research and pilot programs, it falls short in terms of implementing laws available for protecting an abused woman. The state scored very low on reproductive rights, presenting many obstacles to a woman's right to choose, and no community property laws.

This state's relatively strong economic climate for women (15th in the country) lifts its ratings. Additionally, although Minnesota has no women in Congress, it does have one of the higher proportions of women in the state legislature and cabinet, and the lieutenant governor is a woman.

Keep a watchful eye—this highly ranked midwestern state, with much to its credit, could benefit from stronger rights policies.

MISSOURI

Rank: 33
Overall FEM Score: 44.79
Employer: 5.97
Economics: 11.63
Governance: 8.24
Rights: 10.85
General: 8.09
Population: 5,117,073
Revenue: $11.6 billion

Missouri's FEM score suggests that perhaps the state's well-meaning motto, "The Welfare of the People Shall be the Supreme Law," may not fully apply to its female citizens.

Missouri's reasonably good economic climate for women and its perfect score with regard to domestic violence laws kept its score from plummeting too far. In this same category of women's rights, Missouri prohibits any state-funded entities from referring women for an abortion, and there is no support for community property laws.

Salaries for women in government are abysmal, with a gender pay disparity of 76 percent, and the family leave programs are a shadow of what they should be. Women make up 22 percent of Missouri's congressional delegation (and a new female Representative has been added for 1997), but only 9 percent of the governor's cabinet.

This "show me" state still needs to show more progress toward women before we recommend St. Louis as a place for vacation or business travel.

NEBRASKA

Rank: 41
Overall FEM Score: 36.60
Employer: 0
Economics: 13.6
Governance: 8.36
Rights: 4.41
General: 11.58
Population: 1,578,385
Revenue: $3.8 billion

Nebraska lies at the bottom of the midwestern barrel. As a state employer, women's salaries are deplorably low at 69 percent those of men. Additionally, there are very few women in state management positions, and its overall performance as an employer is the second lowest in the country. Nebraska has no female representation in government, except for its lieutenant governor.

Women's rights are also subpar: no community property laws, minimal acknowledgment of abortion rights, and scant support for domestic violence deterrents make it the third worst state in this category.

In surprising contrast to these statistics is a relatively healthy private sector for women, with Nebraska scoring third in the region, and a somewhat above-average score in the general climate category. Overall, however, Nebraska needs to give more attention to its women.

NORTH DAKOTA

North Dakota is slightly below average and very variable. The state has an impressive number of women in private industry management, one

Rank: 29
Overall FEM Score: 47.72
Employer: 9.86
Economics: 15.32
Governance: 5.53
Rights: 6.37
General: 10.65
Population: 638,800
Revenue: $2.1 billion

of the highest in the country, has a female lieutenant governor, and maintains a respectable position on domestic violence.

On the negative side, however, North Dakota has a very dismal women's rights record: The state maintains virtually no support for reproductive choice.

Additionally, there is no female representation in Congress, the percentage of female state managers is low (27 percent), and women's salaries are only 76 percent of those men. Although the family leave policies are quite good, North Dakota's support of child care is practically nonexistent at $4.12 per child per year.

OHIO

Rank: 20
Overall FEM Score: 53.08
Employer: 14.55
Economics: 7.59
Governance: 9.47
Rights: 10.85
General: 10.61
Population: 10,847,115
Revenue: $35.6 billion

As was the case for many other states, Ohio has a mixed record. As a state employer, it is way above average—close to 40 percent of its managers are women. Like Missouri, Ohio gets a perfect score in the women's rights category by maintaining zero tolerance for abusive husbands. The state has a female lieutenant governor and gave women the vote before 1919.

Women, however, have below-average opportunities in the economic and governance spheres with relatively few female managers and few women in positions of political leadership in the state. It is also subpar regarding its female representation in Congress and the cabinet, namely, 10 and 17 percent, respectively.

The most negative aspect of Ohio, as with so many other midwestern states, is its treatment of abortion rights and the many legislative obstacles designed to stop women from exercising their reproductive choices.

SOUTH DAKOTA

Rank: 36
Overall FEM Score: 42.95
Employer: 8.64
Economics: 10.45
Governance: 7.16
Rights: 6.37
General: 10.34
Population: 696,004
Revenue: $1.3 billion

South Dakota has wide open plains, but narrow policies. Like the other Dakota, it is not a state that believes in a woman's right to choose. To the contrary, this state supports legislation that would make abortion illegal, and presently demands parental/spousal consent, waiting periods, counseling, and more. Paradoxically, the state's support of child care is one of the lowest nationally at $4.3 per child per year.

There is no female representation in Congress, and the proportion of women in state management is one of the lowest in the country at 26 percent.

On the more positive side, the scores in the economy category are average, with 32 percent women in private industry management. South Dakota also has a female lieutenant governor. Overall, however, these prairies are not pioneering for women.

WISCONSIN

Rank: 35
Overall FEM Score: 44.16
Employer: 7.39
Economics: 6.94
Governance: 8.99
Rights: 11.69
General: 9.16
Population: 4,891,769
Revenue: $17.3 billion

Wisconsin's record is below average in four of the five categories rated. Its most positive score was in the rights category. It is not particularly good on reproductive rights (its record is lower than 19 other states), but it is average in domestic abuse legislation and is one of the few states to have community property laws.

Wisconsin was below average in every other category, both regionally and nationally: State employees are paid 25 percent less than men, there are no women in Congress, and Wisconsin's score in the economy category was 4 percent lower than the regional average, making it the worst state in the Midwest for women in private industry.

Conference planners beware: Wisconsin is not a business-friendly state for women and needs its public and private systems overhauled before it is worthy of your dollar.

Midwestern Region Summary

As a group, the 12 midwestern states are somewhat below average nationally, with the worst women's rights record of all (an average score of 8.2 versus the West's 12.8). The Midwest is average or slightly below average in the other four categories.

The states are quite variable in their rankings, with FEM scores ranging from 34 to 66. Kansas, Illinois, and Minnesota are the three best, and South Dakota, Indiana, and Nebraska, the three worst. Some midwestern states in the so-called "Rust Belt" are home to traditionally male-dominated industries such as automobile manufacture and automotive products. This often yielded lower-than-average ratings in the economic sphere. The biggest negative for women of this region, however, was encountered in the unfavorable laws and policies associated with the exercise of reproductive rights.

Based on the overall ratings, we suggest that if you have to arrange a meeting in the Midwest, choose Chicago or Minneapolis over Detroit or Indianapolis.

THE WEST

Many of the states in this region do not have very large populations, but have been growing rapidly in recent years. They are frequent vacation destinations because of their national parks, physical beauty, and other attractions.

As a group, the western states receive higher-than-average scores in four of the five categories, and states from this region consistently find their way into the top five measure by measure. In fact, four states from this region are among the top five nationally: Washington (1st), California (2nd), Nevada (3rd), and Hawaii (5th). Unfortunately, the western states come in last of the four regions as employers of women.

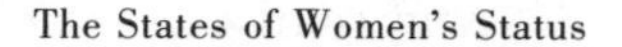

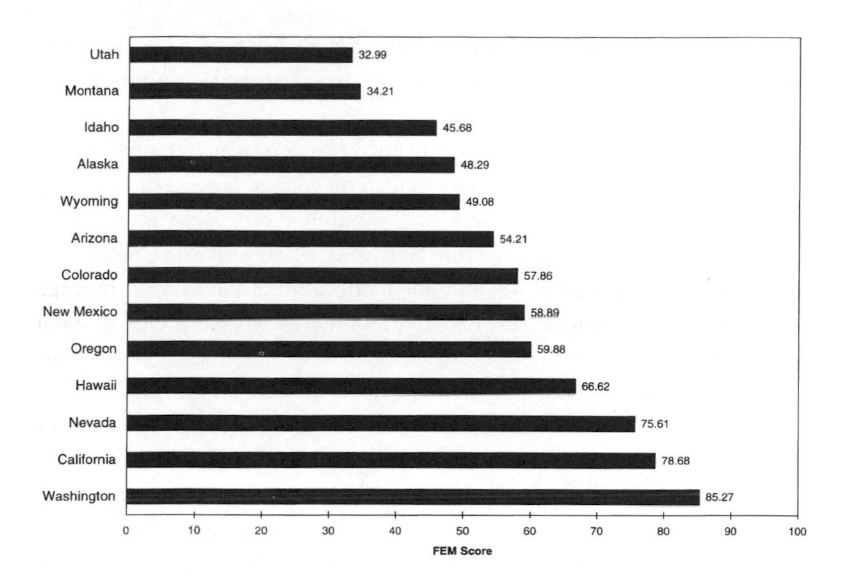

ALASKA

Rank: 25
Overall FEM Score: 48.29
Employer: 7.56
Economics: 9.6
Governance: 5.28
Rights: 10.51
General: 15.35
Population: 950,043
Revenue: $6.3 billion

Alaska's FEM score places it right in the middle of the pack. The state has been an increasingly popular tourist destination in recent years, as rugged and undeveloped places have become more elusive. Tourists flock to Alaska to visit Denali National Park and fish for salmon.

Like other western states, Alaska did not score well as an employer of women. Although the state pays its female employees higher than average salaries relative to men, its percentage of female managers is relatively low.

Alaska's high point came in Category 5, general climate. The state has a female lieutenant governor, a strong history of women's voting rights, and the second highest allocation of money per child for child services nationally (Massachusetts is first). Child services encompasses the amount of money spent preparing children to enter school, including subsidizing child care, Head Start, Block Grants, and inspecting and training staff for child care facilities.

Alaska's low score came in Category 3, women in governance, which includes measures such as the percentage of women serving in Congress and the state legislature as well as the number of female government appointees. Alaska has no women in Congress, a below-average number of women in the state legislature, and about one-third of the average number of female political appointees. Clearly, this leaves Alaska plenty of room for improvement.

ARIZONA

Rank: 19
Overall FEM Score: 54.21
- Employer: 5.14
- Economics: 17.01
- Governance: 10.06
- Rights: 11.69
- General: 10.32

Population: 3,665,228
Revenue: $9.6 billion

Arizona, with its magnificent Grand Canyon and desert landscapes, earned an above-average overall score, with a very good economic climate for women in private industry. The state is above average in both the number of women in private management and the number of women-owned business, ranking in the top ten on both measures.

Arizona is above average in all categories but one, the state as employer. Although the number of women in state management is just slightly below average, female state employees are paid only 75 percent of the rate for men (the national average for state employees is 82 percent). Benefits given to state employees are also subpar.

To the credit of voters in the state, they defeated a ballot initiative several years ago that would have prohibited abortion in the state constitution. However, the state still receives a low score in the abortion measure, requiring parental consent and neither allocating public money to abortion nor legislating policies prohibiting clinic harassment. Arizona makes some points up in Category 4, women's rights, by having community property laws regarding divorce.

CALIFORNIA

Rank: 2
Overall FEM Score: 78.68
  Employer: 11.39
  Economics: 17.51
  Governance: 11.78
  Rights: 20
  General: 17.34
Population: 29,760,021
Revenue: $100.0 billion

California, which ranks second nationally, is a pacesetter in many areas concerning women's progress. It leads the western region in the categories of women's rights and general climate, with many outstanding points.

California really stands out for having two female senators—the only state that can make this boast (but Maine will soon join them)—and about twice the national average of women serving in the House of Representatives. It also ranks high on the list of money spent for child services. An article in *Working Mother* magazine praises the state's free child care referral service, the rich variety of subsidized child care programs, a number of public–private partnerships to expand day care, and a history of continued commitment to quality child care that dates as far back as World War II.

California also excels in the women's rights category. The state has maximum protection for women abused by domestic violence, including such measures as an assumption of custody for women, mandated payment of child, spousal, and other financial support, restraining orders and no-contact provisions that extend to mail and phone calls, and legal eviction of the abuser. It also scores fairly well in the reproductive rights area and has community property laws.

The economic climate for women is a good one: California leads the nation in the percentage of private business managers who are women.

Like most of its neighbors, California's low point came in the state as employer. Although an above-average number of women hold managerial positions within state government, they receive a worse pay differential than women in nonwesterm states.

Overall, California has shown an excellent commitment to women. It earns a strong recommendation as a convention and tourist destination for its excellent FEM score as well as its plethora of beaches, cities, national parks, and, of course, Disneyland (a top scorer in our business rankings).

COLORADO

Rank: 15
Overall FEM Score: 57.86
Employer: 3.11
Economics: 20
Governance: 11.68
Rights: 6.94
General: 14.11
Population: 3,294,394
Revenue: $9.1 billion

Colorado's overall FEM score is above average, but its scores for the individual categories are extremely varied. The state boasts an excellent economic climate for women; in fact, its score in that category was one of the best in the nation. Within private industry, Colorado employers have a very good record of promoting women into managerial positions, and 34 percent of all business owners are women, ranking Colorado second overall in this category. In fact, Colorado's Small Business Administration office backed more loans to women-owned businesses—$61.3 million in fiscal year 1994—than did any other state.

Another strong point for Colorado is in the women in governance category. Its female representation in both Congress and the state capitol are well above average. In fact, the number of women serving in the legislature ranks third nationally.

And although Colorado's child services spending is on the low side of average, it has been recognized by *Working Mother* magazine as a leader, thanks to the initiatives set forth by Governor Roy Romer and his wife, Bea. For example, the state's progressive child care steps include a child care center for 16 businesses in downtown Denver that offers scholarships, and the center is accessible to everyone's children, from mail-room clerks to executives.

Its pattern, however, has valleys as well. Colorado does not do well as an employer, ranking 48th in this category. Although the number of female managers within state government is average, Colorado's female employees are paid only 76 percent of the rate for men earn, and their benefits are abysmal relative to other states. In fact, before federal legislation mandated family leave, Colorado's state government offered its employees only sick or annual leave with no job guarantee. The state's score in the women's rights category was not very strong either, reflecting a lack of community property laws and weak protection for domestic violence victims.

On balance, however, its strong points outweigh the weak ones. So if you're planning a ski vacation to the West, Colorado is a solid choice: it rates below California and New Mexico, but much higher than Wyoming or Utah.

HAWAII

Rank: 5
Overall FEM Score: 66.62
Employer: 9.66
Economics: 20
Governance: 11.9
Rights: 12.35
General: 11.59
Population: 1,108,229
Revenue: $5.2 billion

As if Hawaii needed any more to recommend it than its unique location in the middle of the Pacific, it also has a strong showing in the FEM ratings, ranking fifth nationally. Hawaii places respectably in every category, with one perfect score, three above-average scores, and one average score. Its strong points are the percentage of women-owned business (tops in the country) and strong domestic violence and abortion laws. Hawaii also shows good female representation in the House and has a female lieutenant governor.

Although the amount Hawaii spends on child care and early childhood development is only slightly above average (the range in this category is enormous, from $152 per child per year to 25 cents), the money is spent progressively. For example, the state provides after-school child care in *every* elementary school, eliminating the need for parents to work fewer hours or arrange transportation between school and after-school care.

Where Hawaii has room for improvement is as an employer. Both the percentage of women in state management and the salary differential of female employees are below average. Still, we highly recommend it as a travel destination, both for its physical beauty and for its general treatment of women.

IDAHO

Idaho is both below average nationally and among the bottom of the pack in its western region. The state needs more work as an employer than

Rank: 31
Overall FEM Score: 45.68
Employer: 3.23
Economics: 0.42
Governance: 16.56
Rights: 15.26
General: 10.16
Population: 1,006,749
Revenue: $2.9 billion

even its low-scoring neighbors. It scores way below average in terms of women in management, women's salary rate, and benefits.

Other low points for the state are seen in the economic opportunity category, where Idaho ranks 47th in percentage of women in private management and 41st in the number of women-owned businesses. The state's worst mark is the amount of money spent for child services, namely, 25 cents per child, for which Idaho ranks 50th.

It was a pleasant surprise, then, to discover that Idaho's delegations are well above average in terms of the number of women in Congress and the state capitol. The number of female political appointees is also above average.

Although the state provides excellent protection in domestic abuse situations, women seeking abortions are not nearly as well protected. Idaho scored only above Utah in its restrictive abortion laws, which include waiting periods, mandatory counseling, parental consent, a ban on public funds for abortions, and a lack of policies protecting clinicgoers from violence and harassment.

MONTANA

Rank: 44
Overall FEM Score: 34.21
Employer: 5.51
Economics: 7.65
Governance: 6.82
Rights: 5.67
General: 8.56
Population: 799,065
Revenue: $2.7 billion

Like its neighbor Idaho, Montana is not as progressive as the other states in the region. Its rank, 44th, reflects the state's well-below-average scores in most categories.

Montana's strong points include average scores in the women in private management category, a slightly above-average percentage of female state legislators, and a relatively high score for a history of early women's suffrage. Other than these few highlights, Montana scores consistently below average in all other measures.

Its lowest scores are in child services spending, where it ranks only above Idaho, by spending a mere $3 per child on child care and development, and in restrictive abortion laws.

NEVADA

Rank: 3
Overall FEM Score: 75.61
Employer: 11.72
Economics: 18.37
Governance: 17.83
Rights: 19.05
General: 8.63
Population: 1,201,833
Revenue: $3.9 billion

Nevada's scores range from slightly below average in one category, general climate, to excellent in the other four categories, giving it the third highest FEM score nationally.

Nevada's economic climate is particularly strong for women: 31 percent of the state's business owners and 34 percent of private industry managers are women, both well above average. It was also strong in the women in governance category; Barbara Vucanovich served her seventh term in the House and was the first woman to chair the Appropriations Subcommittee on Military Construction. Unfortunately, she retired in 1996; Nevada's new congressional delegation will be all male. Nevada also has the second highest percentage of female state legislators.

The state really stands out by being a good employer, a feat none of the other western states has accomplished. An above-average number of female managers, above-average pay (female state employees in Nevada earn 83 percent of the rate for men), and average benefits give Nevada a respectable score in this category.

Its rating for general climate was marred by a very low child care allocation, 47th overall. Nonetheless, Nevada is clearly one of the better states in terms of its overall treatment of women and should be considered as a destination for travelers wanting to support a good environment for women.

NEW MEXICO

Rank: 11
Overall FEM Score: 58.89
- Employer: 14.46
- Economics: 11.56
- Governance: 10.43
- Rights: 14.91
- General: 7.54

Population: 1,515,069
Revenue: $5.6 billion

New Mexico's solid performance in our study ranks 11th nationally and makes its tourist destinations, Santa Fe and Taos, good ones for people who support women.

It is just above the national average in terms of women's business opportunity, but not as strong as some of its western neighbors. A few areas require improvement, particularly the absence of female representation in Congress (the national average is 9 percent), and the general climate category, where New Mexico shows another absence of women in high political office, does not have a stellar history of women's voting rights, and spends well below the average for child services.

New Mexico scores much better in the state as employer and women's rights categories. It outperforms all other western states as an employer; in fact its above-average number of women in management and its 87 percent salary differential place the state in eighth place nationally. New Mexico also leads the region in its number of female political appointees (33 percent versus 23 percent nationally).

New Mexico earned its good score in the women's rights category by being one of only nine states with community property laws. That and its strong protection for domestic abuse victims helped the state to overcome a slightly below-average stand on abortion rights.

OREGON

Most people know more about the plight of Oregon's spotted owls than about the state's treatment of women. The good news is that Oregon's economic, political, and legal climates are all hospitable to women, earning the state a place in the top ten. Economic opportunities for women are above average and the state ranks fourth behind Hawaii, Colorado, and Maryland in percentage of business owned by women. In the women in

Rank: 9
Overall FEM Score: 59.88
Employer: 8.75
Economics: 15.73
Governance: 13.13
Rights: 12.35
General: 9.92
Population: 2,843,321
Revenue: $10.0 billion

governance category, Oregon has above-average percentages of women serving in Congress and in the legislature. U. S. Representative Elizabeth Furse, serving her second term, has been a strong voice for women. In addition to making domestic abuse a priority, Furse is a member of both the Pro Choice Task Force and the National Women's Political Caucus. Oregon is also strong in the women's rights category, with very strong policies in favor of abortion.

The lulls in Oregon's performance are as an employer, where women's salary differential is less than the national average and their benefits are also slightly below average, and in the general climate category. Nonetheless, Oregon definitely is a state committed to women's progress.

UTAH

Rank: 46
Overall FEM Score: 32.99
Employer: 0
Economics: 5.88
Governance: 7.13
Rights: 7.98
General: 13.49
Population: 1,722,850
Revenue: $4.9 billion

Utah finished clearly in last place for the region, but ahead of Alabama, Mississippi, West Virginia, and Arkansas among the four regions of the country. However, some of Utah's policies and records are the least progressive of any state, placing it pretty solidly at the bottom of the barrel.

Concerning its good points, Utah has an average number of women-owned businesses, and does have some female politicians and a lieutenant governor. The domestic violence laws are very strong, and the state gave women voting rights long before the federal government did.

As an employer, Utah hits rock bottom. It has the worst numbers of women in management—20 percent of state managers are women as compared with the national average of 31 percent—and the fourth worst salary differential for female government employees. Its numbers are not quite as bad in the private sector, where women occupy 25 percent of managerial positions, compared with the national average of 29 percent. However, it plummets again in the women's rights category, where the state has taken every action possible to restrict a woman's access to an abortion.

Utah is a beautiful state and is a popular destination for skiing and visiting national parks. Based on its record with women, however, we suggest that readers look elsewhere for their vacation travel.

WASHINGTON

Rank: 1
Overall FEM Score: 85.27
Employer: 13.51
Economics: 15.45
Governance: 20
Rights: 19.63
General: 14.68
Population: 4,866,692
Revenue: $17.3 billion

Not only does Washington rank first nationally, its FEM score is significantly higher (by seven points) than even the second place finisher, California. Washington is above average across the board, with only a minor lapse in the state as employer category. It leads the pack in several areas including women in governance and abortion rights.

Washington has an above-average number of women in all parts of government: in the house, the legislature (where it has the highest female percentage of any state), political appointees, and even the Senate. Senator Patty Murray, serving her first term, has identified her policy priorities as children's, family, and women's issues, among others. Her political career began when she garnered support from 12,000 families to save state funding for preschool programs.

Washington excelled in the women's rights category, where it is one of only 11 states to have community property laws, and received the highest possible score in the reproductive rights category (the only state to do so) by having clinic violence and harassment policies, allotting public funds to abortion, banning counseling, parental consent, and forced waiting period laws, and taking an active legislative stand on pro-choice issues and RU-486.

Child services support is another of Washington's strong points. The state has made commendable progress in child care in recent years: it has instituted a strong licensing program and inspection staff, a model support system for family child care providers which offers a resource guide on children's activities, health, and development, and a close relationship with employers to help set up such benefits as flex time. Voters also passed a bond issue to pay for child care facilities at elementary schools.

Washington has shown an exemplary commitment to women and family issues. We wholeheartedly support the selection of Washington for a convention, business trip, or vacation.

WYOMING

Rank: 23
Overall FEM Score: 49.08
Employer: 9.8
Economics: 5.07
Governance: 13.92
Rights: 8.9
General: 11.38
Population: 455,000
Revenue: $2.0 billion

Wyoming's FEM score placed it in the middle of the pack, 23rd overall. Its strengths are in the percentage of women in government and its general climate. Wyoming has the second highest percentage of women in the Congress (only Connecticut scores higher). Although we give the state credit for electing Representative Barbara Cubin in 1994, she is the first woman *ever* to represent Wyoming in the House. The state's percentage of women in the legislature is above average, although its record for appointing women to political positions is below average.

Wyoming's performance in the general rights category was quite varied. Its low per-child spending was offset by its history of women's suffrage: Wyoming allowed women to vote earlier than any other state.

As a state employer, Wyoming was just below the national average. It was weakest in economic opportunity, where it ranked 50th in percentage of female managers in the private sector. That measure was countered somewhat by an above-average percentage of women-owned businesses.

There are many worse states in which to spend your travel and business dollars—and also many better ones.

Western Region Summary

According to the measures studied, the West scored well. In fact, on four of the five categories, the West outscored the other three regions: It is more hospitable to women with respect to economic climate, more willing to elect and appoint women to political offices, more committed to women's rights, and more progressive in terms of its general climate.

However, on the fifth category, the state as employer, the West places last. Female state employees there hold fewer management positions, earn less money, and have access to fewer benefits. Even Washington, California, Nevada, and Hawaii—which rank in the top five nationally on overall FEM score—show some room for improvement.

When the two subregions of the West, the Pacific states and the mountain states, are examined separately, the five Pacific states score higher than the eight mountain states. The average FEM score for the Pacific states (Alaska, Hawaii, Oregon, Washington, and California) is 62.9, versus 54.1 for the mountain states.

If your business needs or vacation travel leads you to this region of the country, we recommend that you consider Seattle, San Francisco, or Las Vegas rather than Salt Lake City. If you wish to ski, we suggest Lake Tahoe or Mammoth in California, Taos (in New Mexico), or any of the Colorado resorts instead of any ski areas in Utah or Wyoming.

COMPARISON OF REGIONS

As noted, the western states clearly outscored all other regions of the country. The four regions are ranked as follows:

RANKING	REGION	FEM SCORE
1	West	57.48
2	Northeast	53.63
3	Midwest	48.98
4	South	42.47

Next we show the scores for each category by region of the country.

Average Category Scores By Region

REGION	I EMPLOYER	II ECONOMIC	III GOVERNMENT	IV RIGHTS	V GENERAL
West	7.88	12.87	12.19	12.76	11.78
Northeast	8.85	10.82	12.14	10.52	11.30
Midwest	9.14	10.92	9.81	8.28	10.83
South	13.02	6.52	6.88	8.75	7.30

As seen, the western states outscored the other regions on four of the five categories. The southern states shine only in the government as employer category, where the overall rankings of the four regions are reversed.

SUMMARIES OF INDIVIDUAL STATES

Below we provide a breakdown of the rankings of each state by category.

STATE	I STATE AS EMPLOYER	II ECONOMIC CLIMATE	III WOMEN IN GOVERNMENT	IV CIVIL RIGHTS	V GENERAL CLIMATE
Alabama	10	45	50	37	48
Alaska	37	25	40	22†	3
Arizona	45	6	24	17†	24
Arkansas	31	46	33	44†	32
California	19	5	18	1	1
Colorado	48	1	19	39†	6
Connecticut	6	16†	5	33	29
Delaware	23	13	44	50	35
Florida	24	12	12	24†	44
Georgia	5	30	29	22†	47
Hawaii	29	2	17	13†	14
Idaho	47	48	6	4	26
Illinois	39	18	16	15†	7
Indiana	22	33	32	47†	38
Iowa	9	27	37	19	22
Kansas	14	22	2	29†	4

STATE	I STATE AS EMPLOYER	II ECONOMIC CLIMATE	III WOMEN IN GOVERNMENT	IV CIVIL RIGHTS	V GENERAL CLIMATE
Kentucky	15	42	43	44†	42
Louisiana	2	49	47	12	43
Maine	20	39	7	6	12
Maryland	21	3	3	27	19
Massachusetts	41	8	8	36	10
Michigan	18	24	22	49	18
Minnesota	26	15	14	31	17
Mississippi	4	50	49	34†	49
Missouri	42	20	31	20†	37
Montana	44	31	38	43	33
Nebraska	50	16†	30	47†	15
Nevada	17	4	4	3	31
New Hampshire	46	29	20	10†	39
New Jersey	32	19	45	10†	2
New Mexico	8	21	23	5	41
New York	43	7	21	8	9
North Carolina	1	38	10†	7	46
North Dakota	25	11	39	41†	20
Ohio	7	32	25	20†	21
Oklahoma	3	40	41	32	13
Oregon	33	9	13	13†	27
Pennsylvania	35	35	27†	26	34
Rhode Island	30	28	15	38	25
South Carolina	13	47	48	15†	45
South Dakota	34	23	35	41†	23
Tennessee	11	43	46	28	36
Texas	40	36	26	9	11
Utah	49	37	36	34†	8
Vermont	28	26	10†	39†	28
Virginia	36	14	34	24†	50
Washington	12	10	1	2	5

STATE	I STATE AS EMPLOYER	II ECONOMIC CLIMATE	III WOMEN IN GOVERNMENT	IV CIVIL RIGHTS	V GENERAL CLIMATE
West Virginia	16	44	42	44†	40
Wisconsin	38	34	27†	17†	30
Wyoming	27	41	9	29†	16

†Tied with another state in this category.

CONCLUSIONS

Considerable sums of money are spent on domestic travel for business and leisure. Obviously, there are times when the trip's destination is dictated by its purpose: if you want to see Yosemite, you'll have to go to California; if your factory is in Detroit, then you must go to Michigan. On the other hand, there are many times when travelers have a choice. If it's skiing you're after, there are slopes in various states that would love to have your tourism dollars. If you are planning a convention, there are many cities with diverse facilities to choose from. It is in these situations that we hope you will consult the ratings in this chapter.

The region that consistently tops the others is the West, home of four of the top five nationally ranked states (Washington, California, Nevada, and Hawaii). As a group, the 13 states in the western region outscore the other regions in four of our five categories. If you are planning a trip to this region, however, we recommend choosing a state other than Utah or Montana, both of which score in the bottom ten nationally. For skiing, California, New Mexico, and Colorado are the best choices; for beautiful parks and mountains, Washington, California, and Hawaii; and for easily accessible airports and hotels for business travel and conferences, Washington, California, and Nevada.

The next strongest region is the Northeast, which boasts New York and Connecticut, two top-ten-ranked states. Also in this region is Massachusetts, which, although not in the top ten, distinguished itself by spending more than any other state (in fact, almost 60 percent more than the state with the next highest spending) on children's services. Another strong point about the region is that none of its constituent states score near the bottom on any FEM category. If it's a city you're seeking, Manhattan is a good choice; both New York and Connecticut showcase many beautiful beaches and lakes.

The states in the Midwest score consistently in the middle of the pack (their average FEM score is 49). The state with the highest ranking in the region is Kansas (6th place nationally), followed by Illinois (12th place). Some states in the region revealed various strengths: Kansas had excellent female representation in both the House and Senate (Illinois also has a female senator); North Dakota has the second highest percentage of female managers in the private sector; and Illinois and Ohio both have outstanding domestic violence laws. At the other end of the scale are Nebraska and Indiana, both in the bottom ten. If your travel plans take you to the Midwest, we recommend that you see the sights and cities in Kansas or Illinois, not the other states in this region.

With a few exceptions—particularly Maryland—the South had a distinctive scoring pattern. For the most part, the states in the South are excellent employers: their female employees are more frequently managers, earn more relative to men, and receive more family benefits than state employees in other regions. However, the southern states are consistently poor in the economic climate and women in governance categories. Women in these states hold fewer private sector management jobs, own fewer businesses, and are elected less frequently to political office than women in other regions. The South also has within its borders the four lowest-scoring states in the country: Alabama (50th), Mississippi (49th), West Virginia (48th), and Arkansas (47th).

The South does have one thing going for it, namely, Maryland. Our fourth-place finisher, this is the star of the southern states. It has a strong economic climate for women, better female representation in government and more progressive reproductive rights laws than any other state in the South. It also has a female U. S. Senator, a distinction shared only by Texas and Louisiana in the South. If you are planning a trip to the South, Maryland stands out as a clear choice.

Clearly, some uniform trends emerge in comparisons of geographical regions, but interesting differences nonetheless exist (such as Maryland). We suggest that you look at the category that is of most interest to you before making your travel plans. We hope that you use these ratings, and that you inform the appropriate state officials as to why you have chosen to visit or to go elsewhere.

Chapter Eight

Globe-Trotting Women

Too often, the great decisions are originated and given form in bodies made up wholly of men ... whatever of special value women have to offer is shunted aside without expression.

Eleanor Roosevelt, 1952

Social science affirms that a women's place in society makes the level of civilization.

Elizabeth Cady Stanton, 1898

Several years ago, the Associated Press reported that two teenage girls in Algeria were killed because they didn't have veils on their faces. Twenty-eight other Algerian women were murdered earlier for the same reason, but no one has yet been arrested or prosecuted for any of these crimes—nor are they expected to be.

In Colorado, a few months earlier, a woman leaving her night job was accosted and raped, but managed to escape and hitch a ride. A good Samaritan drove her toward the police station, with the attacker in pursuit. He caught up with them on the lawn of the police station, shot the driver, kidnapped the woman, and then murdered her. The suspect was caught soon after and imprisoned.

From these two stories, you might conclude that Algeria and Colorado are both unsafe places for women. But there are some important differences in the two scenarios. Although violence against women is omnipresent, the context and consequences differ considerably. In the Colorado incident, the crime was committed by a single individual, and public outcry placed great pressure on police to find the perpetrator, which they did within several days. In Algeria, the murders were committed by a radical religious group whose purpose, apparently, was to frighten women into obedience. The perpetrators were not apprehended, in part because many support their beliefs. In Algeria, unveiled women are likely to be viewed as military targets (Bennoune, 1995).

In fact, the Algerian murders occurred within a sociopolitical climate where women have little power and men have extraordinary control over women's lives—a standard in many Islamic countries. Consequences for offenses ranging from dress code violations to political protest in Islamic countries can be ominous. For example, Dr. Tulima Nasrin, a Bangladeshi writer who wrote a novella considered blasphemous by Moslems, spoke out against religiously mandated discrimination toward women in a newspaper interview, and was forced to flee her country because of death threats (*Daily Camera*, February 17, 1995). In another case in Tehran, Shuida Krami, a 22-year-old who signed a letter calling for the removal of discriminatory measures against women, was stabbed and killed (*Rocky Mountain News*, August 4, 1994).

Condoned physical violence is not the only problem faced by women in some countries. Laws and customs, often under the guise of religion or tribal tradition, dictate brutal treatment of women, ranging from genital mutilation in girls to women being forced to have multiple children because birth control is inaccessible. Or, in China's case, sometimes being forced to terminate wanted pregnancies (Human Rights in China, 1995).

This chapter addresses the issue of what women can do to improve the condition of their sisters globally. Their problems are overwhelming in scope, and there are many historical factors that make change difficult. Many of these practices have been accepted for millennia, and are enshrined within religious frameworks. Because women have not held positions of power, they have had much less influence in political decision-making than their numbers would suggest. Even in democracies where women are well educated and economically well off, their voices have often not been heard. In the United States, for example, foreign affairs are not conducted with women in mind. The authors are hard-pressed to recall a single instance when the treatment of women has influenced U. S. international policy decision-making.

But beyond the need to sensitize our public officials to pay more attention to women's concerns, is there something more we can do? Fortunately, the answer is yes.

There is an economic option that has not been widely used, but which can have important effects: we can examine, country by country, our spending on international travel and imports, focusing on the countries' treatment of women. Americans spend over $43 billion on international travel, a considerable proportion of which is spent by women.[1] We urge you to carefully consider the destination of your next trip. If your international tourist dollars are spent in countries that treat women well, and not in those that don't, your money can have a positive effect on women on a global level.

The same principle applies to purchases of imported products. Look carefully at labels to learn where the products were made and avoid buying any from countries with poor ratings. Dollars speak loudly all over the world. When your spending decisions are based on a consideration of the status of women, you can directly affect policies.

In the previous chapter, we demonstrated the extent to which variation exists in the treatment of women in the United States. Countries differ much more than do our 50 states. In some countries, women lack such simple freedoms as the right to vote, to travel, or to learn to read. Sometimes, just being born female can be extremely hazardous. Population figures from a number of countries in both developed and underdeveloped areas suggest that sex ratios differ from expectation by a wide margin. About 100 million women appear to be "missing," many as the result of female infanticide, abandonment, and higher abortion rates of female fetuses (Kohoni, 1994).

This chapter examines a number of factors that influence women's status in 35 countries,[2] with a view toward helping you make sensible spending decisions. We examined and rated a variety of government policies and practices that affect the status of women within each of the 35 countries. A FEM rating scale, comparable to those used in previous chapters, was then established and applied to our sample.

As was the case in the other chapters in this book, we chose five general categories to explore.

The first category evaluated each country's general legal rights of women. Several indices were employed here. Did women have the right to vote, and if so, how long have they had it? Could they own property in their own name? Were they permitted to work? Interestingly, all 35 countries we rated legally permitted paid employment, although the recent Islamic government that took over in Afghanistan has prohibited women from working, revealing how quickly rights may be taken away. Finally, how much freedom did women have relative to men in their country[3]: were they permitted to dress as they wished, to drive a car, or to travel outside their city without a male companion?

The second category looked at family roles and rights. Here, we considered each country's laws and policies that affect women as mothers, wives, and ex-wives. A variety of indices were used, including whether women had equal legal status in entering into or dissolving a marriage, whether they could control the number of children they had, and whether they had the option to end an unwanted pregnancy. The husband's legal responsibilities were also considered. Was child support mandated after divorce? What, if anything, happened to husbands who physically abused their wives? How adequately did the country's laws protect the victim from further abuse?

The third category evaluated the reproductive health status of women. Here, we examined maternal mortality rates, as well as several other factors that have been shown to relate to women's family role stress, such as fertility rate and household size.

The fourth category rated the overall economic and educational climate for women. This category included the adequacy of women's economic opportunities and whether they had legal protection against workplace discrimination. It also examined how much women earn relative to their male counterparts. Because women earn less than men in all countries, we assessed the size of this disparity, relative to other countries. Thus, a high rating on this index indicates a smaller difference in earning between women and men—not equality. A third measure was the percentage of female managers in private industry, or in government positions for noncapitalist countries (based on data from Antal and Izraeli, 1993). As no country has yet attained gender parity, we rated the extent of the gender differential in positions of authority. A certain level of post-secondary school education is normally needed to obtain any kind of middle- or high-level management position. Thus, we also included the percentage of university students who were women.

The final category, general climate, explored a variety of measures that reflected how positive or negative the overall atmosphere was for women within each country rated. We included in this category whether and when each country ratified the United Nations Convention on The Elimination of All Forms of Discrimination Against Women, which was adopted by the United Nations General Assembly in 1979 (as of this writing, the United States has signed, but has not ratified this document). A second measure was whether or not there are national laws outlawing discrimination against women. A third index was the proportion of women in elected government positions (or serving in the government for non-democratic countries). We also looked at women's average life expectancy, literacy rate, and their access to education relative to men.[4] Each of these factors reflects women's well-being—or lack of it.

The countries rated were chosen either because they attract tourists and business travelers or because our level of trade with them is high. The geographical areas covered include: North and Central America, our closest neighbors, South America, Europe, the Middle East, Africa, and the Far East. The United States was also rated so that comparisons with other countries could be made. Information was compiled from a variety of sources

Ratings are provided for each country, and countries are organized by region. Scores for each of the five categories were totaled to provide an overall FEM score for each country. The best score possible is 100. The average for each category is 10; the overall average is 50. Graphs are presented to allow the reader to compare the five different categories easily. In order to compare categories logically, ratings were statistically transformed into standard score units.[5]

Ratings were based on information compiled from a variety of sources.[6]

COUNTRY RANKINGS

Here's how the 35 countries ranked on FEM scoring, based on a scale of 1 to 100:

	COUNTRY	FEM SCORE
Best	1. Finland	73.34
	2. Australia	72.78
	3. Canada	71.54
	4. Denmark	69.53
	5. United States	68.97
	6. Germany	67.48
	7. France	66.23
	8. United Kingdom	66.12
Intermediate	9. Russia	63.59
	10. Italy	63.03
	11. Greece	62.15
	12. Spain	59.62
	13. Singapore	54.14
	14. Panama	53.78
	15. Mexico	52.16
	16. Jamaica	51.57
	17. China	50.94
	18. Israel	50.07
	19. Brazil	48.55
	20. Switzerland	48.02
	21. Japan	47.89
	22. Dominican Republic	47.12
	23. Venezuela	46.98
	24. Argentina	45.71
	25. Ecuador	45.25
	26. Chile	45.13
	27. South Africa	38.62
	28. Turkey	38.31
	29. Guatemala	38.13
Worst	30. Kenya	31.44
	31. India	30.77
	32. Morocco	29.36
	33. Egypt	28.98
	34. Tanzania	14.28
	35. Saudi Arabia	7.38

FAR EAST/PACIFIC

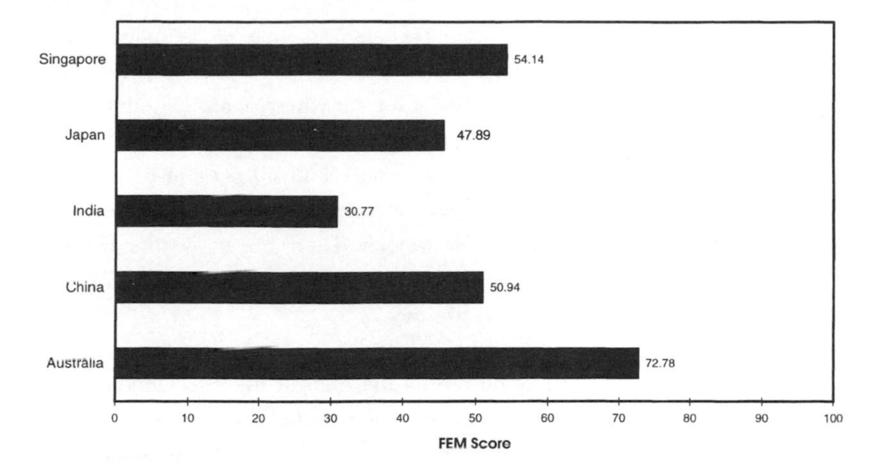

Five countries were rated in this region: Australia, China, India, Japan, and Singapore. They vary enormously in size and type of government. Australia was included on the basis of geography, even though it differs considerably from the others in terms of major population group and historical and cultural influences.

Not surprisingly, Australia is by far the most progressive in its treatment of women. When it is included in the region's averages, the FEM score for the Far East and Pacific is 51.30, or slightly above average; without Australia, 45.93.

AUSTRALIA

Australia, although far from the United States geographically, is not so distant culturally. It received the second highest FEM score of all 35 countries rated, with high scores in all categories.

Its high score in the category of legal rights[7] is partially attributable to its early granting of suffrage to women. Australian women had the vote at the turn of the century, long before American or European women did.

Australia's record in providing economic and educational opportunities for women is also outstanding. Many laws affirm women's equality

Overall FEM Score: 72.78
Legal: 15.91
Family: 14.36
Health: 14.01
Economy: 16.48
General: 12.92
Population: 17,803,000
Revenue: $293.5 billion

in the workplace, and Australia's record with regard to pay equity and women in management is one of the strongest. Australian women are paid 80 percent of the rate for their male counterparts (much higher than the United States's 72 percent), and 30 percent of the nation's managerial positions are occupied by women. There are no gender differences in educational attainment or literacy.

Other high points for Australia include one of the lowest maternal mortality rates in the world, and one of the highest life expectancies for women. Australian women live to an average age of 80, perhaps because of all the rights and opportunities they have, second only to Japan's age of 82.

The only discordant note to this harmonious picture is the relatively low percentage of women—just over 6 percent—in elected government positions. Australian women need to exercise their political options more forcefully to achieve better representation. In all other respects, however, it is one of the world's best places for women.

CHINA

Overall FEM Score: 50.94
Legal: 10.34
Family: 11.43
Health: 10.96
Economy: 6.23
General: 11.98
Population: 1,160,017,000
Revenue: N/A

China is the world's oldest and most populous country. Some of its historical practices, such as foot-binding, have become symbols of how male tastes took precedence over women's well-being. Girls' feet were bound as they grew, a painful way to preserve the daintiness of women's feet and prevent them from doing labor. This deliberate crippling of daughters in more affluent families (poorer families could not afford the "luxury" of women unable to work in the fields) is a particularly graphic symbol of female disempowerment.

Fortunately, China has made great strides in its treatment of women since its revolution, propelling its FEM score to somewhat above average. Its overall score is not based on consistent scores across categories, however. As the accompanying chart demonstrates, China ranks considerably below average in economic and educational opportunities.

Women in China do not have equal opportunity in employment and education. Although China did pass equal rights laws in 1982, they are not well enforced. There are few female managers—only 11 percent—and relatively few women attend universities—only 28 percent.

The general category scores are above average because China was one of the earliest countries to ratify the U. N. convention on abolishing discrimination against women. It also has the highest rate of female participation in government in this region, at 21 percent.

Because of the closed nature of its society, some information concerning women's status is either not known or unverified. Several ominous trends have been noted in the Western press in recent years that appear to have increasingly negative ramifications for its female citizens. In its zeal to control its enormous population, for example, China has instituted some draconian procedures that drastically reduce women's reproductive freedom. The one-child family appears to be a two-edged sword for women. The good side is that girls without brothers will be the recipients of greater parental aspirations; the bad side is that there are more unwanted female babies, reports of female infanticide in rural areas and abandoned female babies appear frequently, and the adoption of female babies by foreign families is on the rise. There appears to be a growing disparity in the number of males and females in the country (Human Rights in China, 1995).

INDIA

India, the other great Asian center of culture, is a much more democratic country than China, but shares with it the need to restrain population growth. India's overall FEM score of 30.77, the lowest of the five countries rated in this region, demonstrates that women do not invariably fare better in democratic countries.

Like China, its huge Asian neighbor, India has historically tolerated horrendous customs involving its female citizens. The practice of purdah

Overall FEM Score: 30.77
Legal: 10.24
Family: 10.42
Health: 2.36
Economy: 5.58
General: 2.17
Population: 883,910,000
Revenue: $240 billion

forbade women from leaving their homes if not accompanied by their husband. Even more horrible was the practice of suttee, in which a wife was linked in every way to the fate of her husband. It went so far as to mandate that if he died before her, she would be cremated with him, even if she was still alive (interestingly, this tradition occurred with greater frequency in more affluent families). Although this practice has been forbidden during this century, there are occasional reports of it still being practiced. Suttee demonstrates the ultimate in women's devaluation.

Getting back to the present, FEM ratings are average for the first two categories, women's legal and family rights, indicating that India's laws with regard to women are reasonably progressive. India is a democracy, and women have had the vote since the country achieved independence in 1948. Within the family, the government strongly supports the use of birth control by the population, although the actual rate of contraception use is below other countries in the region. Government policies with regard to marriage and divorce are mixed. India has very stringent domestic violence laws, but divorce is not as easy for women to obtain as it is in other countries of the region—in part because of religious practices favoring men.

In the employment area, India is somewhat above average in pay, with women earning 72 percent of what men make, but women comprise only 2 percent of managers in private industry. In education, only 28 percent of university students are women, and India's female literacy rate is the lowest in the region (the ratio of female to male readers is 39:100). Girls are also less likely to obtain an education, with only half as many going to high school as boys.

The average life expectancy for women in India is the lowest in the region, 61 years compared with a regional average of almost 75. Indian women are more likely to die during childbirth than in any other country assessed in the region. Household size and fertility rate are larger than for any other Asian country rated.

Although many of these deleterious effects may be related to poverty, this would still not explain the significant variations between men and women. Indira Gandhi certainly symbolized that women could reach the top, but she was quite an exception to the normal lot of Indian women.

JAPAN

Overall FEM Score: 47.89
Legal: 10.73
Family: 13.2
Health: 11.78
Economy: 0
General: 12.49
Population: 124,536,000
Revenue: $2.468 trillion

Japan's overall FEM score of 47.89 is just below the global average. When the category scores are considered, however, a great deal of variation is seen. Japan's ratings are at or above average in all but one category, economic opportunity for women, for which Japan receives a score of zero.

General laws regarding women's rights are average, and its laws and policies concerning women's family roles are particularly good. A considerable majority of the population uses birth control, wives legally have equal rights within a marriage, divorce is not biased in favor of men, and abortions are legal. Domestic violence laws protect women, although studies show that much of such violence goes unreported in Japan. Health statistics and the general climate for women are both pretty good: precollege girls and boys are educated equally, women's literacy rates are equal to men's, and Japanese women can expect to live to age 82—the longest life expectancy of any country in the world.

Because of deep-seated values regarding women's subordination, however, very few of the top government posts go to women. This is equally true in the private sector, where Japan receives *the lowest score of any country rated*. This category is based on women's chances for advancement in employment and higher education. Although equal opportunity laws were passed in 1986, there are no penalties stipulated for lack of compliance. As a result, Japan is alone among industrialized countries in its extremely discrepant pay scales for men and women: women receive only 52 percent of men's earnings. In addition, a very small proportion of women—only 7 percent—hold management positions. Female university graduates who go to work for large multinational companies report having nothing to do at work except to go out and buy snacks or make tea for their male colleagues. When women who do work return home at night, they spend an average of 4½ hours a day on household chores—compared with only 15 minutes for their husbands. And although there is no discrimination against females at lower levels of education, only 24 percent of college students are women, the lowest of any country in the region. This is a very peculiar picture for a democratic, highly industrialized, and wealthy country.

SINGAPORE

Overall FEM Score: 54.14
Legal: 6.95
Family: 14.86
Health: 11.37
Economy: 12.94
General: 8.02
Population: 2,930,000
Revenue: $45.9 billion

Singapore is a small but economically strong island. The annual per capita income is over $15,000. It has in recent years received considerable publicity for its harsh sentencing for crimes generally regarded in the United States as minor, one highly publicized case involving an American teen caned for vandalism. The U. S. State Department also voices concerns about human rights abuses involving censorship, travel restrictions, indefinite confinement, and intimidation of political opposition. These problems potentially affect men and women equally. Despite these problems, Singapore's overall FEM score of 54.14 places it somewhat above average for its treatment of women and the second highest in the region.

Singapore's pattern of scores, like Japan's, is uneven, albeit in a different way. The country receives average scores for women's health and economic opportunities and above-average scores for women's family roles. Its low scores are in the areas of legal rights and general climate for women.

Singaporean women are reasonably well protected with regard to their family roles and in the workplace. Birth control is readily available and used, women are legally regarded as equal partners in a marriage, abortions are available without undue restrictions, and domestic violence laws are protective of women and enforced. Men found guilty of abuse receive mandatory caning and a minimum of 2 years in prison. (Women are never caned, nor are men over 50.) Within the workplace, women were paid 69 percent of what men made in 1991 (slightly higher than the United States for that year), and the proportion of female managers is three times greater in Singapore than it is in Japan (22 percent versus 7 percent). Additionally, women comprise almost half of the college population. Unfortunately, the government does not enforce any type of maternal leave policy. In this regard, Singapore is behind the United States—but not by much, since the U. S. Congress only passed this legislation in 1993.

Health indicators for women are also quite good in Singapore: low maternal mortality rates, low fertility rates, and high life expectancy.

Where Singapore slumps is in the extremely low rate of participation of women in government; at just over 2 percent, it is the lowest of the five countries examined in the region. It is also one of the few countries that has not signed the U. N. nondiscrimination convention.

In summary, despite its questionable record with regard to some human rights abuses, Singapore does not treat its female citizens badly relative to other countries in the Far East. Considerable improvement is still needed, however, in increasing the level of female participation in government.

Far Eastern Summary

The average FEM rating for this region, 51.30, is somewhat above average. As we have seen, however, there is considerable variation among these five countries. Australia, which is not part of the Asian continent, but is geographically and economically closer to Asia than to other continents, receives the highest score. When only Asian countries are considered, the top scorer is Singapore. China and Japan are average, although Japan has abysmally low ratings with regard to women in employment situations (with a negative score of −.31). The low-scoring country of this region is India, with a long history of female devaluation. Although its laws are now very female friendly, its practices have a long way to go to catch up with them. India is particularly poor on all indices of women's health. Moreover, its educational practices are quite discriminatory. If Australia were eliminated from this grouping, the Asian average would be much lower.

It is clear from the scores for the various countries in this region that there is no easily discernible pattern relating women's well-being to the country's form of government. Although Australia (a capitalistic democracy) is best, India is worst, with the same form of government.

In terms of travel, all of these countries are quite far from us and expensive to get to. If you are traveling to this region of the world, we would recommend that you choose Australia or Singapore. If you are considering purchasing products from this region, you might refuse to buy items from India. If you do pursue this course, it would be helpful to write to the country's embassy explaining the reason for your boycott. Japanese products are ubiquitous in the States, so you should consider your buying decisions here very carefully in view of their record of women in industry. As our earlier chapters suggest, their corporate records in the United States are also poor, for the most part.

MIDDLE EAST/NORTH AFRICA

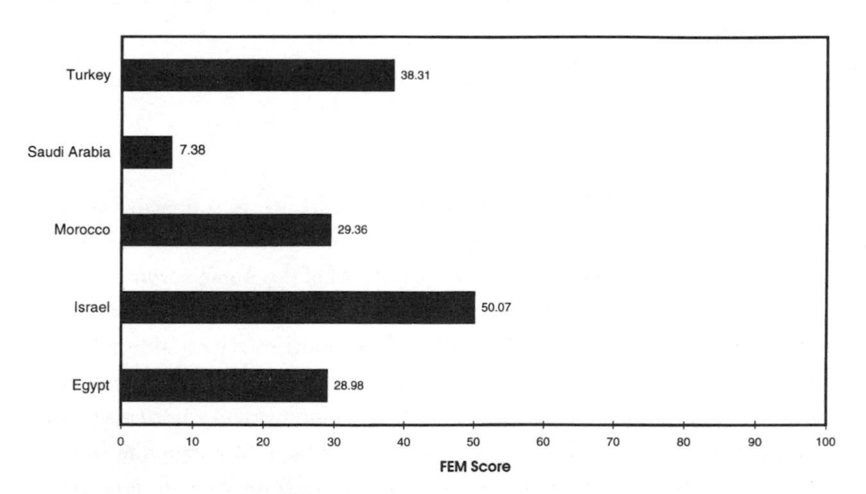

Women in the Middle East and northern Africa are plagued by the strong influence of Islam, a religion that, even in contemporary practice, is very clear about the subordinate role women should hold. Perhaps because of the infiltration of Islamic law into the cultures and laws of the countries rated in this region, it ranks near the bottom of all world regions and includes the worst rated country, Saudi Arabia. Other countries in this region that we rated were Egypt, Israel, Morocco, and Turkey.

EGYPT

Overall FEM Score: 28.98
Legal: 9.66
Family: 5.85
Health: 4.52
Economy: 4.49
General: 4.46
Population: 56,673,000
Revenue: $41.2 billion

Egypt, with a FEM score of 28.98, is not a very attractive place for women. In fact, it is the third worst country in our world ratings. Category 1, the legal category, is the closest Egypt came to an average score, as women there do have some legal rights. Egypt does poorly in the family role category because it has no legal statement of equality concerning women. There are also no domestic violence laws protecting women, and Egyptian women cannot freely obtain a divorce as the husband has the power to refuse it. In 1979, there was some action to correct these inequities, but the legislation was declared unconstitutional in 1985 because the rights granted to women conflicted with the laws of Islam, the state religion.

Women's health status in Egypt is also terrible. Despite government support of birth control, actual usage is low. The fertility rate is high, maternal mortality rates are among the highest in our sample, and women's life expectancy is the lowest in the region. Female circumcision is still practiced.

Egyptian women do not fare much better in the employment or educational areas. They have poor employment opportunities, few laws that protect them, and limited educational options. Egypt reported that 14 percent of its managers are women, as are only a third of its university population. Accounting for the country's poor showing on the last category, general climate, was a very poor literacy rate for women (ratio of female to male readers, 32:100) and the very low proportion of high government positions held by women (1.5 percent).

Egypt is a country that relies heavily on foreign tourism, and admittedly, it does have sights that are quite unique. However, it has a very long way to go in improving the status of women, and we would urge that you go elsewhere.

ISRAEL

Overall FEM Score: 50.07
Legal: 8.8
Family: 9.49
Health: 12.32
Economy: 8.86
General: 10.6
Population: 5,423,000
Revenue: $47.4 billion

Although Israel only achieved an average FEM rating of 50.07, it is by far the best country in this region with regard to treatment of women.

The family rights area is mixed. Although religious restrictions are present for divorce, Israel has very stringent rules against domestic violence. It has the lowest fertility rate in the region and women have good access to birth control. Israel's best category is health. Its maternal mortality rate of 8 per 100,000 births is the best in the region, and one of the best of any country rated. Average life expectancy for Israeli women, at 78.4 years, is the highest in the region, and similar to that for European and North American countries. Israel also has the highest regional literacy rate for women (88 percent) and the highest percentage of women in universities (48 percent).

Where Israel does not compare favorably with other countries in our sample is in the employment rating. Although it has equal employment laws, it has the highest reported pay disparity between men and women in the region—women earn only 57 percent of what men earn—and only 14 percent of managers are women. And despite having had a strong female prime minister, only 8.3 percent of elected positions were occupied by women at the time of writing.

Despite these drawbacks, it is clear that if you wish to visit the Middle East, Israel is the place to go.

MOROCCO

Morocco's FEM score was the fourth worst in our sample. Ratings for four of the five categories are substantially below average, the only exception being the workplace. A respectable score in this category is based on a questionnaire on which the Moroccan embassy indicated that 25 percent of the country's managers are women and that female employees have a salary differential of 72 percent (equal to U. S. pay equity figures). We have no

Overall FEM Score: 29.36
Legal: 7.23
Family: 3.41
Health: 4.89
Economy: 12.1
General: 1.73
Population: 26,069,000
Revenue: $28.1 billion

independent means for verifying this information, however. We would note that it is very much out of sync with other measures of women's well-being in this country.

Even if the good reports of employment advancement are accepted at face value, Morocco still gets a terrible overall FEM score. Women's travel is greatly restricted; they need the permission of their husbands or fathers to travel out of the country. Even a divorced woman needs her father's permission to obtain a passport. Family law offers little protection to women with regard to marriage, divorce, spousal abuse, or child support. Many of these restrictions seem attributable to Islam being the official religion. For example, if a husband discovers his wife committing adultery, he can kill her without penalty (not surprisingly, women do not have the same right when the situation is reversed). Some legal changes favorable to women were passed in 1993, but the U. S. State Department describes them as "cosmetic."

In addition, maternal mortality is very high, there are practically no women in government positions, and the country has the worst record of educating girls of any country we rated in this region.

As exciting as Fez and Marrakech are, we would advise that you make other plans.

SAUDI ARABIA

Overall FEM Score: 7.38
Legal: −6.80
Family: 2.03
Health: 4.7
Economy: 7.02
General: 0.45
Population: 17,119,000
Revenue: $111 billion

In many respects, Saudi Arabia is in a world of its own, and its treatment of women is no exception. Every scale needs an anchor, and Saudi Arabia serves that purpose in our study: it gets by far the lowest score of any country rated.

It receives the lowest scores of any country in three of the five categories.

Because its score in the legal rights category is so far removed from that of other countries, it is actually a negative number (−6.8). As many of you may recall from information presented on television during the Persian Gulf War of 1991, Saudi Arabia provides few, if any, basic rights to its female citizens. They must follow a dress code in covering their face and hair, they cannot drive a car or ride a bicycle, they cannot travel out of the country unless they are accompanied by a man, and they cannot be present in situations with a man who is not related to them. They cannot even travel within the country unless they have permission from a close male relative. They also cannot participate in government. The actions of a few brave women fighting for the right to drive were harshly repressed: following the 1991 demonstration in which 47 women took to the wheels of their cars, all of the women were detained and denounced in religious settings as "corrupters of society." According to a follow-up in *Ms.* magazine (November/December 1991), the female drivers lost their jobs and travel papers, and their families were threatened. Although there is a pervasive abuse of human rights for everyone in the country—for example, no freedom of speech, religion, assembly, or freedom from unwarranted arrest—restrictions on women are greater.

The only category that even approaches average is the economic climate category. That score is based on employment data provided by the Saudi government suggesting that 17 percent of managers are women and that female employees receive a salary differential of 72 percent. If this information is accurate (as in the case of Morocco, we have no independent means of verifying it), it is probably because the country's stringent sex segregation requires that some women occupy management positions in all-female institutions, such as hospitals and schools for women.

It is often assumed that impoverished conditions are associated with poor treatment of women. Although this may be true in some instances, Saudi Arabia presents an interesting exception. Its per capita income places it among the wealthiest countries in the world. Despite this, however, only half as many women as men complete secondary school, only a third as many women as men are literate, and the country has one of the highest fertility rates in the world, with women averaging 7.2 children each. Many occupations are off limits to women, such as engineering, architecture, and journalism.

Saudi Arabian women have no legal protection with regard to their family roles. They are not regarded as equal within the marriage, they cannot divorce without their husband's consent, and they have no legal protection for either domestic abuse or child support if the marriage does not last. In fact, they have no clear rights to property or child custody in the case of divorce. Islamic law also discriminates against women's inheritance: women often receive only half of what their families leave to them, with the other half going to male relatives.

It would be wonderful to be able to help women in this theocratic country gain some basic freedoms, but several factors make this difficult to accomplish. Saudi Arabia's revenues do not rely on tourism, and it is not even clear that Western tourists are welcomed. Their major export is, of course, oil. As consumers, however, we have no way of ascertaining whether our oil purchases come from there or elsewhere. Because of their dominance in oil production, our government has treated them as a staunch ally. We protected them in the Gulf War because we are dependent on their oil, despite their horrendous treatment of women. Clearly, this is one instance where political action may be our only recourse to redress things.

TURKEY

Overall FEM Score: 38.31
- Legal: 11.8
- Family: 9.28
- Health: 7.25
- Economy: 4.57
- General: 5.41

Population: 60,227,000
Revenue: $219 billion

Turkey has the second best record toward women of any of the countries rated in the Middle East, although its overall score still places it below the world average. Turkey scored above average in legal rights and just below average in family rights. Helping women's role in the family is the government's strong advocacy of birth control, and citizens' easy access to it. Unfortunately, however, child support and protection against domestic violence are not legal rights of Turkish women.

Other category ratings are considerably below average. The work world is not particularly favorable to women. Only 3 percent of Turkish managers are women, as are only 31 percent of university students. Although Turkey recently had a female prime minister, very few elected government positions are occupied by women.

Turkey exports to the United States a number of products, including many leather and copper items. It also is the locale for many interesting archaeological sites from antiquity. We would recommend caution before spending consumer or travel dollars here.

Middle East Summary

In terms of our rating system, countries in the Middle East and North Africa are some of the worst places for women. The average FEM score for this region is 30.82, just slightly better than our worst region (see the next section). The top score in the Middle East is Israel's, although its rating is only average relative to the other 35 countries examined. The second best is Turkey, with a score of 38.31. Morocco is a distant third, with a FEM score of 29.36, followed closely by Egypt. The Middle East has the dubious distinction of having the lowest rated country in our sample, Saudi Arabia, with an atrocious 7.38 FEM score that is practically off the bottom of the scale.

Women in the Islamic countries rated have few freedoms and very little in the way of government policy to protect them in either family or employment roles. They do not enjoy equal rights in any area, including education, and they have little if any input into the political process. The educational differentials are particularly striking in Saudi Arabia because it is such an affluent country. Another pattern exhibited for this region is a very high fertility rate. Overall, women in these countries appear to have very little control over most aspects of their lives.

Of the four Islamic countries rated, Turkey differentiates itself by giving women some say in the political process, and in having a government that strongly supports the use of birth control. Nevertheless, very few women there receive a university education, and they have hardly made a dent in the managerial ranks. Egypt and Morocco have very high rates of maternal mortality and low rates of female literacy. Three of these four Islamic countries receive the lowest rankings of our entire sample.

In terms of consumer behavior, the reader can choose not to visit or buy from most of these countries. Turkey and Egypt export many products to the United States, and Morocco exports somewhat less. Saudi Arabia's main export, oil, is not labeled at the consumer level, so it is difficult to take any consumer-related action without political pressure to remedy this situation. If you do wish to travel to this region of the world, we would advise that you go to Israel, and write the others to explain why you are not spending your tourist dollars there.

AFRICA

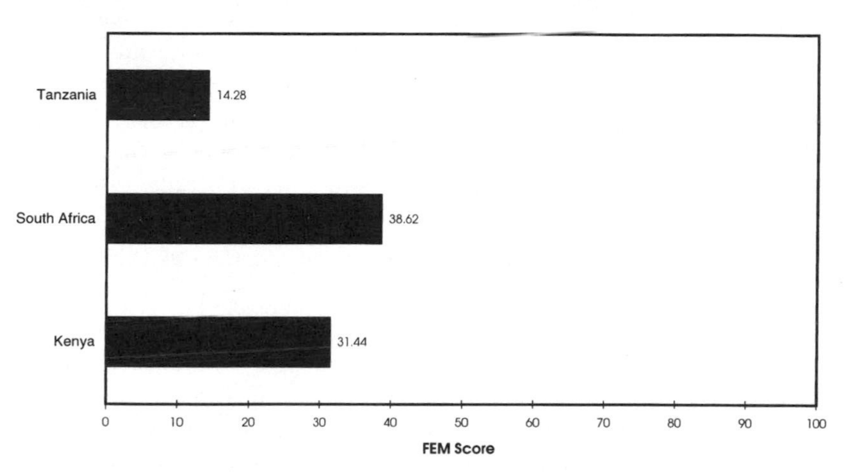

The three countries rated in Africa—Kenya, Tanzania, and South Africa—were chosen because they are common African destinations for tourists and U. S. business dollars. Not surprisingly, South Africa, which is the most developed of the three, scored best on the FEM scale. However, as a region, these countries scored at the bottom of the world order. Tanzania and Kenya were among the bottom six countries. And not a single category score in any country came in at or above average; in most cases it was just a case of how far below average the score was.

It is difficult to view the circumstances of African women from a Western perspective. Although in many cases the countries do have laws or policies that look out for women's interests, there exist conflicting tribal laws that do not. Many women may be caught between these sets of rules.

KENYA

Overall FEM Score: 31.44
Legal: 8.98
Family: 5.41
Health: 2.93
Economy: 8.51
General: 5.61
Population: 28,113,000
Revenue: $8.3 billion

In the February 1994 U. S. State Department report on the state of human rights, the entry for Kenya reads "Domestic violence against women is a serious and widespread problem. Wife beating is common, and rape is widespread." In Kenya, as in other African nations, tribal laws often conflict with government policies. For example, a woman may legally go to court with a domestic violence complaint, but most courts will not do anything because traditional culture permits a man to discipline his wife by physical means, including genital mutilation. Many tribal laws go so far as to prohibit women with claims of rape or domestic abuse from taking these charges outside of the clan. Kenyan women have no right to initiate a divorce, nor are they considered equal partners in a marriage under any existing set of laws.

Nevertheless, Kenya's highest score came in the legal rights category. Women have the right to own property and to travel freely, and working women may take maternity leave. Women may vote, though they have only been able to do so since 1963. The ratio of female to male literacy is 58 to 100. An above-average percentage of managers are women (17 percent versus the African average of 12 percent), as are an above-average number of university students, although that number is still only 29 percent.

In other areas, Kenya scored very low, its worst score coming in the health category. Kenyan women have average access to birth control by African standards, yet the average woman has almost seven children. The maternal mortality rate is 168 deaths per 100,000 births, well above the world standard.

Kenya had one plus and one minus worth mentioning in the general climate category. On the bright side, Kenya is the only African country rated that has signed and ratified the U. N. treaty to end discrimination against women. However, the government has turned a blind eye to the practice of genital mutilation.

For a country whose economy is heavily dependent on tourism dollars, Kenya could do much more to bring its women up to world standards.

SOUTH AFRICA

Overall FEM Score: 38.62
Legal: 5.67
Family: 8.83
Health: 7.72
Economy: 8.55
General: 7.85
Population: 40,435,000
Revenue: $115 billion

South Africa rated the best of the three countries examined in the region. Its FEM score was 38.62, below average in our global sample. Before the recent upheaval of the political system that gave black people basic rights, South Africa was the target of great international criticism and sanctions for its policy of apartheid. Women (white women at least) in South Africa had some of those basic rights before blacks. They could vote in 1979, for example, whereas blacks only got to vote for the first time in April 1994. However, it seems that the country's women, like its black citizens, stand to benefit from the recent turn of events there and the ensuing tide of human rights improvements.

For example, South Africa has recently passed two acts promoting women's rights. The first targets equality between men and women, eliminating the last vestiges of a husband's power to control his wife's property and finances. It removes legal differences in matters of citizenship and repeals prohibitions against women in high-risk occupations, such as mining. The second act aims to prevent domestic violence by simplifying arrest and injunction procedures. It both empowers judges in domestic violence cases and enables convictions on marital rape charges.

Despite this promising progress, South Africa still had several low points. The percentage of women in government stands at a measly 3 percent, well below average even for Africa. Although South African women have fewer children than in neighboring countries, they still average almost five children each. Estimates of illiteracy among women range from one-fourth to one-third of the population. A South African employment act states that no employer shall permit pregnant employees to work from 4 weeks before through 8 weeks after the birth, but while the woman is on leave, the employer is required neither to pay her salary nor to guarantee her job.

TANZANIA

Overall FEM Score: 14.28
Legal: 0.9
Family: 4.27
Health: 1.39
Economy: 6.18
General: 1.54
Population: 25,635,000
Revenue: $7.2 billion

Tanzania's score places it at the bottom of the African countries and second only to Saudi Arabia as the worst country rated. The Tanzanian constitution prohibits discrimination based on nationality, tribe, religion, color, or even life-style. It does not, however, mention women. That is pretty typical of how things are for women in Tanzania.

Tanzania received one of its worst ratings in the area of women's health. Its maternal mortality rate is a staggering 340 per 100,000 (by comparison, many European countries have fewer than 12 maternal deaths per 100,000). A woman's life expectancy is 57 years, the lowest in all of Africa. There is some access to birth control, but women still have almost seven children on average. A recent study done for a United Nations' World Population Conference estimated that 24 percent of married women in Tanzania would like to space their children out by at least 2 years, and another 17 percent do not want any more children. Of those who wish to use some form of family planning, however, the study estimates that because of unavailability, only about 10 percent do, and fewer than 7 percent use modern contraception.

The country is repressive in the area of family law. Women are not seen as having equal standing within a marriage, and although violence against women is still widespread, women have little or no legal recourse. The law does not provide for either divorce or child support in any way that benefits women.

Women are denied equal access to an education, and as a result their literacy rates, ability to participate in the government, and their economic opportunities are all severely limited. Only 41 percent of Tanzanian women have the same opportunity as men to attend high school, and only 18 percent of the university population is made up of women. Only 54 percent of Tanzanian women are literate.

Because of an economy centered around agriculture—90 percent of the employment is in this sector—very few "working women" are actually paid for their work; one source estimates that only 3 percent of the country's working women receive any wages. On a more positive note, large growth in the public service sector of the economy is offering women some rays of hope. There are laws mandating equality in government hiring and pay, perhaps contributing to a female salary differential of 92 percent—an impressive figure anywhere in the world. Women in these civil service positions are entitled to 56 days of paid maternity leave.

African Summary

Many of the basic rights and privileges of women in Africa seem to be dogged by a long tradition of patriarchal tribal cultures. With the exception of South Africa, the governments in countries in this region seem to be doing little to promote women's interests, leading to a regional FEM score of 28.11. Neither Tanzania nor Kenya has laws protecting women from widespread discrimination in their family lives, and Kenya has no laws protecting women in employment situations. Both countries are highly protective of their wildlife to maintain tourism, but need to pay more attention to treatment of their female citizens.

EUROPE

Ten countries in Europe were rated: Denmark, Finland, France, Germany, Greece, Italy, Russia, Spain, Switzerland, and the United Kingdom. This region was the most civilized and progressive of any in the world in its treatment of women.

In some categories, the European countries showed very little or no variation between each other. For example, all European women have certain basic rights: to own property, to vote, to be considered an equal partner in marriage, to divorce, and to travel freely. European women average 1.8 children, and they are entitled to child support in the case of

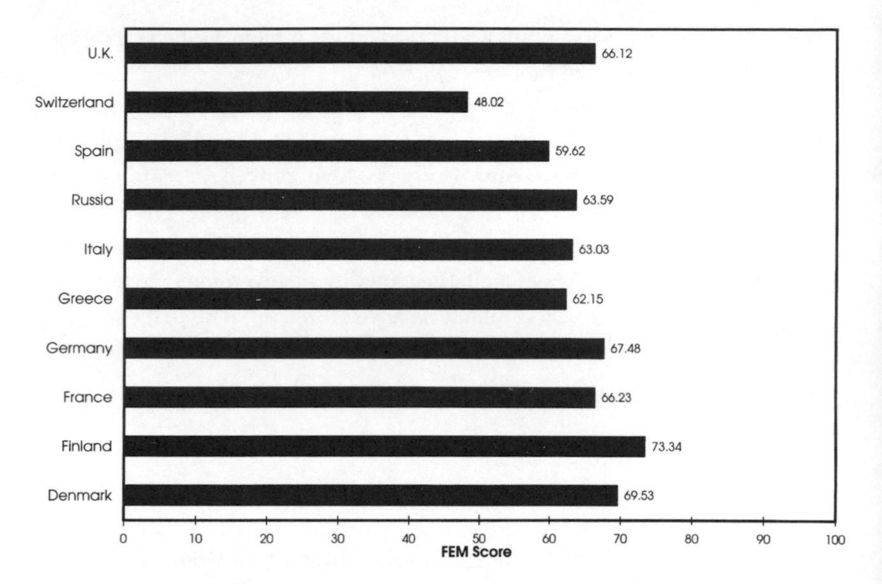

divorce. The legal age for marriage in most of Europe is 18, with the exception of Russia, where women may marry at 16. And more than 90 percent of European women are as literate as men.

The differences, therefore, are elsewhere. Areas in which there was considerable variation include female representation in government and professional managerial positions, when women first received the right to vote, and life expectancy.

DENMARK

Overall FEM Score: 69.53
Legal: 13.65
Family: 14.54
Health: 14.77
Economy: 12.26
General: 14.31
Population: 5,212,000
Revenue: $94.2 billion

Denmark has a very good record regarding treatment of women, scoring well above average and fourth in the world. In the legal rights area, Danish women have had suffrage since 1915, are entitled to 14 weeks of maternity leave, and have broad property rights. The woman's role in the family is also strongly protected legally, including an absolute right to a first trimester abortion and community property standards in a divorce.

Women in Denmark have a salary differential of 82 percent (the European average is 75 percent); however the number of women holding jobs in management is below average (14 percent versus 18 percent for the region). Danish women are well represented in government (33 percent versus 19 percent for the region).

Denmark, along with its Nordic cousin, Finland, seems to be solidly on the right track in its treatment of women.

FINLAND

Overall FEM Score: 73.34
Legal: 14.52
Family: 14.32
Health: 14.43
Economy: 13.58
General: 16.49
Population: 5,212,000
Revenue: $79.4 billion

Finland wins the FEM gold star for outscoring all other countries rated.

In 1906, Finnish women became the first in Europe—and the second in the world, after New Zealand—to get the right to vote and to run for office. Finland also had the world's best ranking in the women in government category, with 43 percent of government positions held by women. In addition, 7 of the government's 17 minister positions are held by women, and the country has the world's only female minister of defense.

Economic opportunities for women in Finland are among the world's best, with women receiving an above-average salary differential and holding an above-average number of upper- and middle-level positions. Perhaps women's success in the Finnish workplace reflects the country's extremely progressive family policies or the fact that half of the university students are women. Parents may take up to *11 months of paid leave* to care for children and the mother has the right to care for her children for *up to 3 years without losing her job*. The state also provides day care for all children under 3 years of age, which may be taken in the form of an allowance toward a private day care facility.

If you're hankering for a European vacation, we highly recommend Finland.

FRANCE

Overall FEM Score: 66.23
Legal: 10.83
Family: 15.16
Health: 14.39
Economy: 14.36
General: 11.49
Population: 57,850,000
Revenue: $1.08 trillion

France scored above average in all five categories, including a first-place European finish in the family roles and rights category.

The country's lowest score was in the area of general legal rights, despite a strong national maternity leave policy that gives mothers 16 to 18 weeks of paid leave. Part of the reason for its score in this category was a late granting of women's suffrage, in 1944 (although it should be recalled that this was only one of five measures in this category).

The high rating for French women's family roles came from several factors: excellent access to birth control (which is subsidized by the government) and very strong domestic violence protection.

Thanks to a 1983 French law prohibiting any discrimination against women, direct or indirect, and guaranteeing women the same access to employment that men have, women have good economic opportunities and a healthy economic climate in which to work. Women's earnings in France compare most favorably to their male colleagues: 89 percent (the highest differential in the region; the European average is 75 percent). Additional legislation, passed in 1992, protects women from sexual harassment in the workplace. There are more women than men in the country's higher education system.

France also did very well in the women's health category, boosted by the second longest life expectancy average in Europe—almost 81 years.

Despite this progress, French women still hold fewer management positions than their European sisters and have a dismally low level of representation in government (6 percent versus the regional average of 19 percent).

GERMANY

Germany's overall FEM score was slightly higher than France's, related in part to its very high score in the women's health category.

Overall FEM Score: 67.48
Legal: 13.26
Family: 14.24
Health: 14.82
Economy: 11.36
General: 13.8
Population: 81,373,000
Revenue: $1.398 trillion

Germany has one of the lowest fertility rates in all of Europe; at an average of 1.5 children per family, Germany's population is below replacement value. Although this issue has been the subject of some concern among Germans, low fertility rates are generally positive because it implies that women are in control of their reproduction.[8] As a result, Germany also has an extremely small average household size.

Germany also treats women fairly well with regard to family roles and rights. The country offers respectable maternity leave, good prenatal care, and free contraception with a doctor's prescription. One of the results of reunification has been negative for women: abortion, once freely available in East Germany, is now illegal (but supposedly overlooked).

Germany's lowest score came in the economic and educational climate category. The country's salary differential was only 67 percent, tying it with Switzerland for the region's worst pay equity.

GREECE

Overall FEM Score: 62.15
Legal: 10.05
Family: 14.18
Health: 13.98
Economy: 11.87
General: 12.07
Population: 10,350,000
Revenue: $82.9 billion

Although Greece's FEM score was well above average in world terms, it was in the bottom third of the European countries rated, probably because the country only recently began paying attention to women's issues. Prior to 1946, the Greek Civil Code, reflecting Roman and Byzantine traditions, gave the husband very substantial power over his wife. These terms were mitigated slightly in 1946, but were not actually revised until 1983.

Since then, women have made substantial progress, both in terms of family rights and reproductive health protection. Women have good access to birth control, although only 59 percent of Greek women use it, compared to an average of 71 percent in Europe. Women are now considered equal within a marriage and have equal obligations to support the family if their means allow. They are entitled to 12 weeks of paid maternity leave and then may return for a shorter workday until their children are 4 years old.

Women have also made progress in the Greek workplace. Laws mandate pay equality and no discrimination, and at 78 percent of what men make, women's salaries are above average for Europe. The number of women in management positions and enrolled in universities, however, is slightly below average.

One real blemish on Greece's record is in the small number of women represented in the government: a record-low 4 percent (the average in Europe is 19 percent). On a somewhat redeeming note, the country does have a female minister of culture.

ITALY

Overall FEM Score: 63.03
Legal: 8.98
Family: 12.45
Health: 14.32
Economy: 15.96
General: 11.32
Population: 57,110,000
Revenue: $1.012 trillion

Italy's scores showed considerable variation in the different categories. It rated fairly low—below average—in the legal rights category for two reasons: late granting of suffrage and very poor maternity leave benefits compared with the rest of Europe.

Other lows concern birth control, domestic violence laws, and marriage roles. Perhaps because of the influential tradition of Catholicism in Italy, and the church's stand against birth control, the most common form of Italian contraception is withdrawal. Italy also lost points for failing to give women equal standing within a marriage and failing to provide legal protection for domestic violence victims.

Despite scoring very respectably on most measures used to assess economic opportunity, Italian women fill only 10 percent of management positions, which is just about half the average. Scores for some measures of health status were very high, including life expectancy and a lower-than-average (for Europe) maternal mortality rate.

RUSSIA

Overall FEM Score: 63.59
Legal: 10.98
Family: 11.27
Health: 11.73
Economy: 14.21
General: 15.4
Population: 148,366,000
Revenue: N/A

The 20th century history of Russia has been significantly different from other countries in Europe and has left women there in a role that is familiar, but perhaps more pronounced, to working women around the world. Until the recent end of Communism in the former Soviet Union, all citizens—women included—were ruled by the credo, "He who does not work shall not eat." This added a mandatory career to the lives of women, who, like their sisters around the world, bear a primary responsibility for raising, feeding, and caring for their families. Because of this expectation, Russian women have been a significant part of the economy for many decades.

This has not helped their family situations, however. One of Russia's lowest scores was in the family rights and roles category. Russian women have very little access to birth control compared with their European neighbors, a situation one national population report described as "deplorable." The country provides no legal or other protection for abused women. Russian women have the highest rate of death during childbirth of any European country rated (48 deaths per 100,000 births, versus a regional average of 12). And, women in Russia have a significantly shorter life expectancy rate than other European women (75, four years less than the average).

Russian women face a mixed bag in the area of employment opportunities. Although they have traditionally been required to work, and Russian women represent more than half of the country's work force, they make up a disproportionate share of workers on arduous night shifts and doing manual labor. On the bright side, however, Russia boasts the highest number of female managers: 30 percent compared with the European average of 18 percent. Women also account for half of the students enrolled in universities.

Russia received its best marks in the general climate category. It was the earliest European signer of the United Nations treaty regarding the treatment of women[9] and has the second highest proportion of women serving in the government—34 percent—although this number may be declining with the advent of capitalism.

SPAIN

Overall FEM Score: 59.62
Legal: 12.09
Family: 11.54
Health: 13.58
Economy: 9.12
General: 13.29
Population: 39,167,000
Revenue: $514.9 billion

Spain, which received the second lowest score of the European countries, has a bit of catching up to do relative to the rest of the continent. Its lowest point was in the area of economic and educational opportunity for women. Spain posts an abysmal record for promoting women into management positions: a scant 6 percent hold these jobs, which is less than one-third the European average. Spain also has the lowest number of literate women relative to men and a below-average pay scale.

In the area of birth control, women have had good access since 1978, when contraceptives were legalized. However, relatively few women take advantage of them—only 59 percent, which is the lowest number in the region. Laws regarding a woman's right to an abortion are much more limited in Spain than in other countries. As a result, perhaps, the average size of a household in Spain is 3.8 people, the largest in the region.

SWITZERLAND

Despite a long history of civilized traditions and democracy, Switzerland has by far the lowest FEM rating in Europe. In fact, surprisingly, women in much poorer and less well-educated countries such as Mexico, Panama, and Jamaica fared better than Switzerland in the FEM scoring. Whereas all of its neighbors ranked in the top 12 countries in the world, Switzerland placed an unimpressive 20th of the 35 countries.

Switzerland scored very low in two categories, legal rights and eco-

Overall FEM Score: 48.02
Legal: 6.45
Family: 12.26
Health: 13.81
Economy: 5.01
General: 10.49
Population: 7,005,000
Revenue: $152.3 billion

nomic conditions for women. It did not give women the right to vote until 1971, by far the last European nation to do so. It also has inferior maternity leave policies for workers. In the economic climate category, Switzerland received three more last-place scores. Women receive just 67 percent of what men are paid, which ties Switzerland with Germany for last place in that category. And Swiss businesses, along with those in Spain, give the fewest management positions to women: a low 6 percent. Switzerland was also in last place for its percentage of women in college, a scant 35 percent.

Switzerland did score well in the area of health with one of the lowest maternal mortality rates in Europe and the longest life expectancy for women (81).

One could almost assume, based on the country's ratings, that Switzerland has made some conscious effort not to get caught up in the movement toward equality that has been taking place for decades in all of its neighboring countries. Like the cheese named after the country, the Swiss government's policy toward women is clearly full of holes.

UNITED KINGDOM (INCLUDING NORTHERN IRELAND)

Overall FEM Score: 66.12
Legal: 13.36
Family: 14.44
Health: 14.36
Economy: 12
General: 11.96
Population: 58,191,000
Revenue: $920.6 billion

The United Kingdom's FEM score fell in the middle of the European pack. Women do fine in the legal and family rights categories. The government mandates 14 weeks of job-guaranteed maternity leave and also pays social security benefits to all single parents. Contraception is widely available—with the exception of abortion, which is not legal in Northern Ireland—and English women use contraception more than do the women in any other European country.

The work world in the United Kingdom still has its ups and downs, leading to the country's lowest score, which comes in the economic climate category. Women's salary differential is less than the European average and women are still found primarily in clerical, personal services, and secretarial work. They make up only 40 percent of the university population, the second lowest rate in the region. However, they do hold 22 percent of the managerial positions, second only to Russia. And the country has adopted Opportunity 2000, the first national initiative of its kind, which aims to foster corporate involvement in improving women's issues.

One area in which England could use substantial improvement is that of female representation in government. With the exception of female ministers of employment and health (and Margaret Thatcher), England has a poor record in this area: women represent only 6 percent of the government.

European Summary

As a region, Europe's FEM score was 63.91, tops in the world.

Without exception, women have the same basic rights as men: to vote, to own property, to work, to dissolve a marriage, and to make choices about reproduction and family size. In absolute terms, the European workplace could do more promoting of women into management and equally paid positions; relative to the rest of the world, however, Europe has an impressive record. European women also benefit from very progressive maternal leave policies, which in many cases provide for paid time off for the first several months and unpaid but job-guaranteed extensions that can last up to 4 years.

Finland and Denmark scored the highest of the European countries rated, Spain was near the bottom, and Switzerland was clearly the big loser. Our suggestion would be to visit the Nordic fjords instead of the Swiss mountain lakes, or to see the Alps from the French or Italian angle, rather than the Swiss. If you aren't traveling to Europe, you can still show support by choosing your purchases carefully. All countries in Europe export substantial numbers of products to our shores. Yes, Switzerland makes good chocolates and watches, but so do other countries. Look at the labels before you buy.

NORTH AND CENTRAL AMERICA

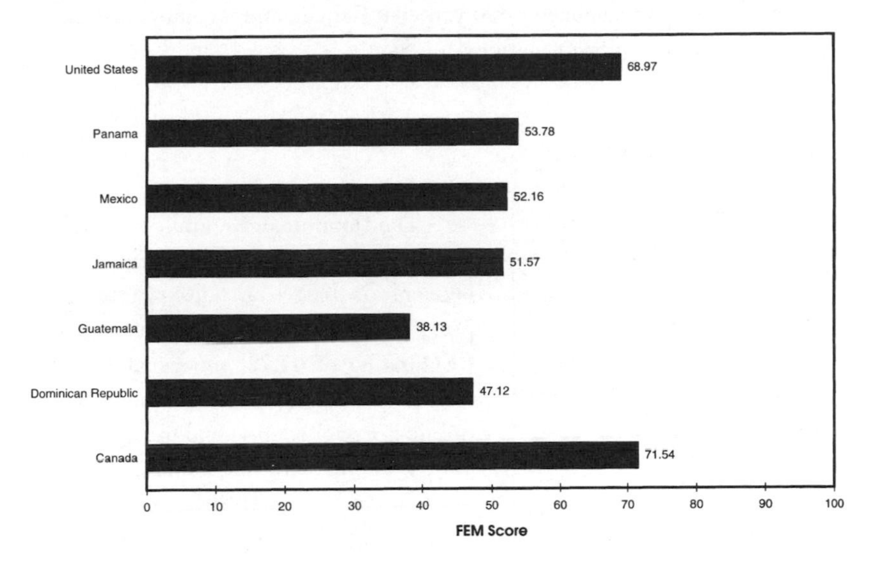

The countries in this section are very varied in economic level and size. We have included the United States with its geographic neighbors for the purpose of comparison.

CANADA

Overall FEM Score: 71.54
Legal: 13.36
Family: 14.74
Health: 13.12
Economy: 16.8
General: 13.52
Population: 29,248,000
Revenue: $537.1 billion

Our northern neighbor has an overall FEM score of 71.54, third highest of any of the countries that we rated. All category scores are well above average for our sample.

Canada makes a particularly strong showing in the employment indices, with a number two ranking overall, because it has one of the world's highest proportions of women in management positions, 35 percent. Its salary differential of 71 percent is reasonably good.

Another notable statistic for Canada is the average female life expectancy of 80.7 years, higher than ours by almost a year. On the basis of this record, we would recommend that you visit Canada and its many interesting cities and magnificent national parks.

DOMINICAN REPUBLIC

Overall FEM Score: 47.12
Legal: 11.22
Family: 6.92
Health: 7.78
Economy: 10.35
General: 10.85
Population: 7,608,000
Revenue: $8.4 billion

The Dominican Republic is a multiparty constitutional democracy whose economy is based on tourism and exports of sugar. It receives an overall FEM rating of 47.12, somewhat below the world average. The Dominican Republic is a poor country and the majority of its workers, including women, receive the minimum wage of $61 a month.

Ratings for three of the categories are in the average range. The Dominican Republic gets low scores on women's rights in the family and women's health. Although discrimination against women is prohibited by law, the gender discrepancies in pay and advancement suggest that they are hardly enforced (and this is documented in a 1994 U. S. State Department Report). Access to birth control is limited, as is child support following divorce. Domestic violence is a problem, and although there are some laws protecting women, they are not very stringent. Women are not permitted to end unwanted pregnancies, and the maternal mortality rate, at 66 deaths per 100,000 births, is much higher than in Canada or the United States. Fertility rates are the second highest in the region, and household size is the highest. Life expectancy is not great either; it is the second lowest of the region at just under 70 years.

Interestingly, women have a relatively high rate of government participation at 11.6 percent, but literacy is relatively low (the ratio of male to female readers is 77:100).

The picture for women, then, appears to be a very mixed one. Despite good access to education and political participation, women's health indices are not good. Moreover, the previously referenced U. S. State Department 1994 report on human rights abuses notes that some women are victims of forced prostitution, and that the workers in the export fields, the so-called Export Processing Zones, which are mostly women, are often exploited.

GUATEMALA

Overall FEM Score: 38.13
Legal: 10.73
Family: 5.68
Health: 5.54
Economy: 6.89
General: 9.29
Population: 10,322,000
Revenue: $12.6 billion

Guatemala's overall FEM score of 38.13 is by far the lowest in the region. Guatemala is a small, poor country heavily reliant on agriculture. Its 1985 constitution calls for democracy, but the country has had a state of armed internal conflict for 33 years, and there are multiple human rights abuses ranging from forced disappearances and extrajudicial killings to torture. The present government seems willing, but unable, to control the police and the army in these areas, which affects women as well as men.

One positive feature of the climate for women is that there are currently several women who hold prominent political positions, including three cabinet posts and two Supreme Court judgeships. In the economic sector, the country's constitution contains guarantees of sex equality and prohibits job discrimination and unequal pay, but unfortunately, these laws are not generally enforced. Only 16 percent of managerial positions are occupied by women, the second lowest rate in the region. Women constitute almost half of university enrollments, but only 23 percent of all university graduates are women, which is a low for the region.

Some other lows include a 5.1 percent rate of female participation in government, a life expectancy of just over 67 years, a very high fertility rate (women average 5.7 children), and a maternal mortality of 110 per 100,000 births. Moreover, women are not legally protected in cases of domestic abuse, cannot obtain legal abortions, and do not have much access to birth control. In fact, only 25 percent of women use birth control, according to a recent report by International Planned Parenthood.

Although Guatemala is not a frequent tourist destination, they do export gift items and textiles to the United States. We recommend that you not purchase items made in this country.

JAMAICA

Overall FEM Score: 51.57
Legal: 11.22
Family: 8.25
Health: 9.48
Economy: 10.74
General: 11.88
Population: 2,366,000
Revenue: $3.7 billion

Jamaica is a frequently visited tourist destination in the Caribbean. Its economy depends on both tourism and exports of products such as bauxite, sugar, bananas, and clothing. Its form of government is a constitutional democracy.

Jamaica's FEM score of 51.57 puts it squarely in the average range of both the overall world sample and its geographical region. This average is derived from a relatively weak showing in women's family and health issues, but a relatively strong showing in the general climate category.

The weak showing in the family rights category reflects the relatively low use of birth control, the absence of any constitutional protection of equality in marriage, and the difficulty in obtaining an abortion. Jamaica also has high rates of domestic violence and a lack of legal procedures for dealing with abusers. One study conducted at the University of the West Indies found that 13 percent of all eighth grade girls had been victims of attempted rapes (Department of State, *Country Reports*, February 1994). Violence against adult women is also regarded as high but underreported.

Jamaica's relatively high score on general climate reflects many laws attesting to women's equality, the highest rate of female political participation in the region (11.7 percent), and a reasonably high average female life expectancy for a poor country (almost 76 years). Jamaica also has a good rate of female literacy (90 percent) and of educational involvement.

MEXICO

Mexico, a democracy with almost 90 million people, is the most frequent destination for international travel from the United States. Its overall FEM score is slightly above average in our global sample. It reveals somewhat low ratings in the health and economic areas and better-than-average ratings for family roles and general climate.

Overall FEM Score: 52.16
Legal: 9.95
Family: 13.07
Health: 8.83
Economy: 7.98
General: 12.33
Population: 87,341,000
Revenue: $328 billion

The low health score reflects the high maternal mortality rate of 91 deaths per 100,000 births, the second worst in the region, and a relatively high fertility rate of 3.5 children. Employment opportunities for women are also mediocre. Despite laws on the books mandating gender equality and equal pay, they are not enforced very strongly. Women make up only 15 percent of the country's managerial positions, a low for the region. Perhaps this is partly attributable to a lack of educational opportunities for women, who comprise only 34 percent of the country's university graduates (only Guatemala's score of 23 percent is worse).

On the more positive side, in the family area, the Mexican government strongly supports access to birth control. Abortions are permitted, but only under limited conditions. Women are legally considered to be equal partners in marriage relationships, and child support and appropriate penalties for spousal abuse are legally sanctioned. Domestic assault is infrequently reported, however.

Mexico's high score in the general climate area was related to several factors. It has a relatively high rate of participation of women in politics (almost 11 percent), and the literacy rate for women, relative to men, is also high.

Clearly, Mexico has a long way to go before women play a less subordinate role, but women's groups there are active, and the government seems to be receptive to making more changes. On this basis, we would not rule out travel to this wonderful country.

PANAMA

Panama, a small Central American country of 2½ million people, was a dictatorship for 21 years prior to the formation of a democratic government in 1989. Its overall FEM score of 53.78 places it somewhat above average—not as high as Canada or the United States, but better than Mexico, Guatemala, or Jamaica.

Overall FEM Score: 53.78
Legal: 11.12
Family: 8.65
Health: 9.69
Economy: 12.69
General: 11.63
Population: 2,583,000
Revenue: $6 billion

Panama's lowest category is family roles and rights. Its score reflects the absence of any laws concerning joint property. Because of this, when a woman gets divorced, she is typically left impoverished. Another problem is that girls are permitted to marry at 12 years of age, an age of consent far lower than any other in the region. Legal abortion is not permitted, so women are very limited in their reproductive freedom.

Panama's best category is employment and higher education. Women constitute 56 percent of university graduates. Discrimination against women is prohibited by law. Statistics provided to us by the Panamanian government indicate that women are paid 72 percent of the rate for men (which is the same as U. S. women), and that 23 percent of management positions are filled by women. In contrast, the 1993 U. S. State Department report estimates that only 5 percent of the country's managers are women. It is not clear whether different ways of calculating this percentage were used.

The State Department report also indicates that discrimination laws are rarely enforced. This pattern seems true for many countries, including our own. We did give credit in our ratings, however, for legal protection, even if not enforced very stringently, because it seemed to us that at least this is a start in the right direction. Women seem to fare even worse in countries that do not have such laws.

UNITED STATES

Relative to the other countries in our survey, the United States is clearly one of the better countries for women. Its total FEM score of 68.97 places it fifth in our sample, behind Finland, Australia, Canada, and Denmark.

The United States does very well on all categories except general climate. This is because it has yet to ratify the U. N. Convention to stop discrimination against women (other countries in the region ratified it at least 10 years ago). It also has a lower percentage of women in elected

Overall FEM Score: 68.97
Legal: 13.16
Family: 14.54
Health: 14.29
Economy: 16.59
General: 10.39
Population: 258,233,000
Revenue: $5.951 trillion

federal positions—a measly 8 percent in the Senate and 10.8 percent in the House of Representatives—than all but two other countries in the region, Guatemala and Panama. (In the November 1996 election, U. S. numbers increased. slightly—to 9 percent in the Senate and 11 percent in the House.) And although the average life expectancy of women here, just under 80 years, is clearly good, it is not the best for the richest country in the world.

The United States received top scores on access to birth control, and about 68 percent of women here use it. Within the marriage, women are recognized as equals and have good resources, relative to the rest of the world, for establishing community property[10] and collecting child support. Our government fell down slightly in the area of domestic violence, with laws that are average for the region but do not take any extra strides to protect women. The U. S. score for the family category was fourth highest overall.

In health issues, U.S. women have a maternal mortality rate of 7 per 100,000 births, higher than Canada but generally respectable in world terms. We average 1.8 children and live in houses with an average of 3.1 people. In this category, the United States ranks seventh, behind Germany, Denmark, Finland, France, the United Kingdom, and Italy.

For economic and educational status, the United States ranks second, behind Canada, and is neck and neck with Australia. Women's salary differential was 65 percent in 1991 (the latest year for which we had comparable data on other countries). This figure (which increased to 72 percent in 1994) is based on U. S. Labor Department reports, but other reports placed that figure as low as 61 percent and as high as 69.9 percent. About 41 percent of managers in this country are women (based on 1994 statistics), and half of the university students are women.

For general legal rights, the United States is seventh, behind Australia, Finland, Denmark, Canada, the United Kingdom, and Germany. But as noted above, the country ranked quite poorly with regard to general climate, a most unimpressive 24th of the 35 countries. The 11 countries below us include Guatemala, Kenya, India, Tanzania, and Saudi Arabia, most of them either very poor or subject to a state religion that seriously limits women's freedoms.

So, despite the activity of many groups dedicated to improving women's status in the United States, we have a long way to go just to catch up with our allies in some areas. We can improve things by becoming more rather than less politically active, and by raising, rather than lowering our voices. We can also improve things in private industry by paying attention to the records of the companies reviewed in Chapters Three through Six, and making our purchases accordingly.

North and Central American Summary

The average FEM score for this region is 54.75, which is above average. The scores ranged substantially within the region, which is not surprising, given the vast economic disparities among the countries rated. Because we are rating international travel, we exclude the United States from this discussion.

The best country, by far, is Canada. We recommend that you consider planning your vacation travel to wonderful cities like Montreal, Toronto, or Vancouver. Canada also has excellent ski areas and breathtaking parks. Canadian imports should also be supported. If you prefer warmer weather, your best choices in this region of the world are Panama and Mexico. Mexico is obviously the closest, and although it is not without problems in its treatment of women, it is better than other tropical countries in Central America.

Countries to be avoided include Guatemala, the low scorer of the region, and the Dominican Republic, which was not quite as bad, but still considerably below average. Don't forget to read the labels on items you are considering buying. These countries don't deserve our support.

SOUTH AMERICA

The South American countries rated—Argentina, Brazil, Chile, Ecuador, and Venezuela—all scored somewhat below the world average and did not vary much. Women there are much better off than they are in

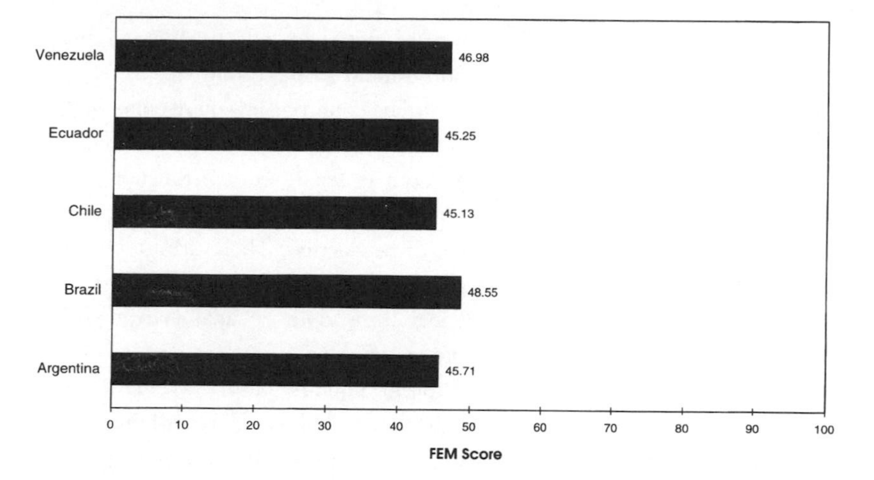

the Middle East: they have basic rights such as voting and free travel. Yet South Americans still have a long way to go before they have the freedoms and equalities enjoyed by women in such countries as Finland and Australia. The average FEM score of the South American countries, 46.32, was well below that of its neighboring region, North and Central America, suggesting that much progress needs to be made.

In all of the South American countries surveyed, women have received some measure of equality and freedom from discrimination as provided by either recent law revisions or the countries' constitutions. In most of the countries, however, historical traditions of male dominance still persist and women have few, if any, methods to challenge inequality. Problems for women are at their worst in the rural segments of South America.

In every country rated in this region, the previously referenced 1994 U. S. State Department report on human rights mentions violence against women as a primary concern. Furthermore, most of these countries have experienced huge population growth, suggesting that family planning services and availability of birth control leave much to be desired everywhere. Abortion is illegal in all five countries rated.

ARGENTINA

Argentina scored just above average in the legal rights category, losing points mainly for being the last country in South America to give

Overall FEM Score: 45.71
Legal: 10.05
Family: 5.12
Health: 11.05
Economy: 7.29
General: 12.2
Population: 31,180,000
Revenue: $112 billion

women the right to vote. Women were not granted suffrage until 1952.

The country's low point came in the area of women's family rights. Although divorce is legal, women are not given equal standing to their husbands while they are married. There are no government programs aimed at preventing or controlling violence against women or domestic abuse.

Despite a constitutional provision for equality of all citizens, women reportedly still find discrimination in the workplace. Women here, as in most places, are overrepresented in low-ranking positions and low-paying fields. And a lack of training opportunities has left Argentinean women with the lowest percentage of women in management of any South American country: just 7 percent, compared with a regional average of 13 percent. Hopefully, Argentina's National Women's Council, which reports directly to the country's president, will have some success as it works toward giving women equal opportunity.

Argentina's highest category score came in the area of general climate. The country has a higher-than-average proportion of women in government for the region, even though the actual number—just under 5 percent—is pretty discouraging in world terms. It should be noted that Argentina is the only country in South America to have had a female head of government, Isabel Perón, during this century (even though she got her start as wife of the president). Argentina has a very high proportion of women in its secondary schools—98 percent—resulting in a 95 percent literacy rate, the highest in South America.

BRAZIL

Women in Brazil seem to be better off than their South American sisters, as the country's FEM score places it first among the five South American countries rated. They have above-average standing in areas of legal rights, economic opportunity, and general climate. Women have also

Overall FEM Score: 48.55
Legal: 11.8
Family: 7.98
Health: 8.12
Economy: 10.18
General: 10.47
Population: 153,792,000
Revenue: $369 billion

had the right to vote since 1934, one of the earliest grantings of this right in South America.

Work opportunities continue to be limited for all of Brazil's citizens, in part because of a problematic economy (with a staggering inflation rate of 2400 percent annually, according to 1993 figures from the U. S. State Department). Such levels of inflation lead to much instability. The same report estimates that 32 million Brazilians receive a less-than-adequate diet every day. Clearly, these problems affect women as much as men.

For women, the country's constitution prohibits discrimination in the workplace based on sex, but inequality continues to be a problem. For example, because women are guaranteed 120 days of paid maternity leave, many women of childbearing age report being forced to show a certificate of sterilization before being hired. Nonetheless, the country received an above-average score for employment opportunities.

Where Brazilian women suffer the most is in family roles and health. Maternal mortality rates are the second highest in the region at 120 deaths per 100,000 births. The average life span of a Brazilian woman is 69, three years less than the regional average. Brazilian women have only marginal access to birth control, and although the percentage of women who take advantage of family planning is above average, the country's fertility rate is also high for the region, with women averaging 3.4 children. The country's fertility rate is quite disparate among socioeconomic groups, with women in the poorer northeast part of the country averaging 5.7 children, whereas more affluent women average 2.9 children, according to the U. S. Agency for International Development. In fact, the population of the country is expected to grow from 156 million to 211 million by the year 2020—or a 35 percent increase.

Within the family, Brazilian women are legally considered equal in marriage and are entitled to some child support in the event of a divorce. In the area of domestic violence, however, Brazilian women receive little help from the laws or the judicial system. For example, in one town studied for the State Department report on human rights, police registered 1500 domestic violence complaints in a year. For the same period, there was not one single conviction.

CHILE

Overall FEM Score: 45.13
Legal: 10.34
Family: 4.5
Health: 9.8
Economy: 9.47
General: 11.02
Population: 14,026,000
Revenue: $34.7 billion

Chile, with below-average scores in four of the five categories, had the worst overall FEM score in the region, and ranked 26th of the 35 countries rated.

Chile's worst score came in the family roles and rights category. Chilean law forbids divorce, so marriages may be terminated only through annulment. If a marriage is annulled, however, it means that legally the couple was never married, and that a woman then has no legal recourse to receive child support. Chile also lost points in this category for allowing girls to marry at age 12. Not surprisingly, there is no governmentally funded recourse available for victims of spousal abuse. On an encouraging note, however, a former female cabinet member has begun, through the National Women's Service, to train police officers in how to assist domestic violence victims. So far, about 5000 officers have been trained.

In the health category, Chile has a relatively low maternal mortality rate—50 deaths per 100,000 births, compared with the regional average of 92—and the lowest fertility rate of any country in South America. The household size, however, remains large at 5.2 people per household.

In the economic field, Chile gets credit for having the highest number of female managers in the region: 18 percent, compared with the average of 13 percent. However, despite the fact that 93 percent of adolescent girls attend secondary school, a below-average 40 percent of university students are women.

Chile also gets credit for having the highest proportion of women in government—almost 8 percent, compared with the average of less than 5 percent—as well as having the longest female life expectancy in the region.

ECUADOR

Overall FEM Score: 45.25
Legal: 12.38
Family: 6.56
Health: 7.56
Economy: 8.59
General: 10.16
Population: 11,221,000
Revenue: $11.8 billion

Conditions for all people in Ecuador are discouraging. About 63 percent of the population lives below the poverty line, 50 percent of children under 5 years old suffer from malnutrition, and the birth rate is one of the highest in Latin America. Not surprisingly, its overall FEM score is just above Chile and very near the bottom for the region.

Ecuador, despite all of its other problems, received above-average scores in both the legal rights and the general categories. Its score in the legal area was bolstered by being the first country in the region to give women the right to vote, in 1928. And as in other countries rated, women in Ecuador are free to own property and travel. Its high score in the general category partially reflects its early signing of the U. N. treaty to end discrimination against women. However, it also received some low scores in the category, including the lowest female life expectancy for the region and the lowest proportion of women in government, just over 1 percent.

Ecuador could use significant improvement in the women's health category. The maternal mortality rate is the highest in the region at 160 deaths per 100,000 births. Perhaps because only 44 percent of women use birth control, the country has the highest fertility rate of any in the region.

VENEZUELA

Given the fact that Venezuela has had a long-standing commitment to democracy, a free press, and well-established unions, it is a little disappointing that it did not do better in our study. Venezuela did score higher than any of its neighbors on the economic climate category, in part because women have better employment opportunities there than in most other South American countries. With the help of active women's rights groups,

Overall FEM Score: 46.98
Legal: 10.53
Family: 5.85
Health: 8.31
Economy: 10.54
General: 11.75
Population: 20,712,000
Revenue: $57.8 billion

women have made advances in such fields as medicine and law, and at 15 percent, the proportion of women in management is above average for the region. Pregnant women on the job get some added protection: in addition to 18 weeks of unpaid leave, women may not be fired either during their pregnancies or for 1 year after.

Venezuelan women have made the least progress in areas of family rights and health issues. On the plus side, the country's Civil Code was revised in the 1980s to afford women and men equality within marriage. On the negative side, the country's laws permit girls to marry when they are 12 years old. Access to and use of birth control are both below average, giving Venezuela the largest household size of any South American country rated (almost 6 people per household). The population of the country tripled between 1950 and 1981, from 5 to 15 million residents.

South American Summary

The countries rated in South America are in varying degrees of commitment to democracy and to women. Although many have granted women workplace equality in their civil codes, women still face long traditions of discrimination and subordination. In the economic sector, there was quite a bit of variation, with Venezuela boasting the best overall climate for women and Argentina the worst.

Every country in this region scored below the world average in the family rights and roles category. Two, Argentina and Chile, do not yet give a woman equal stature within a marriage. All five countries are plagued by a high incidence of domestic violence coupled with a real dearth of resources for victims. Three, Chile, Ecuador, and Venezuela, allow girls to marry at age 12.

On the other hand, every country scored above average in the general climate category. All have signed the U. N. treaty on ending discrimination. About 90 percent of women go as far as men in secondary schools, and the average literacy rate for women in the region is about 89 percent.

If you are heading to South America, we recommend Brazil over Argentina or Chile.

COUNTRY RANKINGS

Below is a summary of each country's ranking in each of the five categories:

Category 1: Legal Rights

	COUNTRY	RANKING
Best	Australia	1
	Finland	2
	Denmark	3
	Canada	4
	United Kingdom	5
	Germany	6
	United States	7
	Ecuador	8
Intermediate	Spain	9
	Turkey	10
	Brazil	11
	Dominican Republic	12
	Jamaica	13
	Panama	14
	Russia	15
	France	16
	Guatemala	17
	Japan	18
	Venezuela	19
	Chile	20
	China	21
	India	22
	Argentina	23
	Greece	24

Category 1: Legal Rights (*continued*)

	COUNTRY	RANKING
	Mexico	25
	Egypt	26
	Italy	27
Worst	Kenya	28
	Israel	29
	Morocco	30
	Singapore	31
	Switzerland	32
	South Africa	33
	Tanzania	34
	Saudi Arabia	35

Category 2: Family Rights and Roles

	COUNTRY	RANKING
Best	France	1
	Singapore	2
	Canada	3
	United States	4
	Denmark	5
	United Kingdom	6
	Australia	7
	Finland	8
Intermediate	Germany	9
	Greece	10
	Japan	11
	Mexico	12
	Italy	13
	Switzerland	14
	Spain	15
	China	16
	Russia	17
	India	18
	Israel	19
	Turkey	20
	South Africa	21

Category 2: Family Rights and Roles (*continued*)

	COUNTRY	RANKING
Intermediate	Panama	22
	Jamaica	23
	Brazil	24
	Dominican Republic	25
	Ecuador	26
	Venezuela	27
Worst	Egypt	28
	Guatemala	29
	Kenya	30
	Argentina	31
	Chile	32
	Tanzania	33
	Morocco	34
	Saudi Arabia	35

Category 3: Women's Health Status

	COUNTRY	RANKING
Best	Germany	1
	Denmark	2
	Finland	3
	France	4
	United Kingdom	5
	Italy	6
	United States	7
	Australia	8
Intermediate	Greece	9
	Switzerland	10
	Spain	11
	Canada	12
	Israel	13
	Japan	14
	Russia	15
	Singapore	16
	Argentina	17
	China	18

Category 3: Women's Health Status (*continued*)

	COUNTRY	RANKING
Intermediate	Chile	19
	Panama	20
	Jamaica	21
	Mexico	22
	Venezuela	23
	Brazil	24
	Dominican Republic	25
	South Africa	26
	Ecuador	27
Worst	Turkey	28
	Guatemala	29
	Morocco	30
	Saudi Arabia	31
	Egypt	32
	Kenya	33
	India	34
	Tanzania	35

Category 4: Educational and Economic Opportunity

	COUNTRY	RANKING
Best	Canada	1
	United States	2
	Australia	3
	Italy	4
	France	5
	Russia	6
	Finland	7
	Singapore	8
Intermediate	Panama	9
	Denmark	10
	Morocco	11
	United Kingdom	12
	Greece	13
	Germany	14
	Jamaica	15

Category 4: Educational and Economic Opportunity (*continued*)

	COUNTRY	RANKING
Intermediate	Venezuela	16
	Dominican Republic	17
	Brazil	18
	Chile	19
	Spain	20
	Israel	21
	Ecuador	22
	South Africa	23
	Kenya	24
	Mexico	25
	Argentina	26
	Saudi Arabia	27
Worst	Guatemala	28
	China	29
	Tanzania	30
	India	31
	Switzerland	32
	Turkey	33
	Egypt	34
	Japan	35

Category 5: General Climate for Women

	COUNTRY	RANK
Best	Finland	1
	Russia	2
	Denmark	3
	Germany	4
	Canada	5
	Spain	6
	Australia	7
	Japan	8
Intermediate	Mexico	9
	Argentina	10
	Greece	11
	China	12

Category 5: General Climate for Women (*continued*)

	COUNTRY	RANK
Intermediate	United Kingdom	13
	Jamaica	14
	Venezuela	15
	Panama	16
	France	17
	Italy	18
	Chile	19
	Dominican Republic	20
	Israel	21
	Switzerland	22
	Brazil	23
	United States	24
	Ecuador	25
	Guatemala	26
	Singapore	27
Worst	South Africa	28
	Kenya	29
	Turkey	30
	Egypt	31
	India	32
	Morocco	33
	Tanzania	34
	Saudi Arabia	35

CONCLUSIONS

It is very difficult to step outside the borders of one's country and be a fair judge of another. The ages of countries vary considerably and their cultures vary greatly. Problems of poverty or overpopulation may be hard to comprehend for one coming from a country where such problems are less visible.

Many of the worst rated countries in the world seem inescapably bound by their histories: in some countries in Africa, by tribal laws; in some areas of the Middle East, by Islam; and in parts of Asia, by age-old traditions of women as inferior. In many of these countries, women still do not have even basic rights, such as to travel, to drive, to make reproductive choices, or to be considered as an equal in a marriage.

Even in countries where there are legal protections against discrimination or granting women equal standing as citizens, the laws often can't compete with the history and traditions of a culture. In these countries, women have a unique problem: they may be prevented from taking advantage of what protection there is because of societal practices.

Even making allowances for the above factors, and comparing countries to others in their region instead of elsewhere in the world, there is still much more that each country could do to improve the standing of its female citizens.

This is no less true in the countries that fared better on the FEM scale. In these countries, women have most basic rights, so the differences are most visible in areas of opportunity, in education, employment, and government. In most countries, women make up about half of the population; why then shouldn't they strive to occupy half of the seats in government? Similarly, if women comprise half of the work force, why can't they hope to be fairly represented in the upper ranks of their professions? Why isn't there one country where women earn one dollar for each dollar earned by a man in the same job? Even the best-rated countries could aim for much higher goals. The United States ranked fifth in the world on overall score; yet only 10 percent of our elected representatives are women. Australia is in the top five on economic climate for women, yet women there make up only 30 percent of managerial and other top-level positions.

Our rankings of the countries reveal that even within regions, there are significant differences in the status of women, and that's where the reader comes in. For example, Kenya and Tanzania are geographical neighbors and share much of the same topography and wild animal viewing that attract visitors. Although it still has far to go, Kenya has done more to help the standing of its women than Tanzania has, so why not plan a safari there, or choose Kenyan coffee over Tanzanian? Similarly, all of the countries in Europe offer wonderful sights and cultures, but one, Switzerland, showed itself to be far behind its neighbors in its treatment of women. For this reason, we would recommend traveling almost anywhere in Europe before Switzerland.

Endnotes

CHAPTER ONE

1. Annual data for 1994 are from the Census Bureau's *Current Population Reports*, Series P-60, U. S. Department of Commerce. The reader should also be aware that statistics for men and women are usually given in terms of the median and not the mean. The median is the 50th percentile. The mean is the average and is usually somewhat higher than the median because salaries at the high end pull it up. Thus, for example, the *mean* salaries for men and women in 1989 were $37,051 and $23,874, whereas the median figures for the same year were $30,549 and $20,932; the disparity between the mean figures is approximately 50 percent more. See *Monthly Review, U. S. Department of Labor, Bureau of Labor Statistics, July 1994*, p. 6. One additional problem is that the Census Bureau figures combine government and private industry salaries, and the latter are less favorable for women.

2. From Lawrence Mischel and Jared Bernstein, *The State of Working America 1994–95*, Washington, DC: Economic Policy Institute, 1994. As The Institute for Women's Policy Research notes, teenagers make up only 15 percent of the minimum wage work force, contrary to popular belief. The majority of those working for minimum wages are adult women.
3. Census Bureau, U. S. Department of Commerce, *Current Population Reports* (Table 5), unpublished data.
4. GDP figures presented for all countries are based on 1990 dollars and exchange rates, so that they may be better compared, and are given in *The World Almanac and Book of Facts*, 1996, p. 200. New York: Funk & Wagnalls. U. S. figures given by the Department of Census are somewhat higher. Without this correction, the 1995 GDP was $6977.4 billion (Bureau of Economic Analysis, U. S. Department of Commerce).
5. Susan Faludi, *Backlash*, New York: Crown, 1991.
6. Total earning capacity for women estimated at $960 billion, based on their 46 percent participation in the work force, and their average earnings of 72 percent of men's salaries.
7. Natania Jansz and Miranda Davies (eds.), *Women Travel, A Real Guide Special*, New York: Prentice Hall, 1990.
8. With regard to RU-486, the authors are pleased to note that as of this writing, the Population Council (a nonprofit group) has obtained the marketing rights and the FDA has finally approved clinical testing of this drug so that it may be available to U. S. women in the future.
9. "Women in Management: The Spare Sex," an article in *The Economist*, March 28, 1992, pp. 17–20, has a good discussion of management possibilities. Most recent statistics for 1995 are from the Federal Glass Ceiling Commission's *Good for Business: Making Full Use of the Nation's Capital, The Environmental Scan* put out by the U. S. Department of Labor in 1995.
10. Sue Alpern, "A History of Women in Management." In E. A. Fageson (Ed.), *Women in Management. Trends, Issues and Challenges in Managerial Diversity*. Newbury Park, CA: Sage, 1993.
11. See *Beyond the Glass Ceiling and The Sticky Floor*, Vol. 1, No. 4, put out by the National Council for Research on Women, 1996.
12. Recent estimates indicate that labor participation rates of women will exceed 60 percent by the turn of the century. See *Facts on Working Women*, No. 92-1, January 1992, U. S. Department of Labor, Women's Bureau.

CHAPTER TWO

1. Information about percentages of managers and professionals comes from either EEO-1 forms provided (1994 was the latest available—1993 and earlier were used when a later one was not available) by the Office of Federal Contract

Compliance Programs (which federal contractors have to provide) or from the companies directly. Percentage of female managers does not distinguish among levels of managers. The number of women in senior management, defined as vice president or above, used in Category 2, was obtained either from the company or from other sources (e.g., annual reports, SEC reports). The percentage of senior female managers is extremely small. Therefore, percentages typically reported on the EEO-1 form represent primarily junior- and middle-level managers. Some researchers have argued that the figures corporations provide for female managers are very inflated, because of the broadness of the Equal Employment Opportunity office's reporting requirements. Thus, secretaries are often listed as managers, and there may have been a general proliferation of managerial titles with no significant status or authority. For discussion of this, see Jerry A. Jacobs, "Women's Entry into Management," Chapter 6, in J. A. Jacobs (ed.), *Gender Equality at Work*, Newbury Park, CA: Sage, 1995.

2. Based on *Job Patterns for Minorities and Women in Private Industry, 1994*, U. S. Equal Employment Opportunity Commission, 1995.
3. How much lower would depend on the actual distribution of scores since all ratings were based on the frequency of scores actually obtained. Thus, a raw score of 0 would not necessarily translate into a rating of 0 unless it was very unusual. Raw scores of 0 more normally were transformed into low positive numbers for the category.
4. J. Fierman, "Why Women Don't Hit The Top," *Fortune*, 122, pp. 40–62, 1990, and U. S. Department of Labor, *A Report on the Glass Ceiling Initiative*, Washington, DC: Government Printing Office, 1991. In 1990, women made up less than 1 percent of Fortune 500 managers, but more recent figures (1995) show an increase to about 3 percent.
5. "Women in Management: The Spare Sex," *The Economist*, March 28, 1992.
6. Figure from Steve Lawrence, "Introduction," *The Foundation Grants Index 1995*. Given the multiple problems faced by women (e.g., domestic violence, breast cancer, poverty), the allocation is woefully inadequate. According to a report prepared in 1995 by The National Council for Research on Women (*Who Benefits, Who Decides*, edited by Mary Ellen S. Capek and Susan A. Hallgarth), funding for women has increased only one-tenth of one percent since 1989 and so-called "generic" program funding does not meet the needs of girls and women.
7. Cited in Tamar Lewin, "Working Women Say Bias Persists," *The New York Times*, October 15, 1994.
8. *Femlin* is a contraction of *feminist* and *gremlin*.
9. Another possible area that could have been included in this factor is the way in which the company's advertising depicts women. We decided not to include this, however, because it is subject to rapid change and is already in the public domain—you can see the ads yourself on TV or in magazines—to a much greater degree than the factors we do rate.

10. There were a few rare exceptions with scores somewhat higher than 20, which generally meant that the company was off the continuum—statistically above the 99th percentile.
11. We were able to obtain this comparative salary data for states, but not, unfortunately, for most individual companies.
12. Data based on elected positions were current as of August 1996 in the ratings. Where significant changes occurred after November, they are mentioned in Chapter 7. Other information was based on 1995 data, from a National Directory put out by the National Women's Political Caucus every 2 years.
13. *The World's Women: 1970–1990 Trends and Statistics*, Social Statistics and Indicators, Series K, No. 8, New York: United Nations, 1991.
14. Interestingly, a recent survey conducted by The 9 to 5 National Association of Working Women (Revitalizing the EEOC, 1995) found considerable general discontent with how discrimination complaints were handled. These included insensitivity and long delays in response. Sixty-five percent of the survey's respondents rated their contact with the agency as negative. Part of this may reflect the underfunded nature of this agency.
15. This is based on both correspondence and phone calls with various officials at the Department of Labor. They base it on their interpretation of several court decisions. A reading of some of these decisions suggests considerable variability, but the authors did not have the considerable financial resources that would be required for a legal challenge.

CHAPTER THREE

1. This story is based on the experience of the first author, Phyllis Katz.
2. Based on studies by Reskin and Rass (1995) in the United States, and by Boyd, Mulvihill, and Myles in Canada (1995).
3. Institute for Women's Policy Research, "The Wage Gap: Women's and Men's Earnings," from Census Bureau, U. S. Department of Commerce, *Current Population Reports*, 1994.
4. Gary Powell, *Women and Men in Management*, 1993.
5. A. M. Morrison, R. P. White, & E. Van Welsor. *Breaking the Glass Ceiling*. Reading, MA: Addison–Wesley, 1987.
6. Institute for Women's Policy Research, "Earnings Differences by Educational Level," from Census Bureau, U. S. Department of Commerce, *Current Population Reports*, unpublished data (1995).
7. Catalyst noted that in March 1994, 52 percent of the Fortune 500 companies had a woman on their corporate boards, up from 49 percent the previous year. There has been a steady increase since then—according to *Working Woman*, the most recent study found a significant jump by August 1996 to 81 percent,

but total proportion of seats occupied by women remains low, as it is still rare for a company to have more than one woman on a board.

8. For category score distributions, the average was 10 and the standard deviation was 4. Thus, companies that were way off the continuum, i.e., more than 2½ standard deviations from the mean, could receive scores as high or higher than 23. This occurred in only five instances (out of more than 2000 scores).
9. We have made every effort for this list to be as up-to-date as possible in our final revision of the book. Given the rate of acquisitions, mergers, and spin-offs in the corporate world, however, it is possible that some of the parent companies listed may have changed.
10. There were a few exceptions. If the company was recently acquired, and we had no information on the new parent company, we used the data available. In some instances, the scores for the parent company were lower than for the recently acquired company, and we tried to use the company data in these instances, particularly if it was not clear that much would be changed.

CHAPTER SEVEN

1. These figures were based on 1995 data, obtained in a personal communication with an economist at the Travel Industry Association of America in Washington, D. C.
2. One exception occurred with regard to the stand taken by many women's organizations during the 1980s that recommended not traveling to States that failed to ratify the Equal Rights Amendment (before it was defeated).
3. *Romer v. Evans*, 116 S. Ct. 1620 (1996).
4. *Job Patterns for Minorities and Women in Private Industry*, 1990, U. S. Equal Employment Opportunity Commission.
5. Interestingly, the Constitution precludes amendments to some of its provisions. For example, before 1808, no amendments about slavery were permitted. Also, every state shall always have two senators.
6. The Equal Rights Amendment fell three states short of passage.
7. The category of sex was added to The Civil Rights Act as a joke (Powell, 1993), but as they say, she who laughs last ...
8. Earlier versions of this bill contained the concept of "comparable worth" for equal pay, but the bill that was passed referred to equal work for equal pay.
9. Figures are from *Job Patterns for Minorities and Women in State and Local Government, 1993*, Washington, DC: U. S. Equal Employment Opportunity Commission, 1994. These are not published on an annual basis, and were the most recent data available during the writing of this chapter.
10. In the ratings, we used the percentage of cabinet-level appointments within each of the 43 states that used an appointment system. The most recent figures available were from 1994 and were compiled by the Center for American

Women in Politics. High electoral positions were rated on the basis of how many women occupied congressional and gubernatorial positions as of August 1996, but subsequent significant changes are noted.

11. As was true for the corporate FEM scores, the state scores were transformed statistically into what are called standard scores. The essence of this transformation is that it takes into consideration the actual average and spread of scores obtained, and changes them so you can compare different metrics. In our case, an average for each category is 10, and the standard deviation is 7. These category scores were summed to form the overall FEM score, and this distribution had an average of 50, and a standard deviation of 13.
12. A complete list of all data sources is contained in Appendix One, together with a description of how ratings were assigned to all measures.
13. This information comes from the previously mentioned *Job Patterns for Minorities and Women in State and Local Government, 1993*, U. S. Equal Employment Opportunity Commission, 1994, and M. Lord and M. King (eds.), *The State Reference Guide to Work–Family Progress for State Employees*, put out by the Families and Work Institute, 1991, in New York. For each measure, we used the most current data available at the time of writing.
14. Sources for this information included the U. S. Equal Employment Opportunity Commission's *Job Patterns for Minorities and Women in Private Industry, 1993*, and *Women-Owned Businesses, 1987*, put out by the U. S. Department of Commerce, which were the latest we could locate when the ratings were compiled. In addition, we tried very hard to obtain gender disparities in wages for each state, but found that no one had these figures.
15. As of January 1997, the new Congress has 52 women Representatives (out of 435) including a nonvoting Representative from the District of Columbia (a net gain of 3) and 9 female Senators (out of 100—a net gain of 1). This gain does not change the overall percentages substantially, although it will affect several individual state ratings. California will have a net gain of 2 congresswomen; Arkansas, Indiana, Kentucky, Missouri, New York, Oregon, and Texas will have 1 more. In contrast, the following states will have 1 less woman in the Congressional delegation: Arizona, Illinois (going from one woman out of 20 to none), Kansas, Nevada, Washington, and Utah. Several other notable gains for women occurred. New Hampshire elected its first woman govenor, Louisiana elected its first female Senator, and Maine will be the second state with 2 female Senators.
16. These indices were based on numerical ratings made by other organizations. Reproductive choice ratings of states were obtained from *Who Decides? A State-by-State Review of Abortion Action and Reproductive Rights, 5th Edition*, 1995, put out by the National Abortion Rights League. Community property information was obtained from *A National Survey of State Laws* by Richard Leiter (ed.), Detroit: Gale Research, Inc., 1993. Finally, ratings of spousal abuse laws were based on Barbara J. Hart's article in *Juvenile and Family*

Court Journal, 1992, Vol. 43, No. 4. Many other measures could not be included. Interested readers are urged to use the relevant references to follow up on those we did not include.

17. We recognize that an argument can be made not to base scores on history, namely, that history does not reflect current status. We felt, however, that states should receive a small bonus in giving women the right to vote before they had to, as this does reflect a long-term favorable climate.
18. We hope that you follow up your travel decisions by writing a letter to the state you chose, and the ones you didn't, delineating the reasons for your choice.
19. This correlation was 0.62, which is reasonably high.

CHAPTER EIGHT

1. This figure was obtained from *Statistical Abstracts of the U.S., 1994*, 114th edition, U. S. Department of Commerce, projected 1994 figures, p. 264. We were not able to find a gender breakdown for 1994, but in 1991, women constituted 40 percent of all travelers (business and pleasure), so an estimate that U.S. women spend over $17 billion would not be out of line.
2. The 35 countries are ones that are frequently visited and/or major exporters of products to the United States.
3. If men were subjected to the same limitations, it was not counted as a negative for women within this category.
4. Some of these may, at first glance, overlap with other categories. We believe, however, that average life expectancy data are more global than health characteristics associated with reproductive health, and gender differentials reflect more than the level of medical care. Similarly, differential literacy rates and access to grade school education are more global and affect more people than the index of universal attendance included in the third category. This was our reasoning for putting these in the more general category.
5. Specific ways in which scores were determined for each measure are discussed in Appendix One. All percentages and rating scales were transformed into standard scores so that different metrics could be compared. Standard scores essentially "correct" for the fact that distributions differ in their averages and degree of variability. For individual categories, standard score means were set at 10, with a standard deviation of 4. For the total scores, the mean was set at 50, with a standard deviation of 20.
6. Appendix One contains a list of all sources used for this chapter. In most instances, ratings are based on 1991–1993 data which were the latest we could find from existing sources. In addition to many United Nations publications, we also surveyed countries' embassies in Washington for each country rated.
7. It should be noted that these rights are more applicable to the white majority than the aboriginal minority.

8. China could be an exception to this assumption insofar as fertility choices are enforced by the government.
9. Ratification of this treaty does not, of course, mean that women are treated equally within the country, only that the principles embodied in the treaty have been endorsed by the government. We regard this as a first step, comparable to equality laws passed within countries that are not fully equal in practice. One of the issues that can be raised with regard to Russia and other countries with totalitarian histories and/or current practices is whether the absence of general human rights (for both genders) should be considered in country ratings. Although this is of legitimate concern to our readers, there are many other institutions that monitor such rights (e.g., Amnesty International, government agencies), and their findings are given widespread publicity. For this reason, ratings in this book are focused on that portion of women's rights that are either unique (e.g., reproductive rights) or discrepant from their male counterparts.
10. It should be noted, however, that there is variation among states along these dimensions. For a further breakdown, see Chapter Seven.

Buying Guide

Products Names and Companies by Category

Many of the companies we rate are enormous conglomerates with divisions, subdivisons, and diversified product lines. Just knowing the company, and how it fared on the FEM scale, isn't always enough. The ratings in this guide, in a simplified form, correspond to the FEM scores for each company discussed in earlier chapters.

In this Buying Guide, we have sorted the companies according to their more familiar product names. Please keep in mind that mergers and acquisitions change these data daily. The products are organized into chapter divisions and industries to correspond to Chapters 3, 4, 5, and 6. We trust this presentation will make them more useful and more accessible as you go to the grocery store, drug store, or mall.

FOOD

Baking Products

BRAND NAME	PARENT COMPANY	FEM	RATING
Duncan Hines	Procter & Gamble	64.75	●
Hershey's	Hershey Foods	63.70	●
Baker's Joy	Alberto-Culver	59.99	●
Betty Crocker, Gold Medal, Bisquick	General Mills	54.04	◐
PET Evaporated Milk	PET†	53.99	◐
Pillsbury, Bundt	Grand Metropolitan	53.73	◐
Fleischmann's Yeast, Baker's, Calumet baking powder	Philip Morris	50.72	◐
Baking Powder, RapidRise	RJR Nabisco	50.43	◐
Arm & Hammer Baking Soda	Church & Dwight	48.68	◐
Grandma's Molasses	Cadbury Schweppes	46.34	◐
Argo, Karo syrup	CPC	45.34	◐
Toll House Morsels	Nestlé	44.44	○
Red Star yeast	Universal Foods	39.89	○

Bread

BRAND NAME	PARENT COMPANY	FEM	RATING
Pepperidge Farm	Campbell Soup	58.93	●
Country Hearth	Grand Metropolitan	53.73	◐
Grant's Farm, Iron Kids	Anheuser-Busch	51.42	◐
Lender's Bagels	Philip Morris	50.72	◐
Rainbow Bread, Colonial, Bran'nola, Thomas', Levy's	CPC	45.34	◐
Wonder, Home Pride, Bread du Jour	Ralston Purina	45.03	◐

*Did not have scores for all five categories

†PET was acquired by Grand Metropolitan in early 1995.

Rating Key: ○ = Female friendly company; buy products when possible ◑ = Average; buy with good conscience ● = Below average; avoid buying these products

Breakfast Foods

BRAND NAME	PARENT COMPANY	FEM	RATING
Cheerios, Wheaties, Total, Kix, Cocoa Puffs	General Mills	54.04	◐
Quaker Oats, Cap'n Crunch, Life, 100% Natural, Quaker Oat Squares	Quaker Oats	53.40	◐
Cream of Wheat	RJR Nabisco	50.43	◐
Post Raisin Bran, Grape Nuts, Fruit & Fibre, Honeycomb, Nabisco Shredded Wheat	Philip Morris	50.72	◐
Corn Flakes, Rice Krispies, Frosted Flakes, All-Bran, Raisin Bran, Special K, Pop-Tarts, Eggo, Healthy Choice	Kellogg	46.43	◐
Bisquick	General Mills	54.04	◐
Downyflake Waffles	PET†	53.99	◐
Hungry Jack	Grand Metropolitan	53.73	◐
Aunt Jemima	Quaker Oats	53.40	◐
Vermont Maid	RJR Nabisco	50.43	◐
Log Cabin, Country Kitchen	Philip Morris	50.72	◐
Golden Griddle	CPC	45.34	◐
MacDonald's Syrup	Borden	45.29	◐
Mrs. Butterworth's	Unilever	42.54	○

Candy and Chewing Gum

BRAND NAME	PARENT COMPANY	FEM	RATING
Hershey's, Reese's, Mounds, Almond Joy, Kit Kat, York, Twizzlers	Hershey Foods	63.70	●
Godiva chocolates	Campbell Soup	58.93	●
Junior Mints	Warner-Lambert	57.81	●
Toblerone, Kraft Caramels, marshmallows, Callard & Bowser toffees	Philip Morris	50.72	◐
Cadbury candy	Cadbury Schweppes	46.34	◐
Nestlé Crunch, Oh, Henry!, Sno-Caps, Baby Ruth, Butterfinger	Nestlé	44.44	○

M&Ms, Milky Way, Snickers, Twix, Starburst	Mars	43.23	○
Big Red, Juicy Fruit, Hubba Bubba, Doublemint, Freedent	Wm. Wrigley, Jr.	58.20	●
Certs, Clorets, Trident, Chiclets, Dentyne, Beeman's, Bubblicious	Warner-Lambert	57.81	●
Life Savers, Bubble Yum, Breath Savers, Care*Free	RJR Nabisco	50.43	◐

Canned Goods

BRAND NAME	PARENT COMPANY	FEM	RATING
Franco-American, Swanson, Campbell's	Campbell Soup	58.93	●
Underwood, B&M Baked Beans, Progresso	PET†	53.99	◐
Green Giant	Grand Metropolitan	53.73	◐
Van Camp's beans	Quaker Oats	53.40	◐
Del Monte	RJR Nabisco	50.43	◐
Chef Boyardee, Dennison's Chili, Luck's Beans, Ro*Tel	American Home Products	50.01	◐
Dole	Dole Food	46.57	◐
Mott's	Cadbury Schweppes	46.34	◐
Libby's	Nestlé	44.44	○
Hormel Chili, Dinty Moore, Top Shelf fried onions, Chi-Chi's, Mary Kitchen	Hormel Foods	43.12	○
Durkee	Reckitt & Colman	38.46	○

Cheese/Yogurt/Dairy/Margarine

BRAND NAME	PARENT COMPANY	FEM	RATING
Yoplait, Colombo	General Mills	54.04	◐
Kraft, Cheez Whiz, Velveeta, Philadelphia cream cheese, Light n' Lively, Polly-O, Breakstone's, Breyer's yogurt, Sealtest	Philip Morris	50.72	◐
Fleischmann's, Blue Bonnet, Parkay, Snackwell's	RJR Nabisco	50.43	◐
Whitney's yogurt	Kellogg	46.43	◐
Mazola	CPC	45.34	◐

†PET was acquired by Grand Metropolitan in early 1995.

Rating Key: ○ = Female friendly company; buy products when possible ◐ = Average; buy with good conscience ● = Below average; avoid buying these products

Cheese/Yogurt/Dairy/Margarine

BRAND NAME	PARENT COMPANY	FEM	RATING
Mountain High, Viva, Borden, Eagle Brand, Meadow Gold	Borden	45.29	◐
County Line cheeses	ConAgra	44.69	○
Carnation Coffee-mate	Nestlé	44.44	○
Farm Fresh milk	Pilgrim's Pride*	42.58	○
I Can't Believe It's Not Butter, Country Crock, Promise	Unilever	42.54	○

Condiments

BRAND NAME	PARENT COMPANY	FEM	RATING
Crisco, Puritan oils	Procter & Gamble	64.45	●
PAM cooking spray	American Home Products	50.01	◐
Clover oils	Chiquita Brands	46.86	◐
Mazola	CPC	45.34	◐
Wesson oils	ConAgra	44.69	○
Bertolli olive oils	Unilever	42.54	○
Heinz	H.J. Heinz	47.77	◐
Hunt's, Healthy Choice	ConAgra	44.69	○
Miracle Whip, Kraft mayonnaise	Philip Morris	50.72	◐
Best Foods/Hellmann's mayonnaise	CPC	45.34	◐
Grey Poupon	RJR Nabisco	50.43	◐
Gulden's mustard	American Home Products	50.01	◐
French's, Colman's mustards	Reckitt & Colman	38.46	○
Open Pit barbecue sauce	Campbell Soup	58.93	●
Bull's Eye, Kraft barbecue, Sauceworks	Philip Morris	50.72	◐
A-1 Steak Sauce	RJR Nabisco	50.43	◐

BRAND NAME	PARENT COMPANY	FEM	RATING
KC Masterpiece, Kitchen Bouquet	Clorox	49.23	◐
Old Smokehouse	Hormel Foods	43.12	○
Vlasic pickles	Campbell Soup	58.93	●
Reese capers, mint jelly, artichoke hearts, Progresso cooking wine	PET[†]	53.99	◐
Golden Dipt batter	McCormick	52.05	◐
Claussen pickles, sauerkraut	Philip Morris	50.72	◐
Regina cooking wine	RJR Nabisco	50.43	◐
Knorr	CPC	45.34	◐
ReaLemon, ReaLime	Borden	45.29	◐
Maggi	Nestlé	44.44	○

Cookies/Desserts

BRAND NAME	PARENT COMPANY	FEM	RATING
Grandma's cookies	PepsiCo	63.45	●
Pepperidge Farm	Campbell Soup	58.93	●
Sara Lee cheesecake, coffee cakes	Sara Lee	55.67	●
Pet-Ritz pie shells, Aunt Fanny's pecan twirls	PET[†]	53.99	◐
Jell-O, D-Zerta, Minute tapioca	Philip Morris	50.72	◐
Stella D'oro, Oreos, Chips Ahoy!, Newton's, Nilla Wafers, Barnum's Animals, Nutter Butter; Honey Maid graham crackers	RJR Nabisco	50.43	◐
Hostess: Twinkies, HoHos, DingDongs, Suzy Q's	Ralston Purina	45.03	◐
Snack Pack, Swiss Miss puddings	ConAgra	44.69	○
Maryland Cookies	Allied Domecq[‡]	42.04	○

Diet Foods

BRAND NAME	PARENT COMPANY	FEM	RATING
NutraSweet, Equal, Simplesse	Monsanto[§]	65.15	●

*Did not have scores for all five categories

[†]PET was acquired by Grand Metropolitan in early 1995.

[‡]Formerly Allied Lyons.

[§]Monsanto is a wholly owned subsidiary of Bayer.

Rating Key: ○ = Female friendly company; buy products when possible ◐ = Average; buy with good conscience ● = Below average; avoid buying these products

Diet Foods

BRAND NAME	PARENT COMPANY	FEM	RATING
Sugar Twin	Alberto-Culver	59.99	●
Weight Watchers	H.J. Heinz	47.77	◐
Sweet Success	Nestlé	44.44	○

Foreign/Specialty Foods

BRAND NAME	PARENT COMPANY	FEM	RATING
Tostitos, Salsa Rio	PepsiCo	63.45	●
Old El Paso, Progresso	PET†	53.99	◐
Near East (rice pilaf, couscous, etc.)	Quaker Oats	53.40	◐
Eagle Salsa	Anheuser-Busch	51.42	◐
Ortega	RJR Nabisco	50.43	◐
Chef Boyardee	American Home Products	50.01	◐
La Choy, Rosarita, Chun King, Patio	ConAgra	44.69	○
House of Tsang	Hormel Foods	43.12	○

Frozen Foods

BRAND NAME	PARENT COMPANY	FEM	RATING
Swanson, Mrs. Paul's	Campbell Soup	58.93	●
Gorton's frozen seafood	General Mills	54.04	◐
Van de Kamp's	PET†	53.99	◐
Totino's, Pappalo's, Green Giant	Grand Metropolitan	53.73	◐
Celeste Pizza	Quaker Oats	53.40	◐
Budget Gourmet, Tombstone Pizza	Philip Morris	50.72	◐
Ore-Ida	H.J. Heinz	47.77	◐

Healthy Choice, Banquet, Kid Cuisine, Morton, Armour Classics, Patio	ConAgra	44.69	○
Stouffer's, Lean Cuisine	Nestlé	44.44	○
Hormel, Mrs. Peterson's	Hormel Foods	43.12	○
Gourmet Selection, Tasty Bird	Tyson Foods	42.87	○
Birds Eye	Unilever	42.54	○

Ice Cream/Whipped Cream

BRAND NAME	PARENT COMPANY	FEM	RATING
Ben & Jerry's	Ben & Jerry's Homemade	64.96	●
Colombo frozen yogurt	General Mills	54.04	◐
Häagen-Dazs	Grand Metropolitan	53.73	◐
Frusen Glädje, Sealtest	Philip Morris	50.72	◐
Healthy Choice	ConAgra	44.69	○
Dove Bars	Mars	43.23	○
Breyer's, Klondike, Popsicle, Good Humor	Unilever	42.54	○
Dreyer's	Dreyer's Grand Ice Cream	38.48	○
Dream Whip, Cool Whip	Philip Morris	50.72	◐
Reddi Whip	ConAgra	44.69	○

Jams & Jellies

BRAND NAME	PARENT COMPANY	FEM	RATING
Health Valley	Health Valley Natural Foods	53.38	◐
Kraft	Philip Morris	50.72	◐
Polaner's	American Home Products	50.01	◐
Lite & Fruity	Borden	45.29	◐
Home Brands	ConAgra	44.69	○
Crosse & Blackwell	Nestlé	44.44	○

†PET was acquired by Grand Metropolitan in early 1995.

ey: ○ = Female friendly company; buy products
ssible ◐ = Average; buy with good conscience
ow average; avoid buying these products

Jams & Jellies

BRAND NAME	PARENT COMPANY	FEM	RATING
Welch's	Welch Foods*	40.64	○

Meats

BRAND NAME	PARENT COMPANY	FEM	RATING
Jimmy Dean, Ball Park franks, Hillshire Farms	Sara Lee	55.67	●
Underwood deviled ham	PET†	53.99	◐
Jim Dandy	Grand Metropolitan	53.73	◐
Oscar Mayer, Louis Rich	Philip Morris	50.72	◐
StarKist tuna	H.J. Heinz	47.77	◐
John Morrell	Chiquita Brands	46.86	◐
Swift, Butterball, Hebrew National, Healthy Choice, Brown 'N Serve, Armour, Golden Star, Decker, Webber's	ConAgra	44.69	○
Armour Star canned meats, Lunch Bucket	Dial	43.96	○
SPAM, Little Sizzlers, Jennie-O, Chicken by George, Dubuque, Hormel Black Label	Hormel Foods	43.12	○
IBP	IBP*	42.99	○
Holly Farms, Tyson, Weaver, Tasty Bird	Tyson Foods	42.87	○
Pilgrim's Pride chicken	Pilgrim's Pride*	42.58	○
Smithfield hams	Smithfield Foods	36.19	○

Pasta/Sauce

BRAND NAME	PARENT COMPANY	FEM	RATING
Delmonico, Ronzoni, American Beauty, Light 'N Fluffy, Pastamania, San Giorgio, P&R, Skinner, Mrs. Weiss	Hershey Foods	63.70	●
DiGiorno fresh pasta	Philip Morris	50.72	◐
Mueller's	CPC	45.34	◐

Creamette pasta, Goodman's	Borden	45.29	◐
Contadina refrigerated pasta, sauce	Nestlé	44.44	○
Gooch Pasta	Archer-Daniels-Midland	39.16	○
Newman's Own	Newman's Own*	59.85	●
Prego sauces	Campbell Soup	58.93	●
Progresso	PET†	53.99	◐
Golden Grain	Quaker Oats	53.40	◐
Hunt, Healthy Choice	ConAgra	44.69	○
Ragú	Unilever	42.54	○

Peanut Butter

BRAND NAME	PARENT COMPANY	FEM	RATING
Jif	Procter & Gamble	64.45	●
Reese's	Hershey Foods	63.70	●
Skippy	CPC	45.34	◐
Peter Pan	ConAgra	44.69	○

Prepared Foods

BRAND NAME	PARENT COMPANY	FEM	RATING
Hamburger Helper, Potato Shakers	General Mills	54.04	◐
Rice-A-Roni	Quaker Oats	53.40	◐
Minute Rice, Shake 'N Bake, Stove Top stuffing, Oven Fry	Philip Morris	50.72	◐
Pizza Pockets, Manwich	ConAgra	44.69	○
Appian Way Pizza Mix	Dial	43.96	○
Uncle Ben's	Mars	43.23	○
Chicken Tonight	Unilever	42.54	○

*Did not have scores for all five categories

†PET was acquired by Grand Metropolitan in early 1995.

Rating Key: ○ = Female friendly company; buy products when possible ◑ = Average; buy with good conscience ● = Below average; avoid buying these products

Salad Dressing

BRAND NAME	PARENT COMPANY	FEM	RATING
Newman's Own	Newman's Own*	59.85	●
Marie's	Campbell Soup	58.93	●
Good Seasons, Catalina, Kraft, Seven Seas	Philip Morris	50.72	◐
Hidden Valley Ranch, Take Heart, Salad Crispins	Clorox	49.23	◐
Club Chef, Chef Classic	Chiquita Brands	46.86	◐
Henri, Western	CPC	45.34	◐
Wish-Bone	Unilever	42.54	○

Snacks—Crackers, Bars

BRAND NAME	PARENT COMPANY	FEM	RATING
Nature Valley granola bars	General Mills	54.04	◐
Rice Cakes, Quaker Chewy	Quaker Oats	53.40	◐
Club House	McCormick	52.05	◐
Handi-Snacks	Philip Morris	50.72	◐
Ritz, Stoned Wheat Thins, Triscuit, Premium, Better Cheddars, Wheatsworth	RJR Nabisco	50.43	◐
Kudos	Mars	43.23	○
Keebler, Carr's, Zesta, Club	United Biscuits Holdings*	39.61	○

Snacks—Popcorn, Chips

BRAND NAME	PARENT COMPANY	FEM	RATING
Smartfood	PepsiCo	63.45	●
Newman's Own	Newman's Own*	59.85	●
Betty Crocker Pop Secret popcorn	General Mills	54.04	◐
Jiffy pop, Crunch 'n Munch	American Home Products	50.01	◐
Cracker Jack	Borden	45.29	◐

Snacks—Popcorn, Chips

BRAND NAME	PARENT COMPANY	FEM	RATING
Orville Redenbacher	ConAgra	44.69	○
Pringles chips	Procter & Gamble	64.45	●
Fritos, Chee-tos, Doritos, Rold Gold, Sun Chips, Ruffles, Lay's, Tostitos, Santitas	PepsiCo	63.45	●
Bugles	General Mills	54.04	◐
Eagle Snacks	Anheuser-Busch	51.42	◐
Planters	RJR Nabisco	50.43	◐
Clover	Chiquita Brands	46.86	◐
Clover Club, Cracker Jack, Wise, Red Seal potato chips	Borden	45.29	◐
O'Boisie's	United Biscuit Holdings*	39.61	○
Fruit Roll-ups, Gushers	General Mills	54.04	◐
Fruit & Fitness	Health Valley Natural Foods	53.38	◐

Soup

BRAND NAME	PARENT COMPANY	FEM	RATING
Healthy Request, Campbell's Chunky Soup, Swanson, Soup Starters	Campbell Soup	58.93	●
Progresso	PET†	53.99	◐
Great American	H.J. Heinz	47.77	◐
Knorr	CPC	45.34	◐
Soup Starter, Wyler's	Borden	45.29	◐
Herb-Ox	Hormel Foods	43.12	○
Lipton's Cup-a-Soup	Unilever	42.54	○

Spices/Seasonings

BRAND NAME	PARENT COMPANY	FEM	RATING
Mrs. Dash	Alberto-Culver	59.99	●

*Did not have scores for all five categories

†PET was acquired by Grand Metropolitan in early 1995.

Rating Key: ○ = Female friendly company; buy products when possible ◐ = Average; buy with good conscience ● = Below average; avoid buying these products

Spices/Seasonings

BRAND NAME	PARENT COMPANY	FEM	RATING
Ac'cent	PET†	53.99	◐
Schillings, McCormick, Old Bay	McCormick	52.05	◐
Lawry's	Unilever	42.54	○

NONALCOHOLIC BEVERAGES

Bottled Water

BRAND NAME	PARENT COMPANY	FEM	RATING
$H_2Oh!$	PepsiCo	63.45	●
Deer Park	Clorox	49.23	◐
Vitter, Perrier water	Nestlé	44.44	○

Iced/Cold Drinks

BRAND NAME	PARENT COMPANY	FEM	RATING
Sunny Delight, Hawaiian Punch	Procter & Gamble	64.45	●
All Sport, Lipton Original	PepsiCo	63.45	●
Newman's Own Lemonade	Newman's Own*	59.85	●
Gatorade, Snapple	Quaker Oats	53.40	◐
Kool-Aid, Tang, Country Time lemonade, Crystal Light	Philip Morris	50.72	◐
Hi-C, Five Alive	Coca-Cola Company	47.71	◐
Nestea	Nestlé	44.44	○

Hot Beverages

BRAND NAME	PARENT COMPANY	FEM	RATING
Hershey Hot Cocoa	Hershey Foods	63.70	●
Swiss Miss	ConAgra	44.69	○

Carnation, Nestlé Quik	Nestlé	44.44	○
Folgers	Procter & Gamble	64.45	●
Maxwell House, Sanka, Yuban, Maxim, General Foods International Coffees	Philip Morris	50.72	◐
Taster's Choice, Hills Bros., Nescafé	Nestlé	44.44	○
Celestial Seasonings	Celestial Seasonings	62.02	●
Crystal Light Tea	Philip Morris	50.72	◐
Lipton	Unilever	42.54	○
Tetley	Allied Domecq‡	42.04	○

Juice/Juice Drinks

BRAND NAME	PARENT COMPANY	FEM	RATING
Ocean Spray Splash	PepsiCo	63.45	●
V8	Campbell Soup	58.93	●
Capri Sun	Philip Morris	50.72	◐
Minute Maid	Coca-Cola	47.71	◐
Naked Juice, Chiquita	Chiquita Brands	46.86	◐
Dole	Dole Food	46.57	◐
Red Cheek, Mott's	Cadbury Schweppes	46.34	◐
Libby's	Nestlé	44.44	○
Tropicana	Seagram*	44.20	○
Welch's	Welch Foods*	40.64	○

Mixers

BRAND NAME	PARENT COMPANY	FEM	RATING
Rose's Lime Juice, Grenadine, Mr. & Mrs. T, Clamato	Cadbury Schweppes	46.34	◐
Tropicana Mixers	Seagram*	44.20	○

*Did not have scores for all five categories

†PET was acquired by Grand Metropolitan in early 1995.

‡Formerly Allied Lyons.

Rating Key: ○ = Female friendly company; buy products when possible ◐ = Average; buy with good conscience ● = Below average; avoid buying these products

Soda

BRAND NAME	PARENT COMPANY	FEM	RATING
Mountain Dew, Slice, Mug, Pepsi	PepsiCo	63.45	●
Sprite, Mello Yellow, Fanta, Tab, Mr. Pibb, Fresca, Coke	Coca-Cola Company	47.71	◐
Schweppes, Canada Dry, Sunkist, A&W, Crush, Hires, 7Up,† Dr Pepper,† Welch's sodas, Squirt	Cadbury Schweppes	46.34	◐
Royal Crown Cola	Triarc	40.89	○

ALCOHOLIC BEVERAGES

Beer

BRAND NAME	PARENT COMPANY	FEM	RATING
Budweiser, Michelob	Anheuser-Busch	51.42	◐
Miller, Löwenbräu, Meister Bräu, Milwaukee's Best, Molson, Icehouse, Foster's	Philip Morris	50.72	◐
Coors	Adolph Coors	48.96	◐
Carlsburg, Tetley Bitter	Allied Domecq‡	42.04	○
Harp, Guinness Stout	Guinness*	40.07	○

Liquor

BRAND NAME	PARENT COMPANY	FEM	RATING
Smirnoff, Popov, Christian Brothers, Black Velvet, Club, J&B, Baileys Irish Cream, G.W. Archer's, Bombay Gin, Cinzano	Grand Metropolitan	53.73	◐
7 Crown, Crown Royal, VO, Chivas Regal, Four Roses, Passport, Meyer's Rum	Seagram*	44.20	○
Jim Beam, Old Crow, DeKuyper, Ronrico, Calvert, Gilbey's	American Brands	43.10	○
Ballantine's, Courvoisier, Beefeater, Canadian Club, Kahlúa	Allied Domecq‡	42.04	○
Gordon's gin, Tanqueray, Dewar's, Johnnie Walker, VAT69, Bell's	Guinness*	40.07	○

Wine/Sparkling Wine

BRAND NAME	PARENT COMPANY	FEM	RATING
Inglenook, Almaden, Lancers, Glen Ellen, Christian Bros., La Piat d'Or	Grand Metropolitan	53.73	◐
Beringer Wines	Nestlé	44.44	○
Mumm	Seagram*	44.20	○
Dom Perignon, Moët, Simi, Hennessey	Guinness*	40.07	○

CIGARETTES

BRAND NAME	PARENT COMPANY	FEM	RATING
Marlboro, Benson & Hedges, Parliament, Virginia Slims, Merit	Philip Morris	50.72	◐
Camel, Now, Doral, Salem, Winston	RJR Nabisco	50.43	◐
duMaurier, Players	Imasaco	41.03	○

GROCERY STORES

BRAND NAME	PARENT COMPANY	FEM	RATING
King Soopers, Kroger, Dillon	Kroger	60.33	●
A&P, Waldbaums, Food Emporium, Dominion, Super Fresh, Farmer Jack, Kohl's, Food Bazaar, Miracle, Food Mart, Big Star	Great Atlantic & Pacific Tea*	49.55	◐
Giant Foods	Giant Foods	49.11	◐
Jewell/Osco Foods, Albertson's	Albertson's	47.85	◐
Safeway	Safeway	46.91	◐
7-Eleven	Southland	44.73	○
Publix	Publix Super Markets*	43.83	○
Circle K	Investcorp*	42.67	●
Jewel Food, Star Market, Acme Market	American Stores	42.40	○
Winn-Dixie, Marketplace, Buddies	Winn-Dixie Stores*	42.38	○
Stop & Shop	Stop & Shop	41.79	○

*Did not have scores for all five categories.

†Dr Pepper/7Up was acquired by Cadbury Schweppes.

‡Formerly Allied Lyons.

Rating Key: ○ = Female friendly company; buy products when possible ◐ = Average; buy with good conscience ● = Below average; avoid buying these products

GROCERY STORES

BRAND NAME	PARENT COMPANY	FEM	RATING
Food Lion	Food Lion*	39.67	○
Grand Union	Grand Union*	39.67	○

COSMETICS AND PERSONAL CARE PRODUCTS

Bar Soaps

BRAND NAME	PARENT COMPANY	FEM	RATING
Ivory, Coast, Zest, Safeguard, Camay	Procter & Gamble	64.45	●
Softsoap, Irish Spring, Palmolive	Colgate-Palmolive	55.80	●
Dial, Tone, Pure & Natural	Dial	43.96	○
Caress, Dove, Shield, Lux, Lifebuoy	Unilever	42.54	○

Cleansers

BRAND NAME	PARENT COMPANY	FEM	RATING
Clean & Clear, RoC, Purpose	Johnson & Johnson	81.81	●
Clinique	Estée Lauder	65.50	●
CBiothem	Cosmair	64.86	●
Noxzema, Clearasil, Oil of Olay	Procter & Gamble	64.45	●
Gloria Roberts	Johnson Products	61.14	●
Oxy, Clear by Design	SmithKline Beecham	60.94	●
Sea Breeze, Fostex	Bristol-Myers Squibb	59.62	●
Stridex, pHisoderm	Eastman Kodak	58.54	●
Neutrogena	Neutrogena†	49.75	◐
Yardley	Maybelline*	47.05	◐
Jafra	Gillette	44.05	○
Ponds, Vaseline	Unilever	42.54	○

Deodorants/Antiperspirants

BRAND NAME	PARENT COMPANY	FEM	RATING
Shower to Shower	Johnson & Johnson	81.81	●
Old Spice, Secret, Sure	Procter & Gamble	64.45	●
Ban, Tickle, Mum	Bristol-Myers Squibb	59.62	●
Tussy	Eastman Kodak	58.54	●
Speedstick, Mennen	Colgate-Palmolive	55.80	●
Arm & Hammer	Church & Dwight	48.68	◐
Dry Idea, Right Guard, Soft & Dri	Gillette	44.05	○
Dial	Dial	43.96	○
Fabergé Power Stick	Unilever	42.54	○
Mitchum	Revlon*	39.59	○
Lady's Choice, Arrid	Carter-Wallace	39.18	○

Fragrances

BRAND NAME	PARENT COMPANY	FEM	RATING
Aramis	Estée Lauder	65.50	●
Anais, Anais; Cacharel; Giorgio Armani; Gloria Vanderbilt; Guy Laroche	Cosmair	64.86	●
Old Spice, Hugo Boss	Procter & Gamble	64.45	●
Lady Stetson	Pfizer	57.70	●
Calvin Klein, Elizabeth Arden	Unilever	42.54	○

Hair Care

BRAND NAME	PARENT COMPANY	FEM	RATING
Final Net, Nice 'n Easy, Frost & Tip, Loving Care	Bristol-Myers Squibb	59.62	●
Ogilvie	Eastman Kodak	58.54	●

*Did not have scores for all five categories

†Neutrogena was acquired by Johnson & Johnson.

Rating Key: ○ = Female friendly company; buy products when possible ◐ = Average; buy with good conscience ● = Below average; avoid buying these products

Hair Care

BRAND NAME	PARENT COMPANY	FEM	RATING
Goody accessories	Newell	56.76	●
Perma-Soft	Dow Chemical	56.24	●
Epic Waves, Dippity-Do	Gillette	44.05	○

Lotion/Moisturizer

BRAND NAME	PARENT COMPANY	FEM	RATING
Skin-So-Soft	Avon Products	86.73	●
Body Shop	Body Shop	71.60	●
Keri, Moisturel	Bristol-Myers Squibb	59.62	●
Complex 15	Schering-Plough	58.51	●
Lubriderm	Warner-Lambert	57.81	●
Neutrogena	Neutrogena†	49.75	◐
Yardley	Maybelline*	47.05	◐
Vaseline Intensive Care	Unilever	42.54	○

Makeup/Lips

BRAND NAME	PARENT COMPANY	FEM	RATING
Avon	Avon Products	86.73	●
Body Shop	Body Shop	71.60	●
Origins, Prescriptives, Estée Lauder	Estée Lauder	65.50	●
Helene Rubenstein, L'Oréal, Lancôme	Cosmair	64.86	●
Max Factor, Cover Girl	Procter & Gamble	64.45	●
Coty "24" lipsticks	Pfizer	57.70	●
Fashion Fair	Johnson Publishing*	57.34	●

Makeup/Lips

BRAND NAME	PARENT COMPANY	FEM	RATING
Chap Stick	American Home Products	50.01	◐
Helene Curtis	Helene Curtis	47.66	◐
Maybelline	Maybelline*	47.05	◐
Jafra	Gillette	44.05	○
Vaseline, Cutex polish remover, Elizabeth Arden	Unilever	42.54	○
Revlon, Almay, Bill Blass	Revlon*	39.59	○

Shampoo

BRAND NAME	PARENT COMPANY	FEM	RATING
Johnson's Baby Shampoo	Johnson & Johnson	81.81	●
Body Shop	Body Shop	71.60	●
Aveda	Aveda	70.42	●
L'Oréal	Cosmair	64.86	●
Ivory Shampoo, Head & Shoulders, Pantene, Pert, Vidal Sassoon	Procter & Gamble	64.45	●
General Treatment, Ultrasheen, Soft 'n Free	Johnson Products	61.14	●
Alberto VO5	Alberto-Culver	59.99	●
Miss Clairol, Infusium	Bristol-Myers Squibb	59.62	●
Duke, Raveen	Johnson Publishing*	57.34	●
Paul Mitchell	John Paul Mitchell Systems*	56.78	●
Apple Pectin	Dow Chemical	56.24	●
Halsa, Agree	S.C. Johnson & Son	53.74	◐
Denorex	American Home Products	50.01	◐
Neutrogena	Neutrogena[†]	49.75	◐
Redken	Redken Laboratories*	47.91	◐

*Did not have scores for all five categories

[†]Neutrogena was acquired by Johnson & Johnson.

Rating Key: ○ = Female friendly company; buy products when possible ◑ = Average; buy with good conscience ● = Below average; avoid buying these products

Shampoo

BRAND NAME	PARENT COMPANY	FEM	RATING
Selsun blue	Abbott Laboratories	47.76	◐
Suave, Finesse, Salon Selectives	Helene Curtis	47.66	◐
White Rain	Gillette	44.05	○
Breck	Dial	43.96	○
Fabergé	Unilever	42.54	○
Flex	Revlon*	39.59	○

Shaving Needs/Razors

BRAND NAME	PARENT COMPANY	FEM	RATING
Noxzema, Old Spice	Procter & Gamble	64.45	●
Aqua Velva	SmithKline Beecham	60.94	●
Schick, Personal Touch, Wilkinson Sword	Warner-Lambert	57.81	●
Barbasol	Pfizer	57.70	●
Colgate, Mennen	Colgate-Palmolive	55.80	●
Edge	S.C. Johnson & Son	53.74	◐
Atra, Daisy, Good News!, Sensor, Trac, Foamy	Gillette	44.05	○
Brut 33	Unilever	42.54	○
Nair	Carter-Wallace	39.18	○
Neet	Reckitt & Colman	38.46	○

Tampons/Sanitary Pads/Incontinence Products

BRAND NAME	PARENT COMPANY	FEM	RATING
Stayfree, Carefree, Serenity, Sure & Natural, o.b.	Johnson & Johnson	81.81	●
Always, Attends	Procter & Gamble	64.45	●

BRAND NAME	PARENT COMPANY	FEM	RATING
Promise, Poise	Scott Paper	50.44	◐
Tampax	Tambrands	46.54	◐
Kotex, New Freedom, Lightdays, Depend, Poise	Kimberly-Clark	44.79	○

Toothbrushes/Toothpaste/Oral Hygiene

BRAND NAME	PARENT COMPANY	FEM	RATING
Reach, Act	Johnson & Johnson	81.81	●
Crest, Gleem	Procter & Gamble	64.45	●
Aquafresh	SmithKline Beecham	60.94	●
Plax	Pfizer	57.70	●
Colgate, Ultra Brite, Viadent	Colgate-Palmolive	55.80	●
Arm & Hammer Dental Care, PeroxiCare	Church & Dwight	48.68	◐
Oral-B	Gillette	44.05	○
Aim, Close-Up, Pepsodent, Mentadent	Unilever	42.54	○
Pearl Drops	Carter-Wallace	39.18	○
Scope	Procter & Gamble	64.45	●
Listerine	Warner-Lambert	57.81	●
Cepacol	Dow Chemical	56.24	●
Signal	Unilever	42.54	○
Binaca	Reckitt & Colman	38.46	○
Efferdent	Warner-Lambert	57.81	●
Rigident	Carter-Wallace	39.18	○

OVER-THE-COUNTER DRUGS

Birth Control (see also Prescription Drugs)

BRAND NAME	PARENT COMPANY	FEM	RATING
Ortho-Gynol, Gynol II	Johnson & Johnson	81.81	●

*Did not have scores for all five categories

Rating Key: ○ = Female friendly company; buy products when possible ◐ = Average; buy with good conscience ● = Below average; avoid buying these products

Birth Control (see also Prescription Drugs)

BRAND NAME	PARENT COMPANY	FEM	RATING
Today sponge, Semicid	American Home Products	50.01	◐
Trojan, Mentor, Class Act condoms	Carter-Wallace	39.18	○

Cold/Allergy Medicines

BRAND NAME	PARENT COMPANY	FEM	RATING
Sine-Aid, Tylenol Cold	Johnson & Johnson	81.81	●
Vicks, VapoRub, NyQuil, Vicks Formula 44	Procter & Gamble	64.45	●
Contac, Sine-Off	SmithKline Beecham	60.94	●
Comtrex	Bristol-Myers Squibb	59.62	●
Afrin, Chlor-Trimeton, Coricidin, Drixoral	Schering-Plough	58.51	●
Sinutab, Benadryl, Sudafed	Warner-Lambert	57.81	●
Dristan, Robitussin, Primatine, Dimetapp, Advil Cold and Sinus	American Home Products	50.01	◐
AlkaSeltzer Plus, Neo-Synephrine	Bayer†	46.45	◐
Motrin IB Sinus	Pharmacia & Upjohn	42.77	○

Cough Drops/Lozenges

BRAND NAME	PARENT COMPANY	FEM	RATING
Vicks	Procter & Gamble	64.45	●
Luden's	Hershey Foods	63.70	●
N'Ice, Hold, Sucrets	SmithKline Beecham	60.94	●
Fisherman's Friend	Bristol-Myers Squibb	59.62	●
Halls, Soother's	Warner-Lambert	57.81	●

Eye Drops/Contact Lenses and Solutions

BRAND NAME	PARENT COMPANY	FEM	RATING
Acuvue contact lens systems	Johnson & Johnson	81.81	●
Durasoft, FRESHLook	Schering-Plough	58.51	●
Visine	Pfizer	57.70	●
Clear Eyes, Murine drops	Abbott Laboratories	47.76	◐
Alcon Opti-Free	Nestlé	44.44	○

First Aid

BRAND NAME	PARENT COMPANY	FEM	RATING
Band-Aid, Johnson's Swabs	Johnson & Johnson	81.81	●
Solarcaine	Schering-Plough	58.51	●
Benylin, Caladryl, Proxacol Hydrogen Peroxide, Lavacol, Neosporin	Warner-Lambert	57.81	●
Bengay, Desitin	Pfizer	57.70	●
Camphophenique	Dow Chemical	56.24	●
Curad bandages	Colgate-Palmolive	55.80	●
Heet	American Home Products	50.01	◐
Bactine	Bayer†	46.45	◐
Mycitracin, Cortaid	Pharmacia & Upjohn	42.77	○
Q-Tips, Vaseline	Unilever	42.54	○

Hemorrhoid Treatments

BRAND NAME	PARENT COMPANY	FEM	RATING
Pago Hemorrhoid	Bristol-Myers Squibb	59.62	●
Anusol, Tucks	Warner-Lambert	57.81	●
Preparation H	American Home Products	50.01	◐

†Bayer recently acquired Miles Laboratories.

Rating Key: ○ = Female friendly company; buy products when possible ◐ = Average; buy with good conscience ● = Below average; avoid buying these products

Painkillers

BRAND NAME	PARENT COMPANY	FEM	RATING
Tylenol	Johnson & Johnson	81.81	●
Norwich Aspirin, Aleve	Procter & Gamble	64.45	●
Ecotrin	SmithKline Beecham	60.94	●
Bufferin, Nuprin, Excedrin	Bristol-Myers Squibb	59.62	●
St. Joseph's, Aspergum	Schering-Plough	58.51	●
Vanquish Analgesic	Warner-Lambert	57.81	●
Advil, Anacin, Anbesol	American Home Products	50.01	◐
Bayer, Midol, Pandol	Bayer†	46.45	◐
Motrin IB	Pharmacia & Upjohn	42.77	○

Pregnancy Tests

BRAND NAME	PARENT COMPANY	FEM	RATING
Fact PLUS	Johnson & Johnson	81.81	●
ept	Warner-Lambert	57.81	●
Clearblue Easy	American Home Products	50.01	◐
First Response, Answer	Carter-Wallace	39.18	○

Sleeping Pills/Stimulants

BRAND NAME	PARENT COMPANY	FEM	RATING
Vivarin, Sominex	SmithKline Beecham	60.94	●
NoDoz	Bristol-Myers Squibb	59.62	●
Unisom SleepGels	Pfizer	57.70	●
Sleep-Eze	American Home Products	50.01	◐
Miles Nervine Nighttime Sleep Aid	Bayer†	46.45	◐
Dramamine	Pharmacia & Upjohn	42.77	○

Stomach/Indigestion Medicine

BRAND NAME	PARENT COMPANY	FEM	RATING
Mylanta, Imodium A-D	Johnson & Johnson	81.81	●
Pepto-Bismol, Metamucil	Procter & Gamble	64.45	●
Titralac Antacid	3M	61.45	●
Tums	SmithKline Beecham	60.94	●
Correctol, Di-Gel	Schering-Plough	58.51	●
Rolaids, Gelusil, Myadec	Warner-Lambert	57.81	●
Benedictin	Dow Chemical	56.24	●
Phillips' Milk of Magnesia, Alka-Seltzer	Bayer†	46.45	◐
Kaopectate, Surfak	Pharmacia & Upjohn	42.77	○

Suntan Lotion

BRAND NAME	PARENT COMPANY	FEM	RATING
Piz Buin, Sundown	Johnson & Johnson	81.81	●
Bain de Soleil	Procter & Gamble	64.45	●
PreSun	Bristol-Myers Squibb	59.62	●
Coppertone, QT, Shade, Tropical Blend	Schering-Plough	58.51	●
Sea & Ski	Carter-Wallace	39.18	○

Vitamins

BRAND NAME	PARENT COMPANY	FEM	RATING
Geritol	SmithKline Beecham	60.94	●
Theragran-M	Bristol-Myers Squibb	59.62	●
Myadec	Warner-Lambert	57.81	●
OsCal	Dow Chemical	56.24	●

†Bayer recently acquired Miles Laboratories.

Rating Key: ○ = Female friendly company; buy products when possible ◐ = Average; buy with good conscience ● = Below average; avoid buying these products

Vitamins

BRAND NAME	PARENT COMPANY	FEM	RATING
Stuart Prenatal Tabs	American Home Products	50.01	◐
Centrum, Caltrate 600, Stresstabs	American Cyanamid†	49.57	◐
Ensure	Abbott Laboratories	47.76	◐
One-A-Day, Flintstones, Bugs Bunny	Bayer‡	46.45	◐
Unicap	Pharmacia & Upjohn	42.77	○

Wart Removal

BRAND NAME	PARENT COMPANY	FEM	RATING
Wart-Off Wart Remover	Pfizer	57.70	●
Compound W	American Home Products	50.01	◐
Tronolane	Abbott Laboratories	47.76	◐

Yeast Infection Medicines

BRAND NAME	PARENT COMPANY	FEM	RATING
Monistat 7	Johnson & Johnson	81.81	●
Gyne-Lotrimin	Schering-Plough	58.51	●
Gyne Cure	Pfizer	57.70	●
Mycelex	Bayer‡	46.45	◐

PRESCRIPTION MEDICATIONS

Antibiotics

BRAND NAME	PARENT COMPANY	FEM	RATING
Floxin	Johnson & Johnson	81.81	●
Noroxin	Merck	72.61	●

Flagyl	Monsanto (Searle)[§]	65.15	●
Augmentin, Timentin	SmithKline Beecham	60.94	●
Garamycin	Schering-Plough	58.51	●
Vibramycin	Pfizer	57.70	●
Achromycin	American Cyanamid[†]	49.57	◐
Erythromycin	Abbott Laboratories	47.76	◐
Cipro	Bayer	46.45	◐
Zagam	Rhône-Poulenc Rorer	46.21	◐
Tazidime, Vancocin	Eli Lilly & Co.	38.95	○

Arthritis

BRAND NAME	PARENT COMPANY	FEM	RATING
Clinoril, Indocin	Merck	72.61	●
Relafen	SmithKline Beecham	60.94	●
Feldene	Pfizer	57.70	●

Asthma/Allergies

BRAND NAME	PARENT COMPANY	FEM	RATING
Hismanal (antihistamine)	Johnson & Johnson	81.81	●
Vanceril, Proventil (asthma), Claritin (antihistamine), Vancenase AQ (allergies)	Schering-Plough	58.51	●
Nasalide (allergies)	Syntex	56.93	●
Ventolin (asthma), Aldactazide, Beclovent (allergies)	Glaxo Wellcome	56.56	●

Birth Control

BRAND NAME	PARENT COMPANY	FEM	RATING
Ortho-Novum	Johnson & Johnson	81.81	●
Loestrin	Warner-Lambert	57.81	●

[†]American Cyanamid was bought by American Home Products.

[‡]Bayer recently acquired Miles Laboratories.

[§]Monsanto is a wholly owned subsidiary of Bayer.

Rating Key: ○ = Female friendly company; buy products when possible ◐ = Average; buy with good conscience ● = Below average; avoid buying these products

Birth Control

BRAND NAME	PARENT COMPANY	FEM	RATING
Norinyl	Syntex	56.93	●
Norplant, Lo/Ovral	American Home Products	50.01	◐
Depo-Provera	Pharmacia & Upjohn	42.77	○

Cancer

BRAND NAME	PARENT COMPANY	FEM	RATING
Taxol, Cytoxan	Bristol-Myers Squibb	59.62	●
Intron A	Schering-Plough	58.51	●
Roferon-A	Genentech	50.23	◐
Taxotere, Campto	Rhône-Poulenc Rorer	46.21	◐

Circulation/Blood

BRAND NAME	PARENT COMPANY	FEM	RATING
Procrit (blood)	Johnson & Johnson	81.81	●
Aldactone (blood pressure)	Monsanto (Searle)†	65.15	●
Dyazide (blood pressure)	SmithKline Beecham	60.94	●
Ticlid (stroke), Cardene (cardiovascular)	Syntex	56.93	●
Activase (blood clots)	Genentech	50.23	◐
Verelan (blood pressure)	American Home Products	50.01	◐
Hytrin (hypertension)	Abbott Laboratories	47.76	◐
Dilacor XR, Lozol (blood pressure)	Rhône-Poulenc Rorer	46.21	◐

Depression

BRAND NAME	PARENT COMPANY	FEM	RATING
Paxil	SmithKline Beecham	60.94	●

BRAND NAME	PARENT COMPANY	FEM	RATING
Zoloft	Pfizer	57.70	●
Prozac	Eli Lilly & Co.	39.85	○

Diabetes

BRAND NAME	PARENT COMPANY	FEM	RATING
Micronase	Pharmacia & Upjohn	42.77	○
Iletin, Humulin	Eli Lilly & Co.	39.85	○

Diet/Digestion/Ulcers

BRAND NAME	PARENT COMPANY	FEM	RATING
Prepulsid (digestive aid), Imodium	Johnson & Johnson	81.81	●
Pepcid (ulcer)	Merck	72.61	●
Lomotil (diarrhea)	Monsanto (Searle)†	65.15	●
Tagamet (ulcers)	SmithKline Beecham	60.94	●
Zantac (ulcer)	Glaxo Wellcome	56.56	●
Caltrate, Florvite (nutrition supplements)	American Cyanamid‡	49.57	◐
Prevacid	Abbott Laboratories	47.76	◐
Axid (ulcers)	Eli Lilly & Co.	39.85	○

Heart

BRAND NAME	PARENT COMPANY	FEM	RATING
Mevacor (heart)	Merck	72.61	●
Nitrodisc (angina)	Monsanto (Searle)†	65.15	●
Capoten (cardiovascular), Pravachol (cholesterol)	Bristol-Myers Squibb	59.62	●
Procan SR (heart), Accupril (cardiovascular)	Warner-Lambert	57.81	●
Cardura, Procardia (heart)	Pfizer	57.70	●
Activase (heart)	Genentech	50.23	◐

†Monsanto is a wholly owned subsidiary of Bayer.

‡American Cyanamid was bought by American Home Products.

Rating Key: ○ = Female friendly company; buy products when possible ◐ = Average; buy with good conscience ● = Below average; avoid buying these products

Heart

BRAND NAME	PARENT COMPANY	FEM	RATING
Cordarone	American Home Products	50.01	◐

Hormones

BRAND NAME	PARENT COMPANY	FEM	RATING
Protropin (hormone)	Genentech	50.23	◐
Premarin, Premphase	American Home Products	50.01	◐
Ogen (estrogen)	Abbott Laboratories	47.76	◐
Provera (progestin)	Pharmacia & Upjohn	42.77	○
Darvon, Tazidime, Vancocin, DES (estrogen)	Eli Lilly & Co.	39.85	○

Painkillers/Relaxants

BRAND NAME	PARENT COMPANY	FEM	RATING
Anaprox, Naprosyn (anti-inflammatory)	Syntex	56.93	●
RPR 100893	Rhône-Poulenc Rorer	46.21	◐
Motrin (painkiller)	Pharmacia & Upjohn	42.77	○
Darvon (analgesic)	Eli Lilly & Co.	39.85	○
Soma	Carter-Wallace	39.18	○

Miscellaneous

BRAND NAME	PARENT COMPANY	FEM	RATING
Risperdal (schizophrenia)	Johnson & Johnson	81.81	●
Dilantin (anticonvulsant)	Warner-Lambert	57.81	●
Actimmune (immune system disorder)	Genentech	50.23	◐
Diamox (glaucoma), Nilstat (antifungal)	American Cyanamid†	49.57	◐
Penetrex (urinary infections)	Rhône-Poulenc Rorer	46.21	◐

Miscellaneous

BRAND NAME	PARENT COMPANY	FEM	RATING
Felbatol (epilepsy)	Carter-Wallace	39.18	○

Skin

BRAND NAME	PARENT COMPANY	FEM	RATING
Retin-A (skin treatment)	Johnson & Johnson	81.81	●

Sleeping/Sedatives

BRAND NAME	PARENT COMPANY	FEM	RATING
Thorazine (anxiety)	SmithKline Beecham	60.94	●
Halcion, Xanax	Pharmacia & Upjohn	42.77	○
Doral (sedative)	Carter-Wallace	38.18	○

HOUSEHOLD GOODS

Air Fresheners

BRAND NAME	PARENT COMPANY	FEM	RATING
Glade	S.C. Johnson & Son	53.74	◐
Arm & Hammer	Church & Dwight	48.68	◐
Renuzit	Dial	43.96	○
Airwick, Stick Ups	Reckitt & Colman	38.46	○

Barbecue

BRAND NAME	PARENT COMPANY	FEM	RATING
Sterno	Colgate-Palmolive	55.80	●
Match Light, BBQ Bag, Kingsford charcoal	Clorox	49.23	◐

†American Cyanamid was bought by American Home Products.

Rating Key: ○ = Female friendly company; buy products when possible ◐ = Average; buy with good conscience ● = Below average; avoid buying these products

Batteries

BRAND NAME	PARENT COMPANY	FEM	RATING
Kodak	Eastman Kodak	58.54	●
Energizer, Eveready	Ralston Purina	45.03	◐
Panasonic	Matsushita Electric	36.56	○

Bug, Pest Killers

BRAND NAME	PARENT COMPANY	FEM	RATING
d-Con	Eastman Kodak	58.54	●
Muskol Insect Repellent	Schering-Plough	58.51	●
Off!, Raid	S.C. Johnson & Son	53.74	◐
Combat Ant/Roach Traps, Superbait	Clorox	49.23	◐
Cutter	Bayer†	46.45	◐
Black Flag	Reckitt & Colman	38.46	○

Cookware

BRAND NAME	PARENT COMPANY	FEM	RATING
Mirro, Foley Rema, Wearever, Airbake, Cushionaire	Newell	56.76	●
Corelle, Pyrex, Revere Ware	Corning	56.40	●
Baker's Choice baking cups	Reynolds Metals	40.42	○
Farberware	Farberware/Hanson*	40.11	○

Dish Soap/Sponges

BRAND NAME	PARENT COMPANY	FEM	RATING
Cascade, Dawn, Joy, Ivory Liquid	Procter & Gamble	64.45	●
Scotch-Brite, O-Cel-O	3M	61.45	●
Palmolive, Ajax	Colgate-Palmolive	55.80	●

BRAND NAME	PARENT COMPANY	FEM	RATING
S.O.S., Tuffy scrubber	Clorox	49.23	◐
Sweetheart, Trend, Dobie	Dial	43.96	○
Sunlight, Dove	Unilever	42.54	○

Film

BRAND NAME	PARENT COMPANY	FEM	RATING
Polaroid	Polaroid	61.88	●
Kodak	Eastman Kodak	58.54	●
Agfa	Bayer†	46.45	◐

Household/Bathroom Cleaners

BRAND NAME	PARENT COMPANY	FEM	RATING
Comet, Mr. Clean, Spic and Span	Procter & Gamble	64.45	●
Lysol, Love My Carpet, Mop & Glo, Resolve, Thompson's, Minwax, Formby's	Eastman Kodak	58.54	●
Fantastik, Glass Plus, Dow Bathroom Cleaner	Dow Chemical	56.24	●
Ajax, Murphy Oil soaps, Handi Wipes, Spray n' Wipe	Colgate-Palmolive	55.80	●
Fuller Brush, Ty-D-Bol, Endust	Sara Lee	55.67	●
Windex, Vanish, Drano, Pledge, Brite, Flavor, Fine Wood	S.C. Johnson & Son	53.74	◐
Formula 409, Pine Sol, Liquid-Plumr, Soft Scrub, Tilex, Tackle	Clorox	49.23	◐
Arm & Hammer	Church & Dwight	48.68	◐
Boraxo, Borateem, Brillo, Parson's Ammonia, Sno Bol, Bruce Floor Care	Dial	43.96	○
Armstrong floor cleaners	Armstrong World	40.54	○
Easy-Off, Carpet Fresh	Reckitt & Colman	38.46	○

Laundry Detergent/Fabric Softeners

BRAND NAME	PARENT COMPANY	FEM	RATING
Ivory Snow, Tide, Cheer, Dash, Bold, Dreft, Era	Procter & Gamble	64.45	●
Spray 'n Wash	Dow Chemical	56.24	●

*Did not have scores for all five categories.
†Bayer recently acquired Miles Laboratories.

Rating Key: ○ = Female friendly company; buy products when possible ◐ = Average; buy with good conscience ● = Below average; avoid buying these products

Laundry Detergent/Fabric Softeners

BRAND NAME	PARENT COMPANY	FEM	RATING
Fab, Dynamo, Fresh Start	Colgate-Palmolive	55.80	●
Shout	S.C. Johnson & Son	53.74	◐
Clorox Bleach, Vivid, Yes, Stain Out	Clorox	49.23	◐
Arm & Hammer detergent, bleach, fabric softener	Church & Dwight	48.68	◐
Purex, Ultra Trend, 20 Mule Team Borax	Dial	43.96	○
All, Wisk, Surf	Unilever	42.54	○
Bounce, Downy	Procter & Gamble	64.45	●
Scotchgard	3M	61.45	●
Cling Free, Delicare	SmithKline Beecham	60.94	●
Kleen Guard, Static Guard	Alberto-Culver	59.99	●
Rinse 'n Soft, Sta Puf	Dial	43.96	○
Snuggle	Unilever	42.54	○
Woolite, Easy-On Starch	Reckitt & Colman	38.46	○

Lawn/Garden Supplies

BRAND NAME	PARENT COMPANY	FEM	RATING
Ortho, Roundup, Lasso lawn products	Monsanto†	65.15	●
Rubbermaid planters, wheelbarrows	Rubbermaid	60.21	●
Treflan	Dow Chemical	56.24	●

Light Bulbs

BRAND NAME	PARENT COMPANY	FEM	RATING
GE	General Electric	61.78	●

Miscellaneous Household Goods

BRAND NAME	PARENT COMPANY	FEM	RATING
EZ Topps, Rubbermaid containers, trash bins, mops, ice cube trays	Rubbermaid	60.21	●
Dr. Scholl's foot products	Schering-Plough	58.51	●
Levelor miniblinds, Anchor Hocking glassware	Newell	56.76	●
Kiwi shoe polish	Sara Lee	55.67	●
Braun appliances	Gillette	44.05	○
Simonize, STP	First Brands	39.00	○

Paper Products

BRAND NAME	PARENT COMPANY	FEM	RATING
Bounty	Procter & Gamble	64.45	●
Wet Ones	Eastman Kodak	58.54	●
Wash 'n Dri	Colgate-Palmolive	55.80	●
Viva, Scott Towels	Scott Paper	50.44	◐
Vanity Fair, Dixie Cups, Brawny	James River of Virginia	48.77	◐
Sparkle	Georgia-Pacific	45.58	◐
Hi-Dri, Little Travelers	Kimberly-Clark	44.79	○

Pet Foods

BRAND NAME	PARENT COMPANY	FEM	RATING
Science Diet, Hill's	Colgate-Palmolive	55.80	●
Pounce	Quaker Oats	53.40	◐
Milk Bone, Butcher's Choice	RJR Nabisco	50.43	◐
9 Lives, Amoré cat food	H.J. Heinz	47.77	◐
Cat Chow, Meow Mix, Tender Vittles	Ralston Purina	45.03	◐
Friskies, Fancy Feast	Nestlé	44.44	○

†Monsanto is a wholly owned subsidiary of Bayer.

Rating Key: ● = Female friendly company; buy products when possible ◐ = Average; buy with good conscience ○ = Below average; avoid buying these products

Pet Foods

BRAND NAME	PARENT COMPANY	FEM	RATING
Crave	Mars	43.23	○
Ken-L Ration, Gaines, Kibbles 'n Bits, Gravy Train, Cycle	Quaker Oats Co.	53.40	◐
Meaty Bone	H.J. Heinz	47.77	◐
Dog Chow, Bonz, Chuck Wagon	Ralston Purina	45.03	◐
Sargeant's, Geisler, Singer	ConAgra	44.69	○
Alpo, Mighty Dog	Nestlé	44.44	○
Kal Kan, Pedigree	Mars	43.23	○

Pet Supplies

BRAND NAME	PARENT COMPANY	FEM	RATING
Litter Green, Fresh Step, Control, Scoop Fresh	Clorox	49.23	◐
Scoop Away, Arm & Hammer cat litter, Pet Fresh Carpet Deodorizer	Church & Dwight	48.68	◐
Fresh 'N Clean, Color Guard flea and tick collars, Ear Rite, Shield	Carter-Wallace	39.18	○
Everclean, Jonny Cat	First Brands	39.00	○

Plastic Wrap/Bags/Foil

BRAND NAME	PARENT COMPANY	FEM	RATING
Hefty bags, Baggies, Cinch Sak	Mobil Oil	56.31	●
Handi-Wrap, Saran Wrap, Ziploc	Dow Chemical	56.24	●
Reynolds Wrap, Cut-Rite wax paper, Sure-Seal, Qwik-Seal	Reynolds Metals	40.42	○
Glad	First Brands	39.00	○

School Supplies (see also Office Supplies)

BRAND NAME	PARENT COMPANY	FEM	RATING
Scotch tape, Post-it Notes	3M	61.45	●

Mead, Cambridge notebooks, Class Mate, Kwik Clip, Envelok, Organizer	Mead	48.66	◐
Elmer's Glue, Krazy Glue	Borden	45.29	◐
Scribe, Kimberly, Chieftan	Kimberly-Clark	44.79	○
Eraser Mate, Liquid Paper, Paper Mate	Gillette	44.05	○

Toilet Paper/Tissues

BRAND NAME	PARENT COMPANY	FEM	RATING
Charmin Ultra	Procter & Gamble	64.45	●
Scott Tissue, Cottonelle	Scott Paper	50.44	◐
Quilted Northern, Marina, Nice 'n Soft, Chelsea	James River of Virginia	48.77	◐
Angel Soft, Coronet	Georgia-Pacific	45.58	◐
Delsey	Kimberly-Clark	44.79	○
Puffs	Procter & Gamble	64.45	●
Scotties tissue	Scott Paper	50.44	◐
Lotus	James River of Virginia	48.77	◐
Kleenex	Kimberly-Clark	44.79	○

RETAIL STORES

Department Stores

BRAND NAME	PARENT COMPANY	FEM	RATING
Mervyn's, Dayton's, Hudson's, Marshall Field	Dayton Hudson	71.71	●
Nordstrom	Nordstrom	68.08	●
Sears	Sears, Roebuck	63.55	●
Marshall's	Melville	61.27	●
The Broadway, Broadway Southwest, Weinstock's, Emporium	Broadway Stores†	59.41	●
J.C. Penney, Thrift Drug	J.C. Penney	53.85	◐

†Broadway Stores was formerly known as Carter Hawley Hale.

Rating Key: ○ = Female friendly company; buy products when possible ◐ = Average; buy with good conscience ● = Below average; avoid buying these products

Department Stores

BRAND NAME	PARENT COMPANY	FEM	RATING
Bloomingdale's, Burdine's, Lazarus, Abraham Straus/Jordan Marsh, Stern's, Rich's/Goldsmith's	Federated Department Stores	52.38	◐
Filene's, Hecht's, Kaufmann's, Robinson-May, Famous-Barr, Meier & Frank, Foley's, Lord & Taylor	May Department Stores	50.51	◐
Neiman Marcus, Bergdorf Goodman, Contempo Casuals	Neiman Marcus Group*	49.73	◐
Macy's, I. Magnin†	R.H. Macy*	43.94	○
Montgomery Ward	Montgomery Ward Holding*	42.58	○

Discount Stores

BRAND NAME	PARENT COMPANY	FEM	RATING
Target	Dayton Hudson	71.71	●
K-Mart	K-Mart	51.89	◐
Woolworth, Woolco	Woolworth	48.40	◐
Wal-Mart, Sam's Club	Wal-Mart Stores	41.67	○

Drugstores

BRAND NAME	PARENT COMPANY	FEM	RATING
CVS, People's Drug Stores	Melville	61.27	●
Thrift Drug, Treasury Drug	J.C. Penney	53.85	◐
Walgreen	Walgreen	47.61	◐
Sav-On, Osco Drug	American Stores	42.40	○

Hardware Stores

BRAND NAME	PARENT COMPANY	FEM	RATING
Target	Dayton Hudson	71.71	●

BRAND NAME	PARENT COMPANY	FEM	RATING
Sears	Sears, Roebuck	63.55	●
Builder's Square	K-Mart	51.89	◐
Montgomery Ward	Montgomery Ward Holding*	42.58	○

Other Stores

BRAND NAME	PARENT COMPANY	FEM	RATING
Linens 'n Things, This End Up	Melville	61.27	●
Pearle Vision	Grand Metropolitan	53.73	◐
Borders-Walden	K-Mart†	51.89	◐
Country General stores	ConAgra	44.69	○
Service Merchandise	Service Merchandise*	43.46	○
PACE Membership Warehouse, Sam's Club	Wal-Mart Stores	41.67	○

CLOTHING

At The Mall

BRAND NAME	PARENT COMPANY	FEM	RATING
Benetton	Benetton Group	68.49	●
Ann Taylor	Ann Taylor	64.38	●
Esprit	Esprit de Corp	63.88	●
Wilson's, Bob's Stores	Melville	61.27	●
The Gap, Banana Republic	The Gap	61.03	●
Coach	Sara Lee	55.67	●
Van Heusen	Phillips-Van Heusen	51.46	◐
The Limited, Lane Bryant, Lerner New York, Victoria's Secret, Express, Abercrombie & Fitch, Henri Bendel, Structure (for men)	The Limited*	48.72	◐
Accessory Lady	Woolworth	48.40	◐

*Did not have scores for all five categories.

†K-Mart had plans to sell 51 percent of its Borders-Walden bookstores subsidiary.

Rating Key: ● = Female friendly company; buy products when possible ◐ = Average; buy with good conscience ○ = Below average; avoid buying these products

At The Mall

BRAND NAME	PARENT COMPANY	FEM	RATING
Laura Ashley	Laura Ashley Holdings*	47.30	◐

Catalogs

BRAND NAME	PARENT COMPANY	FEM	RATING
Lands' End	Lands' End*	45.96	◐

Hosiery

BRAND NAME	PARENT COMPANY	FEM	RATING
Round the Clock, Givenchy, Christian Dior, Anne Klein Collection	Danskin	60.02	●
Hanes, L'eggs, Smooth Illusions	Sara Lee	55.67	●

Jeans

BRAND NAME	PARENT COMPANY	FEM	RATING
Levis, Dockers	Levi Strauss	58.83	●
Lee, Wrangler, Maverick, Riders, Marithé & François Girbaud, Timber Creek, Rustler	VF	53.00	◐
Gitano Jeans	Fruit of the Loom	40.62	○

Lingerie/Underwear

BRAND NAME	PARENT COMPANY	FEM	RATING
Warner's, Valentino Intimo, Olga, Calvin Klein, Fruit of the Loom bras, Van Roalte	Warnaco	58.46	●
Hanes, Bali, Playtex, Just my size, Wonderbra	Sara Lee	55.67	●
Vanity Fair, Barbizon, Vassarette, Silhouette, Eileen West	VF	53.00	◐
Cacique	The Limited*	48.72	◐
Fruit of the Loom underwear	Fruit of the Loom	40.62	○

Other Brand Names

BRAND NAME	PARENT COMPANY	FEM	RATING
Liz Claiborne, Lizsport, Liz & Co., Dana Buchman, First Issue, Realities, Russ	Liz Claiborne	70.91	●
Oshkosh B'Gosh, Boston Traders	Oshkosh B'Gosh	61.15	●
Custom Collection	Danskin	60.02	●
Chaps by Ralph Lauren, Christian Dior, Hathaway	Warnaco	58.46	●
Champion athletic knitwear	Sara Lee	55.67	●
Breckenridge, Kasper, Knitivo, Shapely, Victoire, Castlebury Knits	Leslie Fay	53.95	◐
Timber Creek	VF	53.00	◐
Geoffrey Beene, Pickwick, Somerset	Phillips-Van Heusen	51.46	◐
Timberland	Timberland	48.23	◐
Lanz	Lanz*	47.01	◐
Calvin Klein's "CK"	Nine West Group	42.46	○
Gitano	Gitano Group	39.51	○

Sportswear

BRAND NAME	PARENT COMPANY	FEM	RATING
Moving Comfort	Moving Comfort	86.77	●
Danskin, Dance France, Shape Activewear	Danskin	60.02	●
Head, Personal Sportswear	Leslie Fay	53.95	◐
Jantzen, JanSport	VF	53.00	◐
Reebok, Tinley, Above the Rim	Reebok International	50.91	◐
Fruit of the Loom, Wilson, Salem sportswear, Jostens	Fruit of the Loom	40.62	○

*Did not have scores for all five categories.

Rating Key: ○ = Female friendly company; buy products when possible ◐ = Average; buy with good conscience ● = Below average; avoid buying these products

SHOES

Athletic Shoes

BRAND NAME	PARENT COMPANY	FEM	RATING
Nike	Nike	64.83	●
Reebok, AVIA	Reebok	50.91	◐
Adidas, Le Coq Sportif, Pony USA	Adidas USA	48.41	◐
L.A. Gear, L.A. Tech, Catapult	L.A. Gear	47.40	◐
Capezio	Nine West Group	42.46	○
New Balance	New Balance Athletic Shoes	41.32	○

Other Shoes

BRAND NAME	PARENT COMPANY	FEM	RATING
Birkenstock Footprint Sandals	Birkenstock	75.18	●
Sperry Top-Sider, Keds, Grasshoppers	Stride Rite	69.34	●
Thom McAn, Meldisco	Melville	61.27	●
Van Heusen, Geoffrey Beene, Pickwick, Somerset, Bass	Phillips-Van Heusen	51.46	◐
Rockport, Boks, Rockport, AVIA	Reebok	50.91	◐
Timberland	Timberland	48.23	◐
Street Beat, Street Gear, Street Hiker	L.A. Gear	47.40	◐
Air Step, Connie, Life Stride, Naturalizer	Brown Group*	44.09	○
Nine West, Easy Spirit, Amalfi, Pappagallo, Evan-Picone, Bandolino, Calico, Enzo Angiolini, Selby, Westies, 9&Co., Calvin Klein's "CK"	Nine West Group†	42.46	○
Johnston & Murphy, Jarmin, Laredo, Code West, Mitre	Genesco	39.26	○

Shoe Stores

BRAND NAME	PARENT COMPANY	FEM	RATING
FootAction, Thom McAn	Melville	61.27	●

Shoe Stores

BRAND NAME	PARENT COMPANY	FEM	RATING
Payless Shoesource	May Department Stores	50.51	◐
Foot Locker, Lady Foot Locker, Kinney	Woolworth	48.40	◐
Famous Footwear	Brown Group*	44.09	○

BABY PRODUCTS

Accessories

BRAND NAME	PARENT COMPANY	FEM	RATING
Gerry, Soft Gate	Huffy	49.50	◐
Gerber bedding	Gerber Products‡	47.85	◐

Bathing Products/Wipes

BRAND NAME	PARENT COMPANY	FEM	RATING
Johnson's Baby Shampoo, oil, powder, Baby Diaper Rash Relief	Johnson & Johnson	81.81	●
Chubs baby wipes	Eastman Kodak	58.54	●
A&D Medicated Diaper Rash Ointment	Schering-Plough	58.51	●
Desitin	Pfizer	57.70	●
Baby Magic	Colgate-Palmolive	55.80	●
Baby Fresh, Wash a Bye Baby	Scott Paper	50.44	◐
Huggies wipes	Kimberly-Clark	44.79	○

Clothes

BRAND NAME	PARENT COMPANY	FEM	RATING
babyGap	The Gap	61.03	●
Carter's	William Carter	51.93	◐
Gerber play and sleepwear, Onesies	Gerber Products‡	47.85	◐

*Did not have scores for all five categories.

†Nine West recently acquired the shoe divisions of U.S. Shoe.

‡Gerber has been acquired by a Swiss-owned company that is not in our sample.

Rating Key: ● = Female friendly company; buy products when possible ◐ = Average; buy with good conscience ○ = Below average; avoid buying these products

Diapers

BRAND NAME	PARENT COMPANY	FEM	RATING
Pampers, Luvs	Procter & Gamble	64.45	●
Huggies Pull-Ups, Ultratrim	Kimberly-Clark	44.79	○

Food

BRAND NAME	PARENT COMPANY	FEM	RATING
Gerber baby food, Graduates	Gerber Products†	47.85	◐
Heinz	H.J. Heinz	47.77	◐

Formula

BRAND NAME	PARENT COMPANY	FEM	RATING
ProSobee, Enfamil, Lactofree, Pregestimil, Next Step Toddler Formula	Bristol-Myers Squibb	59.62	●
Nursoy, Promil, SMA	American Home Products	50.01	◐
Gerber Baby Formula	Gerber Products†	47.85	◐
Similac, Isomil formula, Pedialyte	Abbott Laboratories	47.76	◐
Carnation infant formula	Nestlé	44.44	○

Miscellaneous Baby Items

BRAND NAME	PARENT COMPANY	FEM	RATING
Band-Aid, Johnson's Baby Sunblock, Healthflow infant feeding system, Pediacare, Children's Tylenol	Johnson & Johnson	81.81	●
Coppertone Kids, Water Babies	Schering-Plough	58.51	●
Hugger, Pur, Tommee Tippee	Hasbro	48.62	◐

Toys

BRAND NAME	PARENT COMPANY	FEM	RATING
Sesame Street books	Children's Television Workshop	76.13	●
Brio toys	Brio	69.74	●
Barbie, Hot Wheels, Disney toys (Little Mermaid, Lion King), Fisher Price, See 'N Say, Li'l Miss, Magic Nursery, PJ Sparkles	Mattel	67.78	●
Circus World, K&K, Kay-Bee toys and stores	Melville	61.27	●
Little Tikes toys	Rubbermaid	60.21	●
Huffy bikes, basketball equipment	Huffy	49.50	◐
Barney, GI Joe, Playskool, Milton Bradley, Play-Doh, Monopoly, Scrabble, Candy Land, Jurassic Park, Batman, Mr. Potato Head, Nerf, Parker Brothers, Tonka, Cabbage Patch kids, Lincoln Logs, Raggedy Ann, Trivial Pursuit	Hasbro	48.62	◐
Toys "R" Us, Babies "R" Us	Toys "R" Us	48.59	◐
Educational aids and kits for school and home	Educational Insights*	45.02	◐
Tyco toys	Tyco Toys	41.86	○

Children's Clothes and Shoes

BRAND NAME	PARENT COMPANY	FEM	RATING
Keds	Stride Rite	69.34	●
Esprit Kids	Esprit de Corp	63.88	●
Sears, Toughskins	Sears, Roebuck	63.55	●
Oshkosh B'Gosh	Oshkosk B'Gosh	61.15	●
GapKids	The Gap	61.03	●
Mousefeathers dresses	Mousefeathers*	57.84	●
Healthtex, Fisher-Price Kids Wear, Wrangler, Lee	VF	53.00	◐
Carter underwear, night and outerwear	William Carter's	51.93	◐
Weebok	Reebok	50.91	◐

*Did not have scores for all five categories.

†Gerber has been acquired by a Swiss-owned company that is not in our sample.

Rating Key: ○ = Female friendly company; buy products when possible ◐ = Average; buy with good conscience ● = Below average; avoid buying these products

Children's Clothes and Shoes

BRAND NAME	PARENT COMPANY	FEM	RATING
The Limited Too	The Limited*	48.72	◐
Kids "R" Us, Babies "R" Us	Toys "R" Us	48.59	◐
Kids Foot Locker, Kids Mart/Little Folks	Woolworth	48.40	◐
Bendable	L.A. Gear	47.40	◐
Lands' End catalog	Lands' End*	45.96	◐
Buster Brown	Brown Group*	44.09	○
Fruit of the Loom	Fruit of the Loom	40.62	○

TRANSPORTATION

Automobiles

BRAND NAME	PARENT COMPANY	FEM	RATING
Chevrolet, Buick, Pontiac, Cadillac, Saturn, Oldsmobile, GMC trucks	General Motors	63.90	●
Toyota, Lexus, 4Runner, Avalon, Land Cruiser, Previa	Toyota	46.33	◐
Volkswagen, Audi	Volkswagen	43.48	○
Chrysler, Dodge, Plymouth, Jeep, Eagle	Chrysler	41.26	○
Mitsubishi	Mitsubishi Motors	39.57	○
Nissan, Infiniti, Pathfinder, Quest, Sentra	Nissan*	37.48	○
Ford, Mercury, Lincoln, Jaguar	Ford Motor	37.22	○
Honda, Acura, Passport, Odyssey	Honda of America*	36.55	○

Auto Supplies

BRAND NAME	PARENT COMPANY	FEM	RATING
Cooper tires	Cooper Tire & Rubber	53.70	◐

Goodyear tires	Goodyear Tire & Rubber	46.69	◐
Midas mufflers, brakes, etc.	Whitman	43.95	○
Champion spark plugs, Anco	Cooper Industries	43.63	○
Motor oil, Champion wiper blades, brakes	Quaker State	43.08	○

Auto Supplies

BRAND NAME	PARENT COMPANY	FEM	RATING
Exhaust pipes, shock absorbers	Arvin	39.34	○
Prestone antifreeze, STP, Simonize	First Brands	39.00	○
Retail stores for parts and service	Pep Boys—Manny, Moe & Jack*	36.59	○

Gasoline Companies

BRAND NAME	PARENT COMPANY	FEM	RATING
Conoco	E.I. DuPont	65.78	●
ARCO	Atlantic Richfield	62.06	●
Mobil	Mobil Oil	56.31	●
Chevron	Chevron	55.86	●
Amoco	Amoco	54.14	◐
Exxon	Exxon	52.49	◐
Texaco	Texaco	48.37	◐
Phillips 66	Phillips Petroleum	48.27	◐
BP	British Petroleum	43.91	○
Shell	Royal Dutch Petroleum/Shell Oil	41.59	○
Sunoco	Sun	41.12	○
Diamond Shamrock	Occidental Petroleum	37.72	○
Citgo	Citgo	34.35	○

*Did not have scores for all five categories.

Rating Key: ○ = Female friendly company; buy products when possible ◑ = Average; buy with good conscience ● = Below average; avoid buying these products

HOUSEHOLD EQUIPMENT

Appliances

BRAND NAME	PARENT COMPANY	FEM	RATING
Kenmore appliances	Sears, Roebuck	63.55	●
Hotpoint dishwashers, ranges	General Electric	61.78	●
Carrier heating, air conditioning	United Technologies	58.62	●
Mixers, can openers, blenders, food processors, toasters, coffee makers, irons	Hamilton Beach/Proctor Silex†	48.37	◐
Snow removers, Lawn-Boy mowers, weed trimmers, outdoor lighting	Toro	46.73	◐
KitchenAid, Roper, Whirlpool	Whirlpool	45.95	◐
Braun juicers, razors, coffee makers, blenders	Gillette	44.05	○
Spacemaker, Dustbuster, Workmate, can openers, coffee makers, toasters, irons, mixers, power tools	Black & Decker	42.05	○
Maytag washers and driers, Jenn-Air, Admiral, Hoover, Magic Chef	Maytag	41.84	○
Toasters and small ovens, clocks, griddles, breadmakers, heaters	Toastmaster	40.56	○
Thermador ovens, Waste King disposals	Masco	39.94	○
Osterizers, Mixmasters, clocks, timers	Sunbeam	39.92	○
Amana, Caloric, Speed Queen	Raytheon	38.51	○
Acme Juicer, Waring blenders	Dynamics of America	37.56	○
Vacuum cleaners	Electrolux*	37.35	○
Vacuum cleaners, irons, shavers	Philips Electronics	37.00	○
Air conditioners	Fedders	36.72	○
Panasonic vacuums, razors, microwaves, dishwashers, washing machines	Matsushita Electric	36.56	○
Fans, In-Sink-Erators, power tools	Emerson Electric	35.44	○
Air conditioners	York	34.52	○
Air conditioners, heaters, microwaves, toasters, dishwashers, vacuums, washer/dryers	Sharp Electronics	34.40	○

Electronic Equipment

BRAND NAME	PARENT COMPANY	FEM	RATING
Phones, answering machines	AT&T	66.18	●
Calculators, computers	Hewlett-Packard	59.66	●
Cellular phones, pagers	Motorola	57.34	●
Stereos, car navigation systems, TVs, CD players, cellular phones, computers	Pioneer Electronics	43.89	○
TVs, phones, camcorders, stereos	Sony	42.90	○
Calculators	Texas Instruments	42.06	○
Cameras, camcorders	Canon USA	38.95	○
VCRs, TVs, stereos	Zenith	37.16	○
TVs, video equipment, car navigation systems	Philips Electronics	37.00	○
Panasonic portable stereos, phones, CD players	Matsushita Electric	36.56	○
VCRs, TVs, portable stereos	Emerson Electric	35.44	○
TVs, VCRs, stereos, CD players, video, electronic personal organizers	Sharp Electronics	34.40	○

FINANCIAL INSTITUTIONS

Banks

BANK NAME	PARENT COMPANY	FEM	RATING
American Express	American Express	64.10	●
Citicorp	Citicorp	59.75	●
Wells Fargo	Wells Fargo	58.95	●
BankAmerica	BankAmerica	57.38	●
Bank of Boston	Bank of Boston	52.99	◐
Great Western	Great Western Financial	52.40	◐
Travelers Bank	Travelers Group	52.38	◐
Chase Manhattan	Chase Manhattan[‡]	51.08	◐

*Did not have scores for all five categories.

†Hamilton Beach/Proctor-Silex is owned by NACCO.

‡Chemical Bank and Chase Manhattan were scheduled to complete their merger in 1996. The new entity will be called Chase Bank.

Rating Key: ○ = Female friendly company; buy products when possible ◐ = Average; buy with good conscience ● = Below average; avoid buying these products

Banks

BANK NAME	PARENT COMPANY	FEM	RATING
First Fidelity	First Fidelity Bancorp	50.60	◐
Mellon Bank	Mellon Bank	45.29	◐
Bank of New York	Bank of New York	45.26	◐

Financial Institutions

SUBSIDIARY NAME (if applicable)	PARENT COMPANY	FEM	RATING
Calvert Social Investment Fund	Calvert Group (Social Investment Fund)	67.09	●
	Working Assets Funding Service	64.68	●
	American Express	64.10	●
GE Capital	General Electric	61.78	●
	Bankers Trust	61.45	●
DRI/McGraw-Hill, Standard & Poor's	McGraw-Hill	58.32	●
	Transamerica	58.31	●
Smith Barney Shearson, Primerica Financial Services	Travelers Group	52.38	◐
ITT Financial Corp.	ITT	51.53	◐
	Charles Schwab	48.60	◐
	IDS Financial	47.91	◐
	PaineWebber Group*	47.01	◐
	Merrill Lynch	46.02	◐
	United Asset Management	45.62	◐
	Dreyfus Service*	45.31	◐
Dean Witter	Dean Witter, Discover	45.24	◐
	Goldman Sachs Group*	43.19	○

Financial Institutions

SUBSIDIARY NAME (if applicable)	PARENT COMPANY	FEM	RATING
	Investcorp*	42.67	○
	American International	42.59	○
	American General	41.88	○
	Lehman Brothers Holding*	41.83	○
	Salomon	41.15	○
	Morgan Stanley Group	40.97	○
	Fidelity Investments*	40.66	○
	Bear Stearns	39.34	○

Insurance

SUBSIDIARY BRAND (if applicable)	PARENT COMPANY	FEM	RATING
Connecticut General Life, Insurance Company of North America	Cigna	69.98	●
	Prudential	63.78	●
GNA, Harcourt General Insurance	General Electric	61.78	●
	John Hancock	61.66	●
	Chubb	61.09	●
	Aetna Life†	59.19	●
Continental Insurance Companies	Continental	58.60	●
	Transamerica	58.31	●
Allstate	Sears/Allstate	55.18	●
	Metropolitan Life	54.41	◐
Traveler's Insurance	Travelers Group	52.38	◐
ITT Hartford	ITT	51.53	◐

*Did not have scores for all five categories.

†Aetna has merged with Traveler's Group. Its score does not reflect his change.

Rating Key: ○ = Female friendly company; buy products when possible ◐ = Average; buy with good conscience ● = Below average; avoid buying these products

Insurance

SUBSIDIARY BRAND (if applicable)	PARENT COMPANY	FEM	RATING
Gerber Life, Grow-up, Young People's	Gerber Products†	47.85	◐
	Equitable*	45.32	◐
	GEICO*	43.46	○
	Paul Revere	43.39	○
American Home Assurance, Lexington, American Life Insurance	American International Group	42.59	○
Franklin Life, VALIC, American General Life & Accident	American General	41.88	○
	US Life	38.16	○

ENTERTAINMENT

Restaurants

RESTAURANT NAME	PARENT COMPANY	FEM	RATING
Pizza Hut, Taco Bell, Kentucky Fried Chicken	PepsiCo	63.45	●
Wendy's	Wendy's	54.10	◐
Olive Garden, Red Lobster, China Coast	General Mills	54.04	◐
Burger King	Grand Metropolitan	53.73	◐
McDonald's	McDonald's	53.62	◐
International House of Pancakes (IHOP)	IHOP	48.73	◐
Denny's, Quincy's, El Pollo Loco	Flagstar	45.40	◐
Domino's	Domino's	44.51	○
Dunkin' Donuts, Baskin Robbins	Allied Domecq	42.04	○
Hardee's	Imasco USA	41.03	○
Dairy Queen, Orange Julius, Karmelkorn, Golden Skillet	International Dairy Queen	40.90	○
Arby's	Triarc Companies	40.89	○

Sporting Goods

BRAND NAME	PARENT COMPANY	FEM	RATING
Bicycles, basketball backboards and accessories	Huffy	49.50	◐
Boats, boat motors, and bowling, fishing, golf, and billiards equipment	Brunswick	44.59	○
Titleist golf equipment (balls, shoes, gloves, and clubs)	American Brands	43.10	○
Head, Tyrolia ski equipment, tennis gear	Head Sports*	42.19	○
Dynamic Gold and Dynalite Gold golf clubs	Black & Decker	42.05	○
Tennis equipment	Prince Manufacturing*	37.47	○

Long Distance

	PARENT COMPANY	FEM	RATING
	Sprint	70.35	●
	AT&T	66.18	●
	MCI Communications	48.97	◐

Women's and Family Magazines

MAGAZINE NAME	PARENT COMPANY	FEM	RATING
Parenting, *Baby Talk*	Time Warner	67.40	●
Ms. Magazine	MacDonald Communications*‡	62.44	●
Vogue, *Vanity Fair*, *Self*, *Glamour*, *GQ*, *Woman*, *Mademoiselle*, *Bride's*, *Allure*	Condé Nast	58.50	●
Ebony, *Jet*, *Essence*, *EM*	Johnson Publishing*	57.34	●
Cosmopolitan, *Redbook*, *Town & Country*, *Harper's Bazaar*, *Esquire*	Hearst*	47.85	◐
McCall's, *Child*, *Parents*, *YM*	Bertelsmann	43.40	○

Cooking, Home, and Travel Magazines

MAGAZINE NAME	PARENT COMPANY	FEM	RATING
Southern Living, *Cooking Light*, *Martha Stewart Living*, *Sunset*, *Hippocrates*	Time Warner	67.40	●

*Did not have scores for all five categories.

†Gerber has been acquired by a Swiss-owned company that is not in our sample.

‡Formerly Lang Communications.

Rating Key: ○ = Female friendly company; buy products when possible ◐ = Average; buy with good conscience ● = Below average; avoid buying these products

Cooking, Home, and Travel Magazines

MAGAZINE NAME	PARENT COMPANY	FEM	RATING
The Family Handyman, Travel Holiday	Reader's Digest	59.58	●
Gourmet, Condé Nast Traveler	Condé Nast	58.50	●
Country Living, Good Housekeeping, House Beautiful, Victoria	Hearst*	47.85	◐
Better Homes & Gardens, Ladies Home Journal, Midwest Living, Country Home, Traditional Home	Meredith	44.04	○
Family Circle, American Homestyle	Bertelsmann	43.40	○

Entertainment Magazines

MAGAZINE NAME	PARENT COMPANY	FEM	RATING
Life, People, Entertainment Weekly, DC Comics	Time Warner	67.40	●
Reader's Digest	Reader's Digest	59.58	●
New Yorker, Parade, Details	Condé Nast	58.50	●
Rolling Stone, US	Wenner Media†	45.80	◐

News and Business Magazines

MAGAZINE NAME	PARENT COMPANY	FEM	RATING
Fortune, Money, Time	Time Warner	67.40	●
Working Woman, Working Mother	MacDonald Communications*	62.44	●
Business Week, BYTE	McGraw-Hill	58.32	●
American Demographics, Barron's, National Business Employment Weekly	Dow Jones	49.71	◐
Newsweek	Washington Post	48.68	◐
SmartMoney	Hearst*	47.85	◐

Sports and Hobbies Magazines

MAGAZINE NAME	PARENT COMPANY	FEM	RATING
Sports Illustrated, *SI for Kids*	Time Warner	67.40	●
Discover	Walt Disney	60.12	●
American Health	Reader's Digest	59.58	●
Field & Stream, *Golf*, *Home Mechanix*, *Ski*, *Skiing*, *Yachting*, *Outdoor Life*, *Popular Science*	Times Mirror	49.94	◐
Popular Mechanics, *Motor Boating & Sailing*, *Sports Afield*	Hearst*	47.85	◐
Cruising World, *Sailing World*, *Golf Digest*, *Golf World*, *Tennis*, *Snow Country*	New York Times	47.05	◐
Golf for Women, *WOOD*	Meredith	44.04	○
Fitness	Bertelsmann	43.40	○

Book Publishers

IMPRINTS	PARENT COMPANY	FEM	RATING
Book-of-the-Month Club, Little, Brown & Co., Time-Life Books, Warner Books, Sunset Books	Time Warner	67.40	●
Hyperion Press	Walt Disney	60.12	●
Reader's Digest Association	Reader's Digest	59.58	●
McGraw-Hill, Osborn Books	McGraw-Hill	58.32	●
Chilton Guides	Capital Cities/ABC‡	57.61	●
Johnson Publishing Co. Book Division	Johnson Publishing*	57.34	●
Harcourt Brace & Co.	Harcourt General	53.86	◐
Houghton Mifflin	Houghton Mifflin	49.95	◐
CRC Handbooks, Jeppesen Sanderson, Wm. C. Brown college texts	Times Mirror	49.94	◐
Random House, Fawcett, Alfred A. Knopf, Crown, Modern Library, Orion, Pantheon, Vintage, Fodor's Travel	Advance Publications	49.13	◐
Avon, Hearst, William Morrow	Hearst*	47.85	◐

*Did not have scores for all five categories.
†Formerly Straight Arrow Publications.
‡Capital Cities/ABC was acquired by Disney.

Rating Key: ○ = Female friendly company; buy products when possible ◐ = Average; buy with good conscience ● = Below average; avoid buying these products

Book Publishers

IMPRINTS	PARENT COMPANY	FEM	RATING
Penguin, Addison-Wesley, Viking, Longman, Signet	Pearson*	43.68	○
Bantam, Doubleday, Dell	Bertelsmann	43.40	○
Golden, Golden Book (kids)	Western Publishing*	42.89	○
Simon & Schuster, Pocket Books, Prentice-Hall, Silver Burdett Ginn, Poseidon Press, Macmillan, Scribner's, The Free Press	Paramount†	41.32	○

Newspapers

NEWSPAPER NAME	PARENT COMPANY	FEM	RATING
USA Today, *Cincinnati Enquirer*, *Courier-Journal* (Louisville, KY), *Des Moines Register*, *Detroit News*, *The Tennessean* (Nashville), *USA Weekend*	Gannett	79.35	●
Charlotte Observer, *Detroit Free Press*, *Miami Herald*, *Philadelphia Daily News*, *Philadelphia Inquirer*, *St. Paul Pioneer Press*, *San Jose Mercury News*	Knight-Ridder	59.07	●
The Fort Worth Star Telegram, *The Kansas City Star*	Capital Cities/ABC‡	57.61	●
Los Angeles Times, *Newsday*, *Baltimore Sun*, *Hartford Courant*	Times Mirror	49.94	◐
Wall Street Journal	Dow Jones	49.71	◐
The (Portland) *Oregonian*, *Plain Dealer* (Cleveland), *Star-Ledger* (Newark), *The Times-Picayune* (New Orleans)	Advance Publications	49.13	◐
Washington Post, *The Herald* (Everett, WA)	Washington Post	48.68	◐
Houston Chronicle, *San Antonio Express-News*, *San Francisco Examiner*, *Seattle Post-Intelligencer*, *Albany Times Union*	Hearst*	47.85	◐
The New York Times, *International Herald Tribune*, *Boston Globe*	New York Times	47.05	◐
The Cincinnati Post, *Commercial Appeal* (Memphis), *Rocky Mountain News* (Denver)	E.W. Scripps*	41.86	○
Garden State Newspapers, *Richmond Times-Dispatch*, *Tampa Tribune*, *Winston-Salem Journal*	Media General	38.35	○

Cable Channels[§]

CHANNEL	PARENT COMPANY	FEM	RATING
HBO, Cinemax, Court TV (55%), Comedy Central (50%), E! (50%), TBS (19.4%), Sega Channel (33%)	Time Warner	67.40	●
Disney Channel	Walt Disney	60.12	●
ESPN (80%), A&E (37.5%), Lifetime (50%)	Capital Cities/ABC[‡]	57.61	●
Encore, STARZ!, Court TV (33%), Discovery (49%), Family Channel (39%), QVC (43%)	Tele-Communications	51.04	◐
CNN, TBS, TNT, Cartoon Network	Turner Broadcasting[§]	50.35	◐
ESPN, A&E, Lifetime (all partial)	Hearst*	47.85	◐
MTV, VH-1, Showtime, Flix, Nickelodeon, The Movie Channel, USA (50%), Comedy Central (50%)	Viacom	44.86	○
QVC	QVC Network	38.51	○

Network TV and Radio Stations

NETWORK STATION	PARENT COMPANY	FEM	RATING
NBC	General Electric	61.78	●
ABC	Capital Cities/ABC[‡]	57.61	●
NPR	National Public Radio*	54.00	◐
PBS	Public Broadcasting System*	53.35	◐
CBS	CBS	51.00	◐

Systems Operators

CABLE SYSTEM OPERATOR	PARENT COMPANY	FEM	RATING
American Televisions and Communications, Warner Cable	Time Warner	67.40	●

*Did not have scores for all five categories.

[†]Paramount was acquired by Viacom.

[‡]Capital Cities/ABC was acquired by Disney.

[§]If a parent company owns less than 10% of a cable network, the percentage is not listed.

Rating Key: ○ = Female friendly company; buy products when possible ◐ = Average; buy with good conscience ● = Below average; avoid buying these products

Systems Operators

CABLE SYSTEM OPERATOR	PARENT COMPANY	FEM	RATING
Comcast	Comcast	54.34	◐
TCI, Communications Services, Heritage, SCI, WestMarc	Tele-Communications	51.04	◐
Community Cablevision (CA), Dimension Cable Services	Times Mirror	49.94	◐
Viacom Cable	Viacom	44.86	○
Meredith	Meredith	44.04	○
Scripps (CO, CA, FL)	E.W. Scripps*	41.86	○

Record Labels

RECORD LABEL	PARENT COMPANY	FEM	RATING
Warner Bros., Atlantic, Elektra	Time Warner	67.40	●
Hollywood Records	Walt Disney	60.12	●
RCA, Arista	Bertelsmann	43.40	○
Columbia, Epic, TriStar Music, Sony Classical	Sony	42.90	○
Motown, A&M, Polygram	Philips Electronics	37.00	○
MCA, Geffen, Uptown, GRP	Matsushita Electric	36.56	○

Theme Parks

PARK NAME	PARENT COMPANY	FEM	RATING
Six Flags Corp.	Time Warner	67.40	●
Walt Disney World, Disneyland, Disney-MGM Studios Theme Park, Epcot Center	Walt Disney	60.12	●
Sea World, Busch Gardens, Cypress Gardens	Anheuser-Busch	51.42	◐
Paramount parks in CA, VA, NC, OH, and Toronto	Paramount†	41.32	○
Universal Studios	Matsushita Electric	36.56	○

Video

VIDEO STORE	PARENT COMPANY	FEM	RATING
Blockbuster Video	Viacom	44.86	○

TRAVEL

Airlines

SUBSIDIARY NAME (if applicable)	PARENT COMPANY	FEM	RATING
	American Airlines	64.49	●
	America West Airlines	49.60	◐
	Southwest Airlines	48.09	◐
	USAir Group	47.48	◐
	United Airlines	45.67	○
Midwest Express	Kimberly Clark	44.79	○
	Alaska Airlines*	42.92	○
	Continental Airlines	40.92	○
	Trans World Airlines	40.26	○
	Delta Airlines	38.49	○
	Northwest Airlines	35.04	○

Hotels

HOTEL NAME	PARENT COMPANY	FEM	RATING
Clarion, Quality Inn, Comfort Inn, Rodeway Inn, Econo Lodge, Sleep Inn, Friendship Inn	Manor Care	66.86	●
Marriott, Courtyard, Residence Inn	Host Marriott	61.13	●
Walt Disney World, Disneyland	Walt Disney	60.12	●
Hyatt, Grand Hyatt, Hyatt Regency, Hawthorne Suites	Hyatt Hotels	54.47	◐

*Did not have scores for all five categories.

†Paramount was acquired by Viacom.

Rating Key: ● = Female friendly company; buy products when possible ◐ = Average; buy with good conscience ○ = Below average; avoid buying these products

Hotels

HOTEL NAME	PARENT COMPANY	FEM	RATING
Stouffer Renaissance, Ramada International	Renaissance Hotels	50.17	◐
Embassy Suites, Hampton Inn, Homewood Suites	Promus Hotel	49.85	◐
Days Inn, Howard Johnson, Ramada (in US), Super 8, Travelodge, Knights Inn	HFS	49.19	◐
Hilton, Hotel Conrad, Waldorf-Astoria	Hilton Hotels	45.84	◐
Westin Hotels	Westin Hotel*	43.97	○
Sheraton Hotels	ITT Sheraton*	40.31	○

Rental Cars

RENTAL CAR COMPANY	PARENT COMPANY	FEM	RATING
Alamo	Alamo Rent A Car*	50.95	◐
Avis	Avis	45.64	◐
Budget	Budget Rent A Car	42.33	○
Dollar, Thrifty, Snappy	Chrysler	41.26	○
Hertz	Hertz	39.29	○

OFFICE EQUIPMENT, SUPPLIES, & FURNITURE

Computers

COMPUTER EQUIPMENT	PARENT COMPANY	FEM	RATING
Computers, ThinkPad	IBM	70.32	●
Laser printers	Xerox	69.78	●
NCR PCs, software	AT&T†	66.18	●
Notes, Lotus 1-2-3, AmiPro word processing, SmartSuite, Soft*Switch, cc:Mail	Lotus‡	62.97	●
Apple computers, PowerBook, Macintosh PCs	Apple Computer	60.24	●

Laser printers, computers	Hewlett-Packard	59.66	●
High-resolution LCD monitors, printers	Tektronix	58.52	●
Computers, modems	Motorola	57.34	●
PCs, LinkWorks software, server systems	Digital Equipment	56.22	●
PCs, notebook computers, Dimension, Latitude, OmniPlex, PowerEdge	Dell Computer	55.62	●
Software for mainframes, PCs	Computer Associates	48.71	◐
NetWare, Quattro Pro	Novell	47.29	◐
VS series computers, software	Wang Laboratories	45.37	◐
Database management, SQL software	Oracle	44.93	○
DOS and Windows operating systems, software	Microsoft	43.96	○
PCs, CD-ROM players	Pioneer Electronics	43.89	○
SPARC workstations, Solaris software	Sun Microsystems	43.31	○
CD-ROM players, computers	Sony	42.90	○
Laptop, portable computers and PCs	Compaq Computer	42.58	○
Notebook PCs, printers	Texas Instruments	42.06	○
Computer chips, software, add-in enhancements	Intel	41.67	○
Paradox, dBase, C++, Delphi software	Borland*	40.73	○
Computers, software, peripherals	Ricoh	40.72	○
Printers, notebook computers	Canon USA	38.95	○
CTOS, software	Unisys	38.69	○
Panasonic computers	Matsushita Electric	36.56	○
Computers, printers	Sharp Electronics	34.40	○

*Did not have scores for all five categories.

†AT&T was one company when we collected this information. It has since divided into three companies. AT&T is still the communications company, NCR makes computers, and Lucent Technologies produces equipment.

‡Lotus was bought by IBM to develop and sell its Notes technology.

Rating Key: ○ = Female friendly company; buy products when possible ◐ = Average; buy with good conscience ● = Below average; avoid buying these products

Electronic Office Equipment

EQUIPMENT	PARENT COMPANY	FEM	RATING
Copiers, printers	IBM U.S.	70.32	●
Copiers, printers	Xerox	69.78	●
Copiers, fax machines, mailing systems	Pitney Bowes	64.12	●
Printers, fax machines, copiers	Hewlett-Packard	59.66	●
Copiers, printers	Eastman Kodak	58.54	●
Printers, calculators	Texas Instruments	42.06	○
Copiers, fax machines, printers	Ricoh	40.72	○
Financial, bar code, document processing equipment	Standard Register	39.04	○
Copiers, fax machines, typewriters, printers	Canon USA	38.95	○
Copiers, fax machines, phones	Matsushita Electric	36.56	○
Copiers, fax machines	Sharp Electronics	34.40	○

Office Supplies

BRAND NAME	PARENT COMPANY	FEM	RATING
Scotch tape, Post-it Notes, diskettes, desk organizers, mailing supplies, transparency film	3M	61.45	●
Eldon desk accessories	Rubbermaid	60.21	●
Major office supply retail store	Staples†	59.66	●
Eberhard Faber markers and pens, Stuart Hall notebooks and stationery, desk organizers	Newell	56.76	●
OfficeMax office supply retail stores	K-Mart	51.89	◐
Paper and office products	Boise Cascade	50.02	◐
Paper: Eureka!, Curtis, Word Pro, Delta Brite	James River of Virginia	48.77	◐

Paper, notebooks	Mead	48.66	◐
Paper, glossy paper	Consolidated Papers	47.29	◐
Pens, other office products	A.T. Cross	46.38	◐
Paper, envelopes	Georgia-Pacific	45.58	◐
Paper products	Alco Standard	45.46	◐
Tabs, labels, markers, adhesives	Avery Dennison	45.36	◐
Neenah paper	Kimberly-Clark	44.79	○
Business and electronic forms, labels, direct mail services	Moore	44.51	○
Flair, Papermate pens, Liquid Paper	Gillette	44.05	○
Office products	International Paper	43.21	○
ACCO fasteners and paper clips, Swingline staplers and staples, Wilson Jones steno books, binders, and columnar pads, Day-Timers	American Brands	43.10	○
Computer City, Incredible Universe, and Radio Shack stores	Tandy*	42.97	○
Major office supply retail store	Office Depot†	41.51	○
Business forms	Standard Register	39.04	○
Packaging, tissue, and kraft paper	Chesapeake	36.57	○

Office Furniture

PRODUCT LINE	PARENT COMPANY	FEM	RATING
Herman Miller, Phoenix Designs, Meridian	Herman Miller	45.25	◐
Gunlocke, Holga, BPI, Ring King Visibles, Remington	Hon*	41.78	○
Globe	Globe Business Furniture*	40.64	○
Anderson Hickey	Anderson Hickey*	38.49	○
O'Sullivan	O'Sullivan*	36.58	○

*Did not have scores for all five categories.

†In mid-1996 Office Depot and Staples announced plans to merge.

Rating Key: ○ = Female friendly company; buy products when possible ◐ = Average; buy with good conscience ● = Below average; avoid buying these products

Mailing Services

PARENT COMPANY	FEM	RATING
Pitney Bowes	64.12	●
Federal Express	46.41	◐
DHL Worldwide Express*	39.45	○
Airborne Express*	38.10	○
United Parcel Service of America	36.34	○

*Did not have scores for all five categories.

Appendix One

Methodology, Scoring, and References

COMPANIES

Procedures and Sources for Gathering Information

We chose our original list of companies from the Fortune 500 and Fortune Service 500. After deleting companies without consumer goods relevant to this book, and culling and adding other companies that did have consumer products, we had a working list. Here is, in general, how we found information regarding those companies.

Questionnaires. We sent questionnaires to between 500 and 600 companies. Four part-time employees, at various times, followed up by phone, and at least one-third were re-sent or faxed on the basis of those phone calls. In some cases, a company representative answered the questions over the phone. In our final sample of 386 companies, we received 124 surveys. When survey information conflicted with other sources, the surveys, as the only firsthand information we had, were always given priority.

After we had filled in what information we could from other sources, each company received a letter showing what information we had for them, and requesting that they correct or update that information. Some did. Others did not. If we never got information in response to questionnaire, fax, or phone call, the company was considered uncooperative.

EEO-1s. Beginning in May 1992, and on four subsequent occasions, EEO-1 reports were requested from the Office of Federal Contract Compliance Programs (OFCCP) on the companies that had federal contracts (about 85 percent of our list). We used FOIA requests, connections, letters, phone calls, and begging. Finally, in early 1994, we began to receive EEO-1s—from 1991, the most recent information then available. The process took over 2 years. Another request was made in 1995, and we received 1994 data for most of the companies. These came in a more timely manner. Much of this was used, especially for the percentage of women in top management and total management, but we updated it based on questionnaires or annual reports whenever possible. We also relied on the Department of Labor for its most recent list of companies in noncompliance with affirmative action laws. Our scores reflect, therefore, that they were cited (a not very common occurrence), and whether they complied within the time frame of our study.

Council for Economic Priorities (CEP). We culled much of our benefits information from CEP reports on companies, as well as information on what sort of advancement programs companies had for women (if any). They also list purchases from women-owned businesses.

The Foundation Center. This was the source for most of the charitable giving information. We also gathered information from various Taft directories.

Kinder, Lydenberg, Domini & Co. (KLD). The corporate rating forms we bought from KLD were an important source of information used for the book. Their information included women/minorities on boards of directors and in important line positions, benefits information, whether the company had been included on any "100 Best"-type lists or received any awards and purchases from women-owned businesses. The reports also had some charitable giving information and occasionally discussed lawsuits concerning women. The forms were also helpful in listing all company products as well as mergers and acquisition data.

Catalyst. Catalyst's reports included much information about boards of directors.

Working Mother Magazine's "100 Best Companies for Working Mothers." Not only did companies receive points for being included on this list, but the article also contained a lot of benefits information. We used the most recent lists available and gave companies additional credit if they were listed at any time during previous years.

Annual Reports. We requested annual reports from all of the companies that sell multiple consumer products, in order to get our products identified with the correct companies. Board of directors and senior executive information was also available from most annual reports.

Other sources.

- *Hoover's Handbook of American Business 1995* (also the 1993 edition). Edited by Gary Hoover, Alta Campbell, and Patrick J. Spain. Austin, TX: The Reference Press, Inc., 1994.
- *Shopping for a Better World.* 1992 and 1994 editions. CEP.
- *The Corporate Reference Guide to Work–Family Programs*. Edited by Ellen Galinsky, Dana E. Friedman, and Carol A. Hernandez. New York: Families and Work Institute, 1991.
- Visits to stores to check products and companies.
- Compact Disclosure data base (for board of directors information).
- Schroeder Wertheim found board of directors data on about 35 companies using their data bases.
- Various articles from *The New York Times* and *Working Women* (for female presidents and CEOs and highest-paid women in corporate America).
- The research department at *The Wall Street Journal*, which provided some information, back articles, and direction.
- We occasionally refer to an amount given to women-owned businesses. This information came from KLD and/or CEP. We originally intended to use this as a measure, but we did not have information for enough companies to justify an entire measure. Instead, we occasionally refer to it in the text.

SOURCES BY CATEGORY

In general, sources are ranked below according to the amount of information obtained from the source (i.e., the top source gave us the most information; the bottom source, the least).

Category 1: Management Opportunities

1, 2. Officials/Managers & Professionals.

a. *The Wall Street Journal* obtained 400 1992 EEO-1s from the Department of Labor and published an article with the top 200 in March 1994. Their article was about women in corporate America. We were able to obtain some of their research.

b. U. S. Department of Labor, Office of Federal Contract Compliance Programs provided us with EEO-1 reports for government contractors for 1991 and 1994.

c. The company itself. These are generally more recent.

d. CEP included some Official/Manager figures on their questionnaires.

e. KLD

3. Female Management/Training Programs.

a. CEP surveys

b. Our surveys

c. *Working Mother* magazine's "100 Best Companies" descriptions.

d. Families and Work Institute

Category 2: Glass Ceiling

4. Board of Directors

a. KLD

b. Catalyst survey

c. CEP surveys

d. Our surveys

e. Annual reports/10-K reports

f. Compact Disclosure

g. DUNS computer search

5. Top Officers

a. CEP surveys

b. Our surveys

c. KLD

d. Directory of Corporate Affiliations

e. Annual reports/10-K reports

f. Compact Disclosure

g. DUNS computer search

6. Highest-Paid Officers

a. KLD

b. CEP surveys
c. Compact Disclosure
d. 10-K or reports from company
e. Our surveys
f. DUNS computer search
g. *Ward's Business Directory of U.S. Private and Public Companies*. Detroit: Gale Research Inc., 1993.
h. *Directory of Corporate Affiliations*. New Providence, NJ: National Register Publishing Co., 1993.

Category 3: Support of Women

7. Charitable Giving
   a. The Foundation Center, New York, New York: from their publications and data
   b. KLD
   c. Our surveys
   d. IRS Form 990-PF (private foundation) reports, which includes lists of organizations that received grants and the amount given to each
   e. Taft
   f. Several companies were reported on a Catalyst brochure as giving to them. These companies were given credit for giving to women's organizations if no other information was available.

8. Survey Cooperation. Based on whether they returned the survey (5 points) and if they gave more than requested (10). If not, 0.

9. Salary Info. From questionnaires.

Category 4: Benefits

10. Benefits.
    a. CEP surveys
    b. Our surveys
    c. Families and Work Institute
    d. KLD
    e. *Working Mother* magazine's "100 Best Companies" (primary information for ten companies for which we had no other source; additional information for others)

Category 5: Femlin Factor

11. Awards received: Inclusion in Domini Social Index (KLD); EVE and Secretary 2000 Awards (DOL).; Catalyst Award

12. KLD Rating. KLD.

13. Listed on *Working Mother* "100 Best." *Working Mother* magazine (for 9 years, through 1995).

14. Noncompliance citation (DOL).

15. Female Founder, CEO, President or Division President
    a. KLD
    b. Magazine, newspaper articles. *Working Woman*, business section of newspapers, etc.

MEASURES USED IN EACH CATEGORY/HOW EACH MEASURE YIELDED RAW SCORES

Category 1: Management Opportunities

Measure 1: Women as Officers and Managers. Number of women as a percentage of total number of officers and managers.

Measure 2: Female Professionals. Number of women as a percentage of total number of employees ranked as professionals.

Measure 3: Female Advancement Programs. The number of a company's formal programs that promote or aid women. For example, could have included recruitment, mentoring, or special training of women.

Category 2: Glass Ceiling

Measure 4: Board of Directors. Percentage of women serving as board members.

Measure 5: Top-Salaried Officers. Percentage of women among the best-paid employees of the company.

Measure 6: Most Senior Officials. Percentage of women among top management as defined by CEP, KLD, and annual reports.

Category 3: Support of Women

Measure 7: Charitable Giving. If we knew the exact percentage a company gave to women's groups, we used that number. If a company gave to United Way, it got credit for 23 percent of the amount given (because United Way targets 23 percent of its donations to women's groups); if a company gave both to United Way and directly to women's groups, it got credit for both. If we knew a company gave to women's groups, but we didn't know how much, we gave it the national average (the average amount of corporate giving targeted toward women's groups for 1994 was 5.3 percent, according to the Foundation Center). If we got no response, had no information or if the company did not give specifically to women or United Way, the company received a zero for this measure.

Measure 8: Cooperation. If the company returned our survey or responded over the telephone, it received a 5. If not, it got a zero. In rare cases, when the company was very cooperative and sent us scads of information—such as corporate handbooks, product listings, or employment policies—it received a 10.

Measure 9: Salary Information. If the company provided salary information (which we requested on the questionnaire) it got a 5 for this category. If it was good news—e.g., that the female employees earn more than the national average of 72 cents to every dollar earned by men—the company received a 10. If we had no information, the company got a zero.

Category 4: Benefits

Measure 10: Benefits. *Leave:* Each company got 1 point for each additional week of job-guaranteed leave it offered above and beyond the 12 weeks guaranteed by the FMLA. It could earn these raw score points for maternity, paternity, child care, and/or dependent care leave. Companies also got an additional 5 points for any sort of paid leave not given as disability pay. A zero for the leave category could mean a no response, a policy that simply complies with the FMLA, or a company that has fewer than 50 employees and thus has no obligation to comply with the FMLA.

Day care: Companies got 20 points for having day care for children of employees on or near site. They got 10 points for securing discounts at neighboring facilities or giving employees substantial subsidies toward day care. They got 5 points for forming a coalition with a group of other companies to form a day-care center or providing sick-child care in emergencies.

Flexible work schedules: Companies got 10 points for each scheduling benefit—job sharing, part-time return, flex time, or work at home—offered either as a formal policy or at all locations, or if the company indicated on a questionnaire that they had those benefits. Five points were awarded for either informal policies or having policies only at some sites.

Family Benefits: Companies got 1 point for each family benefit offered. A "family benefit" could include: after-school or summer programs for kids, pretax set-asides for dependent care, reimbursement of child care costs for nonroutine business travel, paying for sick-child care, gifts for newborns, adoption aid, scholarships for employees' children, floating sick child days, child care resource and referral, elder care resource and referral, full-time work/family coordinator, prenatal education programs, tuition reimbursements or on-site classes, making available breast-feeding rooms, or offering other information in the form of handouts, seminars, or luncheons.

All raw scores were converted to standard scores for the category, with an average of 10 and a standard deviation of 4. This was done for all five categories. Thus, the specific ratings given to each measure were statistically transformed to be comparable, even when the raw scores were not.

Category 5: Femlin

Measure 11: Awards. Companies earned 10 points for being listed on the Domini Social Index. They could also earn 10 points for having won the Department of Labor's EVE award, 10 points for winning the Department of Labor's Secretary 2000 Award, and 10 points for winning the Catalyst Award.

Measure 12: KLD Rating. Companies could earn 2, 1, 0, −1, or −2 points, depending on the ratings they received (how many diamonds) in the "Women" category on KLD's "Strengths/Weaknesses" chart.

Measure 13: Best 100. Companies received 1 point for each year they were listed in *Working Mother* magazine's "100 Best Companies for Working Mothers."

Measure 14: Noncompliance. Companies received a zero if they were on the Department of Labor's list of companies in noncompliance with affirmative action laws. They received a 5 if they were cited but had subsequently corrected their violations. Companies not on the list received a 10.

Measure 15: Female Founder. A company either founded by a woman or run by a female president or CEO received a 10 in this category. A company with one major division or subsidiary run by a woman earned a 5; if two or more major divisions or subsidiaries were run by women, the company received a 10.

HOW RAW NUMBERS WERE STATISTICALLY TRANSFORMED

All corporate measures were standardized using a *z*-score transformation so that all variables would be on the same metric. The mean was chosen to be 10, with a standard deviation of 4. These standard scores were averaged to obtain a total category score for each of the five categories. The scores from each of the five categories were then added together to become the total "FEM score," which had an average of 50 and a standard deviation of 10.

SOURCES BY CATEGORY, STATES (CHAPTER SEVEN)

Fourteen measures were used to determine ratings for each state. Listed below are the measures and the sources from which the information was culled.

Category 1: State as Employer

1, 2. Percentage of Women as Managers and Female Salaries.

- U. S. EEOC. *Job Patterns for Minorities and Women in State and Local Government, 1993* (pp. 222–271). Washington, DC: U. S. Equal Employment Opportunity Commission, 1994.

3. Benefits.

- Lord, Michelle, and King, Margaret (eds.). *The State Reference Guide to Work–Family Programs for State Employees*. New York: Families and Work Institute, 1991.

Category 2: Economic Climate

4. Women in Private Industry Management.

- U. S. EEOC. *Job Patterns for Minorities and Women in Private Industry, 1993* (pp. 37–86). Washington, DC: U.S. Equal Employment Opportunity Commission, 1994.

5. Women-Owned Business.

- U. S. Department of Commerce. "Women-Owned Business." Washington, DC: U. S. Department of Commerce, Economic Census, 1987.

Category 3: Women in Government

6,7. Women in the House of Representatives and Women in State Legislatures.

- National Women's Political Caucus. *National Directory of Women Elected Officials 1995*. Washington, DC: National Women's Political Caucus, 1995.

8. Political Appointments.

- National Women's Political Caucus. *The Appointment of Women: A Survey of Governor's Cabinets* [does not include Florida, Georgia, Indiana, Mississippi, New Hampshire, South Carolina, or Texas because these states have elected or noncabinet governmental structures]. Washington, DC: National Women's Political Caucus, 1990–1992.
- Women Executives in State Government. "Number of women in cabinet level appointed positions." Washington, DC: Council of State Governments, 1995.

Category 4: Civil and Family Rights

9. Community Property.

- Leiter, Richard (ed.). *A National Survey of State Laws*. Detroit: Gale Research, Inc., 1st source edition, 1993.

10. Reproductive Rights.

- NARAL Foundation. *Who Decides? A State-by-State Review of Abortion and Reproductive Rights*. Washington, DC: The NARAL Foundation, 5th edition, 1995.

11. Domestic Violence.

- Hart, Barbara J. "State Codes on Domestic Violence. Analysis, Commentary and Recommendations." In *Juvenile and Family Court Journal*, Vol. 43, No. 4, 1992.

Category 5: General Climate

12. Women in High Political Office.

- National Women's Political Caucus. *National Directory of Women Elected Officials 1995*. Washington, DC: National Women's Political Caucus, 1995.

13. History of Voting Rights.
- Scott, Anne Firor, and Scott, Andrew Mackay. *One Half the People: The Fight for Women's Suffrage* (pp. 166–168). Urbana: University of Illinois Press, 1975.

14. Child Care.
- Adams, Gina, and Sandfort, Jodi. *State Investments in Child Care and Early Childhood Education*. Washington, DC: Children's Defense Fund, 1992.
- Cadden, Vivian. "The 10 Best States For Child Care." *Working Mother*, February 1993.

HOW MEASURES YIELDED RAW SCORES, STATES (CHAPTER SEVEN)

Category 1: State as Employer

Measure 1: Female Managers. Percentage of state-employed managers who are women.

Measure 2: Salary Equity. Salary ratio of male and female state employees; parity would be 100 percent.

Measure 3: Benefits. The number and type of family benefits state employees receive. One point was possible for each of: disability/maternity leave, paternity leave, adoptive parent leave, elder care leave, continuation of healthcare benefits during leave, and job guarantee during extended leaves.

Category 2: Economic Climate

Measure 4: Female Managers. The percentage of management positions in private industry that are held by women.

Measure 5: Women-Owned Business. Percentage of businesses within the state that are owned by women.

Category 3: Women in Government

Measure 6: Women in the House of Representatives. Percentage of a state's congressional representatives who are women.

Measure 7: Women in State Legislature. Percentage of state representatives who are women.

Measure 8: Political Appointments. Percentage of total state cabinet-level appointees who are women. (This does not include Florida, Georgia, Indiana, Mississippi, New Hampshire, South Carolina, and Texas because these states have elected or noncabinet governmental structures. These states did not receive a score on this measure; therefore, their scores for this category are the average of the other two measures.)

Category 4: Civil and Family Rights

Measure 9: Community Property. States with community property divorce laws received credit; states without these laws received no credit.

Measure 10: Reproductive Rights. States could earn 1 point for each of the following policies or programs to enhance women's reproductive freedom: policies banning clinic violence and harassment, allotment of public funds to abortion, pro-choice legislation, support of RU-486, not requiring preabortion counseling, not requiring parental consent for minors, not requiring a waiting period for abortion, and a modification of state laws to reflect *Roe v. Wade*. Thus, the range was from 0 to 8. One point was subtracted from the score of states that have proposed any anti-choice legislation.

Measure 11: Domestic Violence. States earned 1 point for any of the following laws and policies protecting women from domestic violence: easy availability of restraining orders, legal eviction of a spouse, assumption of custody for women, payment of child or spousal support, awarding of attorney's fees, awarding of other standard of living monetary compensation (such as house or car payments), and no-contact provisions that extend to mail and telephones.

Category 5: General Climate

Measure 12: Women in High Office. States with either a female governor, lieutenant governor, or U. S. senator(s) received credit on this measure for each.

Measure 13: History of Voting Rights. Thirty-six states ratified the 19th Amendment, giving women the right to vote. Points were given to those states and extra points were given to states that granted women the vote before 1919. States admitted to the Union after 1919 (Hawaii and Alaska) did not receive a score on this measure. Instead, their category score was the average of the other two measures.

Measure 14: Child Care. The dollar amount spent by the state on child care per child per year.

HOW RAW NUMBERS WERE TRANSFORMED INTO STATISTICS, STATES (CHAPTER SEVEN)

Scores were standardized using a z-score transformation so that all variables would be on the same metric. The mean was chosen to be 10, with a standard deviation of 7, for each of the five categories; thus the possible scores could range from 0 to 20 for each category. The scores for each of the five categories were then added together to yield a "FEM score," with a mean of 50 and a standard deviation of 35.

LITERATURE CONSULTED FOR CHAPTER SEVEN

Blumberg, R. L. (ed.). *Gender, Family and Economy* (pp. 22–23). Newbury Park, CA: Sage Publications, 1991.

Brown, L. K. An historical perspective on the woman manager in the United States. In: *Woman Manager*. Washington, DC: Business and Professional Women's Foundation, 1981.

Business and Professional Women's Foundation. *You Can't Get There from Here: Working Woman and the Glass Ceiling*. Washington, DC: Business and Professional Women's Foundation, April 1992.

Catalyst. *Women in Corporate Management: Model Programs for Development and Mobility*. New York: Catalyst, 1990.

Corson, B., Marlin, A. T., Schorsch, J., Swaninathan A., and Will, R. *Shopping for a Better World*. New York: Council of Economic Priorities, 1989.

The Council of State Governments. *The Book of States* (Volume 30). Lexington, KY: The Council of State Governments, 1994–95.

Faludi, S. *Backlash*. New York: Crown, 1991.

Fuchs, V. R. "Women's Quest for Economic Equality." *The Journal of Economic Perspectives*, Winter, 1989, pp. 25–41.

Galbraith, J. K. *The New Industrial State*. Harmondsworth: Penguin, 1974.

Gist, R. R. *Marketing and Society: A Conceptual Introduction*. New York: Holt, Rinehart & Winston, 1971.

Hart, B. J. *State Codes on Domestic Violence. Analysis, Commentary and Recommendations*, Reno, Nevada: National Council of Juvenile and Family Court Judges, 1991.

Leiter, R. (ed.) *A National Survey of State Laws*. Detroit, London: Gale Research, Inc., 1st source edition, 1993.

Lord, M., and King, M. *The State Reference Guide to Work–Family Programs for State Employees*. New York: Families and Work Institute, 1991.

Lydenberg, S. D., Marlin, A. T., Strub, S. P., and the Council of Economic Priorities. *Rating America's Corporate Conscience*. Reading, MA: Addison–Wesley, 1986.

Morrison, A. M., White, R. P., Van Velsor, E., and The Center for Creative Leadership. *Breaking the Glass Ceiling*. Reading, MA: Addison–Wesley, 1992.

NARAL Foundation. *Who Decides? A State-by-State Review of Abortion and Reproductive Rights 1993*. Washington, DC: The NARAL Foundation, 5th edition, January 1995.

National Women's Political Caucus. *The Appointment of Women: A Survey of Governor's Cabinets*. Washington, DC: National Women's Political Caucus, 1990–1992.

National Women's Political Caucus. *National Directory of Women Elected Officials 1995*. Washington, DC: National Women's Political Caucus, 1995.

Peter, L. J. *The Peter Principle*. New York: William Morrow, 1969.

Robertson, T. D., and Kassarjian, H. H. *Handbook of Consumer Behavior*. Englewood Cliffs, NJ: Prentice–Hall, 1991.

Scott, A. F., and Scott, A. M. *One Half the People: The Fight for Women's Suffrage* (pp. 166–168). Urbana: University of Illinois Press, 1975.

Scott, R. *The Female Consumer*. London: Associated Business Programmes Ltd., 1976.

Smith, N.C. *Morality and the Market*. London: Routledge, 1990.

Stanton, E. C. *Eight Years and More: Reminiscences 1815–1897*. Originally published: T. Fisher Unwin. Printed and bound by Edwards Brothers, Inc., Ann Arbor, MI, 1898.

U. S. Department of Commerce. *Statistical Abstract of the United States 1993*, 113th edition. Washington, DC: U. S. Department of Commerce, 1993.

U. S. Department of Labor. *Facts on Working Women*. Washington, DC: U. S. Department of Labor, Women's Bureau, No. 90-1, June 1990.

U. S. Equal Employment Commission. *Job Patterns for Minorities and Women in State and Local Government, 1993*. Washington, DC: U. S. Equal Employment Opportunity Commission, 1993.

Wight, R. *The Day the Pigs Refused to Be Driven to Market: Advertising and the Consumer Revolution*. London: Hart Davis MacGibbon, 1972.

SOURCES BY CATEGORY, COUNTRIES (CHAPTER EIGHT)

Frequently, the source given under a measure is "questionnaire." This refers to a questionnaire mailed or faxed to each country's embassy in Washington, D.C.

Category 1: Legal Rights

1. Voting Rights.
   - Antal, Arian Berthoin, and Izraeli, Dafna. "A Global Comparison of Women in Management" and "Percentage of Women Elected to Parliament, 1990." In Fagenson, E. (ed.), *Women in Management: Trends, Issues & Challenges in Managerial Diversity*. Newbury Park, CA: Sage Publications, 1993.

2. Property Rights and Employment. Questionnaire.

3. Presence or Absence of Gender Discrimination Laws. Questionnaire.

4. Infringement of General Freedoms.
   - Questionnaire.
   - U. S. Department of State. *Country Reports on Human Rights Practices for 1993*. Submitted to the Committees on Foreign Affairs and Relations. February 1994.

Category 2: Family Roles

5, 6. Access to Birth Control and Use of Birth Control.
   - Questionnaire
   - Freeman, M., and Pohlandt-McCormick, H. *Abortion and Birth Control Country Profiles, 1993*. Regent's Park, London: International Planned Parenthood Federation, 1993.
   - Camp, S. L., and Speidel, J. J. (eds.). *World Access to Birth Control* Washington, DC: Population Crisis Committee, 1987.

7, 8, 9. Divorce Law, Marriage Law, Age of Marriage.
   - Kurian, George Thomas. "Sex Offenses Rate, 1988." In *New Book of World Rankings*. New York: Facts on File, 3rd edition, 1991.

10, 11. Child Support and Domestic Violence Laws. Questionnaire.

12. Abortion Laws.
   - Freeman, M., and Pohlandt-McCormick, H. *Abortion and Birth Control Country Profiles, 1993*. Regent's Park, London: International Planned Parenthood Federation, 1993.

- Henshaw, Stanley K. "Induced Abortion: A World Review, 1990." In *Family Planning Perspectives*, Vol. 22, No. 2, p.77, March/April 1990.
- Camp, S. L., and Speidel, J. J. (eds.). *World Access to Birth Control*. Washington, DC: Population Crisis Committee, 1987.

Category 3: Reproductive Health Status

13, 14, 15. Maternal Mortality Rate, Fertility Rate, Household Size.

- United Nations. *Population and Vital Statistics Report*. Series A, Vol. XLVI, No. 2, April 1994.
- Kurian, George Thomas. "Population Doubling Time, 1988." In *The New Book of World Rankings*. New York: Facts on File, 3rd edition, 1991.

Category 4: Economic and Educational Climate

16, 17, 18. Employment Opportunities, Gender Pay Equity, Women in Management.

- Questionnaire.
- Antal, Arian Berthoin, and Izraeli, Dafna. "A Global Comparison of Women in Management," "Corporate Programs for Women in Management," and "Women in the Labour Force and Women in Management, 1987." In Fagenson, E. (ed.), *Women in Management: Trends, Issues and Challenges in Managerial Diversity*. Newbury Park, CA: Sage Publications, 1993.
- Kurian, George Thomas. "Women in the Labour Force." In *The New Book of World Rankings*. New York: Facts on File, 3rd edition, 1991.
- U.S. Department of Commerce, Economics and Statistics Administration, AID Office of Women in Development. "Relative Wages in Manufacturing Men/Women, 1991" and "Population by Activity Rate and Economic Sector, by Gender,

1980 and 1990." In *Gender and Generation in the World's Labour Force*. Washington, DC: U.S. Department of Commerce, 1991.

19. Women in Universities.
    - Kurian, George Thomas. "Women in University Enrollment Ratio." In *The New Book of World Rankings*. New York: Facts on File, 3rd edition, 1991.

20. Life Expectancy.
    - UN *Population and Vital Statistics Report.*

Category 5: General Climate

21. Literacy Differential.
    - Kurian, George Thomas. "Male and Female Literacy Rate, 1988." In *The New Book of World Rankings*. New York: Facts on File, 3rd edition, 1991.

22. Education Differential.
    - Kurian, George Thomas. "Primary School Enrollment Ratio, 1985." In *The New Book of World Rankings.* New York: Facts on File, 3rd edition, 1991.

23. General Status.
    - Questionnaire.
    - Department of State. *Country Reports on Human Rights Practices for 1993*. Submitted to the Committees on Foreign Affairs and Relations. February 1994.

24. U. N. Treaty Ratification.
    - United Nations. "Convention on the Elimination of All Forms of Discrimination Against Women—List of Participants." 1993.

25. Women in Government.
    - Antal, Arian Berthoin, and Izraeli, Dafna. "A Global Comparison of Women in Management" and "Percentage of Women Elected to Parliament, 1990." In Fagenson, E. (ed.), *Women in Management: Trends, Issues & Challenges in Managerial Diversity*. Newbury Park, CA: Sage Publications, 1993.

HOW MEASURES YIELDED RAW SCORES, COUNTRIES (CHAPTER EIGHT)

Category 1: Legal Rights

Measure 1: Voting Rights. An average was determined for the countries in the sample. Each country earned points based on whether women were granted the right to vote before, during, or after the average year that women received this right in the region. Countries that still prohibit women from voting received a zero.

Measure 2: Property Rights. Countries in which women have the right to own property earn points; countries where women do not have this right received a zero.

Measure 3: Countries received zeros if they had no laws prohibiting gender discrimination, 10 if they did.

Measure 4: Infringement of General Freedoms. Countries where women are free to drive, travel, and dress as they please earned points in this category; if women in a country are restricted in any of these areas, the country received a zero.

Category 2: Family Roles

Measures 5 and 6: Access to Birth Control and Use of Contraception. The scores from these two categories were averaged. Access to birth control was scored in three ways: if the country limited access (if abortion is illegal, for example), the country received a zero; if women in the country have good access to some methods, the country earned an average score; and an above-average score was given to countries in which birth control choices are both varied and accessible. Use of contraception was compared to the average of the countries in the sample.

Measure 7: Divorce Law. Countries in which divorce is either not legal or cannot be initiated by the woman received a zero. If divorce is available, but only on limited grounds, the country earns an average score. Full credit was given to countries in which there are no restrictions on divorce.

Measure 8: Marriage Laws. Maximum credit was given to countries that have laws providing for equality of marriage partners with regard to arranged marriages and initiation of and division of property in a divorce. Partial equality earned partial credit points and no laws providing for equality received a zero.

Measure 9: Age of Marriage. Points were deducted for each year under age 18 that girls are permitted to marry.

Measure 10: Child Support. Countries that mandate men continuing financial support of their children following separation or divorce earned credit; otherwise the country received a zero.

Measure 11: Domestic Violence Laws. This measure was scored by determining the types of domestic violence interventions available to women, where five or more options (such as availability of restraining orders) earned a country the highest possible score. That score was then combined with the country's sexual offense rate relative to the average of the sample.

Measure 12: Reproductive Rights. Countries that severely restrict or prohibit abortion received a zero. Partial credit was given to countries that allow abortion, but on limited grounds. Full credit was awarded to countries in which abortion is available on request.

Category 3: Reproductive Health Status

Measure 13: Maternal Mortality. The percentage of women who die during childbirth.

Measure 14: Fertility Rate. The average number of children that women bear.

Measure 15: Household Size. The average number of people living in each household.

Category 4: Economic and Educational Opportunities

Measure 16: Employment Opportunities. Countries received full credit if they had legislation for both equal opportunity for employment and equality laws regarding pay. They got partial credit if they had either one or the other, and zero if they had neither.

Measure 17: Gender Pay Equity. The average pay differential for men and women in the country (where parity would be 100 percent).

Measure 18: Women in Management. The percentage of management positions in each country that are held by women.

Measure 19: Women in Universities. The percentage of women enrolled in universities (where parity would be 50 percent).

Category 5: General Climate

Measure 20: Female Life Expectancy. Average years of life expected for women.

Measure 21: Literacy Differential. The average number of literate women versus literate men for each country (parity would be 100 percent).

Measure 22: Educational Differential. The average number of years that girls spend in school relative to boys in each country.

Measure 23: General Status. Countries in which there exist government-supported or funded institutions or bureaus that are devoted to furthering the status of women earned points for this measure.

Measure 24: U. N. Treaty Ratification. Countries that signed the United Nations treaty regarding ending all forms of discrimination against women earned credit for each year prior to 1994 that they ratified it.

Measure 25: Women in Government. The average percentage of government positions held by women in each country.

HOW RAW NUMBERS WERE TRANSFORMED INTO STATISTICS, COUNTRIES (CHAPTER EIGHT)

Scores were standardized using a *z*-score transformation so that all variables would be on the same metric. The mean was chosen to be 10, with a standard deviation of 4, for each of the five categories; thus the possible scores could range from 0 to 20 for each category. The scores for each of the five categories were then added together to yield a "FEM Score," with an average of 50 and a standard deviation of 10.

LITERATURE CONSULTED FOR CHAPTER EIGHT

Adler, N. J., and Izraeli, D. N. "Women in Management Worldwide." In N. J. Adler and D. N. Izraeli (eds.), *Woman in Management Worldwide* (pp. 3–16). New York: M. E. Sharpe, Inc.

Al-Mouhandis, Z. *Higher Education for Women in Saudi Arabia* (doctoral dissertation, University of San Francisco). *Dissertation Abstracts International*, Vol. 47, p. 2904A, 1986.

Antal, A. B., and Izraeli, D. N. "A Global Comparison of Women in Management: Women Managers in their Homelands and as Expatriates." In Fagenson, E. (ed.), *Women in Management* (pp. 52–96). Newbury Park, CA: Sage Publications, 1993.

Bennoune, K. E. "The war against women in Algeria." *Ms. Magazine*, pp. 22–23, September/October, 1995.

Camp, S. L., and Speidel, J. J. (eds.). *World Access to Birth Control*. Washington, DC: Population Crisis Committee, 1987.

Council of Economic Advisors. *Economic Indicators, April 1994* (includes data available as of May 3, 1994). Prepared for the Joint Economic Committee. Washington, DC: U. S. Government Printing Office, 1994.

Daily Camera (Boulder, CO). "Exiled author longs for life without fear." February 17, 1995.

Davies, M., and Jansz, N. In M. Davies and N. Jansz (eds.) *Women Travel*. Englewood Cliff, NJ: Prentice–Hall Trade Division, 1990.

Davidson, M. J., and Cooper, C. *Shattering the Glass Ceiling: The Woman Manager*. Paul Chapman Publishing, Ltd., 1992.

Dornbush, R., and Fischer, S. *Macro-Economics*. New York: McGraw–Hill, 1984.

Human Rights in China. *Caught Between Tradition and the State*. New York: Human Rights in China, August 17, 1995.

Institute for Women's Policy Research. "The Wage Gap: Women's and Men's Earnings." In *Briefing Paper*. Washington, DC: Institute for Women's Policy Research, 1994.

International Planned Parenthood Federation. *Country Fact Sheets: Family Planning in the Western Hemisphere*. London: International Planned Parenthood Federation, 1991.

International Planned Parenthood Federation. *Reproductive Rights* (poster). London: International Planned Parenthood Federation, 1991.

International Planned Parenthood Federation. *Abortion Laws in Europe*. London: International Planned Parenthood Federation, November 1993.

International Planned Parenthood Federation. *Country Profiles*. London: International Planned Parenthood Federation, 1993.

Kohoni, N. K. "Where are the Girls?" *Ms. Magazine*, p. 96, July/August 1994.

Kurian, G. T. *The New Book of World Rankings*, New York: Facts on File, 3rd edition, 1991.

National Committee on Pay Equity. *Questions and Answers on Pay Equity*. Washington, DC: National Committee on Pay Equity, 1993.

National Women's Political Caucus. *National Directory of Women Elected Officials, 1995*. Washington, DC: National Women's Political Caucus, 1995.

"Saudi Arabia: Update on Women" (p. 17). *Ms.* Magazine. November/December 1991.

Torabi, B. "Veil of Darkness Descends on Iran's Women." In *The Rocky Mountain News*, August 4, 1994.

"Two Algerian Women Slain for Not Wearing Veils." *Rocky Mountain News* (Denver, Co.). March 31, 1994.

United Nations. *Annual Review of Population Law, 15*, United Nations Population Fund and Harvard Law School, 1991.

United Nations. *The World's Women: 1970–1990 Trends and Statistics*. New York: United Nations, 1991.

United Nations. *Population and Vital Statistics Report*. United Nations Statistical Papers, Series A, Vol. XLVII, No. 1, data available January 1995.

United Nations. *Wistat*. Women's Indicators and Statistic Database (version 2). New York: United Nations, 1992.

United Nations Development Programme. *Human Development Report for 1995*. London: Oxford University Press, 1995.

U. S. Department of Commerce. *Gender and Generation in the World's Labour Force*. Washington, DC: U. S. Department of Commerce, Economics and Statistics, 1991.

U. S. Department of Commerce. *Money Income of Households, Families, and Persons in the United States: 1990–91*. Washington, DC: U. S. Department of Commerce, Bureau of the Census, 1992.

U. S. Department of Commerce. *Statistical Abstracts of the U.S., 1994* (p. 264), Washington, DC: U. S. Department of Commerce, Projected 1994 Figures, 114th edition, 1994.

U. S. Department of Labor. *Country Reports on Human Rights Practices for 1993*. Washington, DC: U. S. Government Printing Office, 1994.

Appendix Two

Category Ratings of Companies

In this section, each company's FEM score is broken down into its constituent category percentile rankings. We do this so that a reader who considers benefits to be a more important measure of a company than, say, its board of directors can locate all such companies excelling in benefits but not necessarily in its board. The categories are as follows:

CATEGORY 1: MANAGEMENT OPPORTUNITIES

This category includes the percentage of women among a company's officers and managers, the percentage of female professionals, and the number of female advancement programs at each company.

CATEGORY 2: GLASS CEILING

This category includes the number of women serving on a company's board of directors and the percentage of top executive positions and top-salaried positions held by women.

CATEGORY 3: SUPPORT OF WOMEN

There are three measures for this category: the percentage of charitable donations targeted toward women and women's groups, whether the company cooperated with our survey, and whether a company provided salary information, and if so, how the latter compared with the national average.

CATEGORY 4: BENEFITS

In this category, companies earned points for the benefits provided to employees. Benefits included extended and/or paid family leave, flex time, day-care centers or assistance, and other family-friendly programs.

CATEGORY 5: THE FEMLIN FACTOR

This category considered whether a company had received prior accolades from such sources as the Domini Social Index, KLD, or *Working Mother* magazine. It also took into account whether companies were in compliance with the EEOC. Lastly, companies with female presidents, CEOs, founders, or division presidents received credit here.

Remember: The scores of each company are relevant only to the scores of other companies.

The categories are presented in a simplified form, which corresponds to the following key:

1 = 80–100th percentile
2 = 60–79th percentile
3 = 40–59th percentile
4 = 20–39th percentile
5 = 0–19th percentile

In-between category scores, such as 3.5 or 4.5, indicate that the scores overlapped 2 percentile categories because of ties in scores in that category.

The company's overall category ranking is also provided, using the same 1–5 rankings used for individual categories.

COMPANY RANK*	†	Categories 1	2	3	4	5
A.T. CROSS 217	3	1	5	4.5	4.5	3.5
ABBOTT LABORATORIES 194	3	2	4	1	3	5
ADIDAS USA 181	3	2	1	4.5	4.5	3.5
ADOLPH COORS 168	3	4	4	1	2	2
ADVANCE PUBLICATIONS 167	3	1	5	4.5	4.5	3.5
AETNA LIFE & CASUALTY 72	1	1	3	3	1	1
AIRBORNE EXPRESS 362	5	5	5	4.5	4.5	2
ALAMO 143	2	3	1	4.5	4.5	3.5
ALASKA AIRLINES DIVISION 280	4	4	2	4.5	4.5	2
ALBERTO-CULVER 64	1	1	1	2	3	2
ALBERTSON'S 187	3	5	3	4.5	1	3.5
ALCO STANDARD 229	3	3	2	3	4.5	2
ALLIED DOMECQ 298	4	3	5	2	4.5	3.5
ALLSTATE (SEARS) 106	2	2	2	3	1	2
AMERICAN AIRLINES 32	1	4	1	1	1	1
AMERICAN BRANDS 276	4	4	4	4.5	2	3.5
AMERICAN CYANAMID (acquired by American Home Products) 161	3	4	4	1	1	5
AMERICAN EXPRESS 36	1	1	2	3	1	1
AMERICAN GENERAL 299	4	2	5	4.5	4.5	3.5
AMERICAN HOME PRODUCTS 153	2	3	1	2	2	3.5
AMERICAN INTERNATIONAL GROUP 289*	4	3	4	4.5	4.5	3.5
AMERICAN STORES 292	4	5	3	2	3	3.5
AMERICA WEST AIRLINES 160	3	4	3	4.5	1	2
AMOCO 110	2	4	2	3	1	1
ANDERSON HICKEY 357	5	4	5	4.5	4.5	3.5
ANHEUSER-BUSCH 136*	2	5	3	3	1	2
ANN TAYLOR 34	1	4	1	2	3	1
APPLE COMPUTER 60	1	2	3	1	1	1
ARCHER-DANIELS-MIDLAND 349	5	5	5	4.5	4.5	2

1 80–100th percentile 2 60–79th percentile 3 40–59th percentile
4 20–39th percentile 5 0–19th percentile
*Company's numerical ranking based on total FEM scores for the final 384 companies. An asterisk next to the ranking indicates a tie between two companies.
†Numbers in this column represent this company's overall category rank.

		Categories				
COMPANY RANK*	†	1	2	3	4	5
ARMSTRONG WORLD 325	5	5	3	3	3	3.5
ARVIN 345	5	5	4	2	3	3.5
AT&T 24	1	1	3	3	1	2
ATLANTIC RICHFIELD 45	1	2	3	1	1	1
AVEDA 11*	1	1	1	1	2	3.5
AVERY DENNISON 231*	4	4	3	2	2	5
AVIS 226	3	2	4	2	2	5
AVON PRODUCTS 1	1	1	1	1	1	1
BANKAMERICA 92	2	1	2	3	2	1
BANKERS TRUST NEW YORK 51	1	2	4	1	1	2
BANK OF BOSTON 126	2	1	2	3	4.5	2
BANK OF NEW YORK 239*	4	1	3	4.5	4.5	3.5
BAYER† 213	3	3	5	2	2	3.5
BEAR STEARNS 344	5	4	4	4.5	4.5	3.5
BEN & JERRY'S HOMEMADE 28	1	5	1	2	1	1
BENETTON GROUP 18	1	1	1	1	3	3.5
BERTELSMANN 269	4	1	5	4.5	4.5	3.5
BIRKENSTOCK SANDALS 6	1	1	1	1	2	3.5
BLACK & DECKER 297	4	5	3	2	3	5
BODY SHOP 9	1	1	1	1	3	1
BOISE CASCADE 152	2	4	3	1	2	5
BORDEN 235	4	4	3	2		5
BORLAND 319	5	3	5	4.5	4.5	3.5
BRIO 16	1	1	1	1	2	3.5
BRISTOL-MYERS SQUIBB 68	1	1	3	1	1	2
BRITISH PETROLEUM (BP) 261	4	3	5	4.5	2	5
BROADWAY STORES (formerly Carter Hawley Hale) 71	1	1	1	4.5	3	3.5
BROWN GROUP 253	4	2	3	4.5	4.5	2
BRUNSWICK 248	4	5	2	1	2	5
BUDGET RENT A CAR 294	4	1	5	4.5	4.5	5
CADBURY SCHWEPPES 218	3	3	1	2	3	5
CALVERT GROUP (Social Investment Fund) 22	1	1	1	1	1	2

Category 1: Management Opportunities Category 2: Glass Ceiling Category 3: Support of Women Category 4: Benefits Category 5: The Femlin Factor

	Categories					
COMPANY RANK*	†	1	2	3	4	5
CAMPBELL SOUP 74*	1	4	3	3	1	1
CANON USA 352	5	4	5	2	4.5	5
CAPITAL CITIES/ABC (acquired by Walt Disney) 90	2	1	4	1	1	1
CARTER-WALLACE 348	5	4	5	4.5	3	3.5
CBS 141	2	2	1	3	4.5	1
CELESTIAL SEASONINGS 46	1	2	1	1	2	2
CHARLES SCHWAB 177*	3	2	2	2	3	3.5
CHASE BANK (merger of Chase Manhattan & Chemical banks) 139	2	2	3	3	2	3.5
CHESAPEAKE 373*	5	5	5	4.5	4.5	3.5
CHEVRON 102	2	3	3	1	1	5
CHILDREN'S TELEVISION WORKSHOP 5	1	1	1	1	2	3.5
CHIQUITA BRANDS 208	3	4	2	2	3	3.5
CHRYSLER 311	5	5	4	4.5	3	2
CHUBB GROUP OF INSURANCE 56	1	1	2	1	2	3.5
CHURCH & DWIGHT 171*	3	3	4	2	2	2
CIGNA 14	1	1	2	1	1	2
CITGO PETROLEUM 384	5	5	5	4.5	4.5	5
CITICORP 66	1	1	4	1	1	1
CLOROX 164	3	2	4	3	2	3.5
COCA-COLA 195	3	4	2	4.5	2	2
COLGATE-PALMOLIVE 103	2	2	2	2	2	2
COMCAST 109	2	1	2	1	3	2
COMPAQ COMPUTER 286	4	4	5	2	3	3.5
COMPUTER ASSOCIATES 171*	3	2	4	2	3	2
CONAGRA 247	4	4	4	2	2	5
CONDÉ NAST PUBLICATIONS 82	2	1	1	4.5	1	3.5
CONSOLIDATED PAPERS 200	3	5	3	1	4.5	2
CONTINENTAL AIRLINES 316	5	5	4	2	4.5	3.5
CONTINENTAL (INSURANCE) 77	1	1	1	2	2	2

1 80–100th percentile 2 60–79th percentile 3 40–59th percentile
4 20–39th percentile 5 0–19th percentile

*Company's numerical ranking based on total FEM scores for the final 384 companies. An asterisk next to the ranking indicates a tie between two companies.

†Numbers in this column represent this company's overall category rank.

COMPANY RANK*	†	Categories 1	2	3	4	5
COOPER 265	4	5	2	1	3	5
COOPER TIRE & RUBBER 118	2	5	5	2	1	5
CORNING 98	2	3	3	3	1	1
COSMAIR 30	1	1	1	2	1	3.5
CPC 233	4	4	3	2	3	2
DANSKIN 63	1	1	1	4.5	4.5	3.5
DAYTON HUDSON 8	1	1	1	1	1	1
DEAN WITTER, DISCOVER 239*	4	5	1	4.5	3	3.5
DELL COMPUTER 104*	2	1	2	2	2	3.5
DELTA AIR LINES 356	5	5	2	4.5	4.5	5
DHL WORLDWIDE EXPRESS 343	5	4	4	4.5	4.5	5
DIAL 258*	4	4	2	4.5	4.5	3.5
DIGITAL EQUIPMENT 100*	2	2	2	2	2	1
DOLE FOOD 211	3	3	2	4.5	2	5
DOMINO'S 249	4	3	5	2	3	3.5
DOW CHEMICAL 100*	2	4	3	2	1	1
DOW JONES 158	3	4	3	4.5	2	2
DREYER'S GRAND ICE CREAM 358	5	5	5	2	4.5	3.5
DREYFUS SERVICE 236	4	2	1	4.5	4.5	3.5
DR. PEPPER/7UP (acquired by Cadbury Schweppes) 330	5	3	5	2	4.5	5
DYNAMICS OF AMERICA 364	5	5	5	4.5	4.5	3.5
E.I. DU PONT 25	1	4	1	2	1	1
E.W. SCRIPPS 302	4	2	4	4.5	4.5	3.5
EASTMAN KODAK 79	2	2	3	2	1	1
EDUCATIONAL INSIGHTS 242	4	2	2	4.5	4.5	3.5
ELECTROLUX 367	5	5	5	4.5	4.5	5
ELI LILLY 336	5	4	4	3	4.5	5
EMERSON ELECTRIC 380	5	5	4	4.5	4.5	5
EQUITABLE 234	4	2	3	3	4.5	2
ESPRIT DE CORP 38	1	1	1	4.5	4.5	1
ESTÉE LAUDER 26	1	1	1	2	4.5	2
EXXON 127	2	4	4	2	1	2

Category 1: Management Opportunities Category 2: Glass Ceiling Category 3: Support of Women Category 4: Benefits Category 5: The Femlin Factor

COMPANY RANK*	†	Categories 1	2	3	4	5
FARBERWARE/HANSON 331	5	3	5	4.5	4.5	3.5
FEDDERS 371	5	5	5	4.5	4.5	3.5
FEDERAL EXPRESS 215	3	4	2	2	3	2
FEDERATED DEPARTMENT STORES 130	2	2	2	1	4.5	3.5
FIDELITY INVESTMENTS 322	5	2	5	3	4.5	3.5
FIRST BRANDS 351	5	5	5	2	3	3.5
FIRST FIDELITY BANCORP 145	2	1	3	2	3	3.5
FLAGSTAR 231*	4	1	3	4.5	4.5	3.5
FOOD LION 337*	5	3	5	4.5	4.5	3.5
FORD MOTOR 368	5	5	3	4.5	3	5
FRUIT OF THE LOOM 324	5	3	5	2	4.5	5
GANNETT 4	1	1	1	1	1	1
THE GAP 57	1	4	1	1	2	2
GEICO 267	4	2	3	3	4.5	3.5
GENENTECH 150	2	2	4	4.5	2	2
GENERAL ELECTRIC 48	1	3	1	1	1	3.5
GENERAL MILLS 112	2	2	2	2	3	1
GENERAL MOTORS 37	1	5	3	1	1	2
GENESCO 347	5	3	5	4.5	4.5	3.5
GEORGIA-PACIFIC 228	3	4	3	2	2	5
GERBER PRODUCTS 190*	3	3	2	3	3	2
GIANT FOODS 166	3	2	5	4.5	2	2
GILLETTE 255	4	3	4	2	3	5
GITANO GROUP 342	5	4	5	3	4.5	3.5
GLAXO WELLCOME 97	2	2	4	4.5	1	2
GLOBE BUSINESS FURNITURE 323	5	4	4	4.5	4.5	3.5
GOLDMAN SACHS GROUP 274	4	2	5	2	4.5	3.5
GOODYEAR TIRE & RUBBER 210	3	5	4	2	1	3.5
GRAND METROPOLITAN 120	2	1	4	2	2	5
GRAND UNION 337*	5	3	5	4.5	4.5	3.5
GREAT ATLANTIC & PACIFIC TEA 162	3	3	1	2	4.5	2

1 80–100th percentile 2 60–79th percentile 3 40–59th percentile
4 20–39th percentile 5 0–19th percentile
*Company's numerical ranking based on total FEM scores for the final 384 companies. An asterisk next to the ranking indicates a tie between two companies.
†Numbers in this column represent this company's overall category rank.

COMPANY RANK*	†	Categories 1	2	3	4	5
GREAT WESTERN FINANCIAL 128	2	1	4	4.5	2	2
GUINNESS 332	5	3	5	4.5	4.5	3.5
H. J. HEINZ 193	3	2	3	4.5	3	2
HAMILTON BEACH/PROCTOR SILEX 183	3	5	5	1	3	5
HARCOURT GENERAL 116	2	1	5	1	3	3.5
HASBRO 177*	3	2	2	3	4.5	2
HEAD SPORTS 295	4	4	2	4.5	4.5	3.5
HEALTH VALLEY NATURAL FOODS 123	2	2	1	4.5	2	3.5
HEARST 190*	3	2	2	2	4.5	3.5
HELENE CURTIS 196	3	2	2	4.5	2	3.5
HERMAN MILLER 238	4	4	4	4.5	2	2
HERSHEY FOODS 40	1	3	2	4.5	1	1
HERTZ 346	5	2	5	4.5	4.5	5
HEWLETT-PACKARD 67	1	2	3	1	1	1
HFS 165	3	1	4	2	3	3.5
HILTON HOTELS 223	3	2	4	2	4.5	3.5
HONDA OF AMERICA 375*	5	5	5	4.5	4.5	3.5
HON 305	4	4	4	2	4.5	3.5
HORMEL FOODS 275	4	5	3	3	2	3.5
HOST MARRIOTT 53	1	1	3	1	1	1
HOUGHTON MIFFLIN 155	3	5	1	2	3	5
HUFFY 163	3	5	1	4.5	3	2
HYATT HOTELS 107	2	1	2	2	2	3.5
IBM 13	1	4	3	2	1	1
IBP 278	4	3	2	3	4.5	5
IDS FINANCIAL SERVICE 188	3	5	1	2	2	2
IHOP 173*	3	2	1	4.5	4.5	3.5
IMASCO USA 314	5	5	2	4.5	4.5	3.5
INTEL 307	4	4	4	3	4.5	2
INTERNATIONAL DAIRY QUEEN 317*	5	5	3	4.5	4.5	2
JOHN PAUL MITCHELL SYSTEMS 96	2	2	1	4.5	4.5	3.5
JOHNSON & JOHNSON 3	1	1	2	1	1	1
INTERNATIONAL PAPER 272	4	4	3	3	2	5

Category 1: Management Opportunities Category 2: Glass Ceiling Category 3: Support of Women Category 4: Benefits Category 5: The Femlin Factor

	Categories					
COMPANY RANK*	†	1	2	3	4	5
INVESTCORP 285	4	2	4	4.5	4.5	3.5
ITT 135	2	3	2	1	3	3.5
ITT SHERATON 328	5	2	5	4.5	4.5	3.5
J.C. PENNEY 117	2	2	2	3	2	1
JAMES RIVER OF VIRGINIA 170	3	4	4	1	1	3.5
JOHN HANCOCK 49	1	2	1	3	1	2
JOHNSON PRODUCTS 55	1	2	1	4.5	4.5	1
JOHNSON PUBLISHING 93	2	2	1	2	4.5	1
KELLOGG 214	3	2	3	4.5	3	3.5
KIMBERLY-CLARK 245	4	4	2	3	3	3.5
KMART 133	2	1	2	2	4.5	1
KNIGHT-RIDDER 73	1	2	1	4.5	2	1
KROGER 59	1	1	2	3	1	1
L.A. GEAR 199	3	1	2	4.5	4.5	3.5
LANDS' END 222	3	2	3	4.5	4.5	2
LANZ 206	3	2	1	4.5	4.5	3.5
LAURA ASHLEY HOLDINGS 202	3	2	2	4.5	4.5	1
LEHMAN BROTHERS HOLDING 303	4	2	4	4.5	4.5	3.5
LESLIE FAY 115	2	5	1	2	1	5
LEVI STRAUSS 76	1	1	1	1	2	3.5
THE LIMITED 173*	3	2	3	4.5	4.5	1
LIZ CLAIBORNE 10	1	1	1	1	2	1
LOTUS DEVELOPMENT 43	1	2	1	2	1	1
MACDONALD COMMUNICATIONS 44	1	1	2	1	4.5	3.5
MANOR CARE 23	1	1	2	1	2	2
MARS 273	4	4	5	1	4.5	3.5
MASCO 333	5	5	3	4.5	4.5	3.5
MATSUSHITA ELECTRIC 375*	5	5	5	4.5	3	5
MATTEL 20	1	2	1	2	1	1
MAYBELLINE 203*	3	2	1	4.5	4.5	3.5
MAY DEPARTMENT STORES 146	2	2	3	3	3	1

1 80–100th percentile 2 60–79th percentile 3 40–59th percentile
4 20–39th percentile 5 0–19th percentile
*Company's numerical ranking based on total FEM scores for the final 384 companies. An asterisk next to the ranking indicates a tie between two companies.
†Numbers in this column represent this company's overall category rank.

COMPANY RANK*	†	Categories 1	2	3	4	5
MAYTAG 301	4	5	4	4.5	3	2
MCCORMICK 131	2	2	3	1	2	3.5
MCDONALD'S 121	2	1	3	4.5	2	1
MCGRAW-HILL 84*	2	1	1	4.5	1	3.5
MCI COMMUNICATIONS 169	3	2	2	3	4.5	2
MEAD 175*	3	4	3	2	2	2
MEDIA GENERAL 360	5	5	5	3	3	3.5
MELLON BANK 237	4	1	5	2	4.5	5
MELVILLE 52	1	1	2	4.5	1	2
MERCK 7	1	2	1	2	1	1
MEREDITH 254	4	4	4	2	4.5	2
MERRILL LYNCH 220	3	2	4	3	3	2
METROPOLITAN LIFE INSURANCE 108	2	1	2	3	1	5
MICROSOFT 256*	4	4	4	1	3	5
3M 50	1	4	3	1	1	1
MITSUBISHI 341	5	3	5	4.5	3	5
MOBIL OIL 99	2	4	4	3	1	3.5
MONSANTO 27	1	4	1	1	1	5
MONTGOMERY WARD HOLDING 288	4	2	4	3	4.5	3.5
MOORE 250	4	2	4	4.5	3	3.5
MORGAN STANLEY GROUP 315	5	4	5	3	4.5	3.5
MOTOROLA 91	2	4	4	2	1	1
MOUSEFEATHERS 88	2	2	1	4.5	4.5	1
MOVING COMFORT 2	1	3	1	1	3	1
NATIONAL PUBLIC RADIO 114	2	2	1	4.5	4.5	3.5
NEIMAN MARCUS GROUP 159	3	2	1	2	4.5	2
NESTLÉ 251	4	3	5	3	2	5
NEUTROGENA (acquired by Johnson & Johnson) 157	3	5	1	4.5	2	2
NEW BALANCE ATHLETIC SHOES 310	5	2	5	4.5	4.5	3.5
NEWELL 95	2	4	5	1	2	3.5
NEWMAN'S OWN 65	1	3	1	2	4.5	3.5
NEW YORK TIMES 203*	3	5	1	4.5	3	1

Category 1: Management Opportunities Category 2: Glass Ceiling Category 3: Support of Women Category 4: Benefits Category 5: The Femlin Factor

COMPANY RANK*	†	Categories 1	2	3	4	5
NIKE 29	1	1	2	1	1	1
NINE WEST GROUP 291	4	5	3	4.5	4.5	1
NISSAN 365	5	5	5	4.5	4.5	3.5
NORDSTROM 19	1	1	1	4.5	1	1
NORTHWEST AIRLINES 381	5	5	5	4.5	4.5	5
NOVELL 201	3	4	1	4.5	4.5	1
NU SKIN 134	2	5	2	1	4.5	3.5
OCCIDENTAL PETROLEUM 363	5	5	4	4.5	4.5	3.5
OFFICE DEPOT (merging with Staples) 309	5	4	4	4.5	4.5	3.5
ORACLE 243	4	4	4	2	2	5
OSHKOSH B'GOSH 54	1	1	3	1	3	2
O'SULLIVAN 373*	5	4	5	4.5	4.5	5
PAINEWEBBER GROUP 205	3	2	1	4.5	4.5	3.5
PARAMOUNT (acquired by Viacom) 138*	2	2	1	2	3	2
PAUL REVERE INSURANCE 270	4	5	2	2	3	2
PENGUIN BOOKS USA (PEARSON) 264	4	2	3	4.5	4.5	3.5
PEP BOYS—MANNY, MOE & JACK 372	5	5	5	4.5	4.5	3.5
PEPSICO 42	1	1	2	1	1	1
PET (acquired by Grand Metropolitan) 113	2	3	3	1	2	3.5
PFIZER 89	2	4	2	1	1	3.5
PHARMACIA & UPJOHN 284	4	5	3	3	2	5
PHILIP MORRIS 144	2	4	2	2	3	1
PHILIPS ELECTRONICS 370	5	5	5	2	3	5
PHILLIPS PETROLEUM 184	3	4	3	2	2	5
PHILLIPS-VAN HEUSEN 136	2	1	1	3	4.5	2
PILGRIM'S PRIDE 287	4	3	3	4.5	4.5	3.5
PIONEER ELECTRONICS 262	4	2	3	4.5	4.5	3.5
PITNEY BOWES 35	1	2	1	1	1	1
POLAROID 47	1	4	3	3	1	1
PRIMERICA 190*	3	2	2	3	3	3.5
PRINCE MANUFACTURING 366	5	3	5	4.5	4.5	5

1 80–100th percentile 2 60–79th percentile 3 40–59th percentile
4 20–39th percentile 5 0–19th percentile

*Company's numerical ranking based on total FEM scores for the final 384 companies. An asterisk next to the ranking indicates a tie between two companies.

†Numbers in this column represent this company's overall category rank.

COMPANY RANK*	†	Categories 1	2	3	4	5
PROCTER & GAMBLE 33	1	2	4	3	1	1
PROMUS HOTEL 156	3	3	2	2	2	3.5
PRUDENTIAL INSURANCE OF NORTH AMERICA 39	1	1	2	1	1	2
PUBLIC BROADCASTING SYSTEM 124	2	2	1	4.5	4.5	3.5
PUBLIX SUPER MARKETS 263	4	3	2	4.5	4.5	3.5
QUAKER OATS 122	2	1	2	3	3	1
QUAKER STATE 277	4	5	1	2	3	5
QVC NETWORK 355	5	4	5	4.5	4.5	3.5
R.H. MACY 260	4	2	2	4.5	4.5	3.5
RALSTON PURINA 241	4	5	3	4.5	2	3.5
RAYTHEON 354	5	5	4	2	4.5	3.5
READER'S DIGEST 70	1	1	1	2	2	3.5
RECKITT & COLMAN 359	5	3	5	4.5	4.5	5
REDKEN LABORATORIES 188*	3	2	2	4.5	4.5	1
REEBOK 142	2	2	4	3	2	2
RENAISSANCE HOTELS 151	2	3	5	1	4.5	3.5
REVLON 340	5	2	5	3	4.5	5
REYNOLDS METALS 327	5	5	4	3	2	5
RHÔNE-POULENC RORER 219	3	5	3	2	1	5
RICOH 320	5	3	5	2	4.5	5
RJR NABISCO 147*	2	3	4	1	2	5
RUBBERMAID 61	1	4	1	1	1	1
S.C. JOHNSON & SON 119	2	3	4	4.5	1	1
SAFEWAY 207	3	2	4	2	3	5
SALOMON 312	5	4	4	3	3	3.5
SARA LEE 104*	2	1	2	2	2	2
SCHERING-PLOUGH 80	2	2	4	2	1	1
SCOTT PAPER 147*	2	4	3	1	2	2
SEAGRAM 252	4	3	4	2	4.5	3.5
SEARS, ROEBUCK 41	1	2	1	4.5	1	1
SERVICE MERCHANDISE 268	4	2	5	4.5	4.5	1
SHARP ELECTRONICS 383	5	5	5	4.5	4.5	5

Category 1: Management Opportunities Category 2: Glass Ceiling Category 3: Support of Women Category 4: Benefits Category 5: The Femlin Factor

	Categories					
COMPANY RANK*	†	1	2	3	4	5
SHELL OIL (ROYAL DUTCH) 308	4	5	5	4.5	2	3.5
SMITHFIELD FOODS 379	5	5	4	4.5	4.5	5
SMITHKLINE BEECHAM 58	1	1	5	2	1	5
SONY 281	4	2	5	3	3	5
SOUTHLAND 246	4	1	4	4.5	4.5	3.5
SOUTHWEST AIRLINES 186	3	4	1	4.5	3	1
SPRINT 11*	1	2	2	1	1	1
STANDARD REGISTER 350	5	4	4	2	4.5	5
STAPLES (merging with Office Depot) 69	1	2	1	1	3	3.5
STOP & SHOP 304	4	2	5	3	4.5	3.5
STRIDE RITE 17	1	2	1	1	1	1
SUNBEAM 334	5	4	4	4.5	4.5	3.5
SUN (SUNOCO) 313	5	5	3	4.5	4.5	2
SUN MICROSYSTEMS 271	4	4	4	2	4.5	3.5
SYNTEX 94	2	2	3	4.5	1	1
TAMBRANDS 212	3	2	1	2	4.5	5
TANDY 279	4	3	4	4.5	4.5	2
TEKTRONIX 81	2	4	2	1	2	2
TELE-COMMUNICATIONS (TCI) 140	2	2	3	1	3	3.5
TEXACO 180	3	4	3	2	2	3.5
TEXAS INSTRUMENTS 296	4	5	4	4.5	2	3.5
TIMBERLAND 185	3	2	4	2	2	3.5
TIMES MIRROR 154	2	2	2	2	4.5	2
TIME WARNER 21	1	2	3	2	1	1
TOASTMASTER 326	5	4	2	4.5	4.5	5
TOPP'S 377	5	5	5	4.5	4.5	3.5
TORO 209	3	5	3	2	2	3.5
TOYOTA 216	3	4	5	2	2	3.5
TOYS "R" US 179	3	1	3	4.5	4.5	1
TRANSAMERICA 84*	2	2	2	1	1	1
TRANS WORLD AIRLINES 329	5	5	2	4.5	4.5	5

1 80–100th percentile   2 60–79th percentile   3 40–59th percentile
4 20–39th percentile   5 0–19th percentile
*Company's numerical ranking based on total FEM scores for the final 384 companies. An asterisk next to the ranking indicates a tie between two companies.
†Numbers in this column represent this company's overall category rank.

	Categories					
COMPANY RANK*	†	1	2	3	4	5
TRAVELERS GROUP 129	2	1	3	2	2	2
TRIARC 317*	5	2	5	4.5	4.5	3.5
TURNER BROADCASTING SYSTEM 149	2	2	4	1	3	2
TYCO TOYS 300	4	5	4	1	4.5	3.5
TYSON FOODS 283	4	4	2	2	4.5	5
UNILEVER 289*	4	3	4	3	3	3.5
UNISYS 353	5	5	3	2	3	5
UNITED AIRLINES 225	3	5	4	2	2	3.5
UNITED ASSET MANAGEMENT 227	3	2	4	4.5	4.5	1
UNITED BISCUIT HOLDINGS 339	5	3	5	4.5	4.5	3.5
UNITED PARCEL SERVICE OF AMERICA 378	5	5	5	3	3	5
UNITED TECHNOLOGIES 78	2	4	3	1	1	3.5
UNIVERSAL FOODS 335	5	5	4	4.5	4.5	3.5
USAIR GROUP 198	3	4	3	1	2	5
USLIFE 361	5	2	5	4.5	4.5	5
VF 125	2	2	1	2	3	2
VIACOM 244	4	1	4	4.5	4.5	3.5
VOLKSWAGEN/AUDI 266	4	5	5	1	2	3.5
WALGREEN 197	3	3	3	1	4.5	2
WAL-MART STORES 306	4	5	3	4.5	3	2
WALT DISNEY 62	1	1	4	1	1	1
WANG LABORATORIES 230	3	4	5	2	3	3.5
WARNACO GROUP 83	2	1	1	2	4.5	1
WARNER-LAMBERT 87	2	1	4	2	1	1
WASHINGTON POST 175*	3	3	2	4.5	4.5	1
WELCH FOODS 321	5	3	5	4.5	4.5	3.5
WELLS FARGO 74*	1	1	2	2	1	1
WENDY'S 111	2	1	2	2	3	2
WENNER MEDIA 224	3	2	1	4.5	4.5	3.5
WESTERN PUBLISHING GROUP 282	4	2	3	4.5	4.5	3.5
WESTIN HOTEL 258*	4	2	2	4.5	4.5	3.5
WHIRLPOOL 221	3	5	3	2	2	3.5

Category 1: Management Opportunities Category 2: Glass Ceiling Category 3: Support of Women Category 4: Benefits Category 5: The Femlin Factor

COMPANY RANK*	†	1	2	3	4	5
		Categories				
WHITMAN 256*	4	5	2	3	2	3.5
WILLIAM CARTER 132	2	1	3	4.5	4.5	5
WINN-DIXIE STORES 293	4	3	5	2	4.5	3.5
WM. WRIGLEY JR. 86	2	4	2	1	1	2
WOOLWORTH 181	3	4	2	2	4.5	1
WORKING ASSETS FUNDING SERVICE 31	1	1	1	4.5	3	3.5
XEROX 15	1	2	2	1	1	1
YORK 382	5	5	4	4.5	4.5	5
ZENITH ELECTRONICS 369	5	5	3	4.5	4.5	3.5

1 80–100th percentile 2 60–79th percentile 3 40–59th percentile
4 20–39th percentile 5 0–19th percentile

*Company's numerical ranking based on total FEM scores for the final 384 companies. An asterisk next to the ranking indicates a tie between two companies.

†Numbers in this column represent this company's overall category rank.

Appendix Three

Organizations that Help Women

AFL-CIO, Standing Committee on Salaried and Professional Women
815 16th St. NW—Room 707, Washington, DC 20036; (202) 638-0320
Works on improving women's workplace conditions

AMERICAN ASSOCIATION FOR HIGHER EDUCATION—WOMEN'S CAUCUS
One Dupont Circle, Ste. 600, Washington, DC 20036; (202) 293-6440
Addresses women's issues in higher education

AMERICAN ASSOCIATION OF UNIVERSITY WOMEN
1111 16th St. NW, Washington, DC 20036; (202) 785-7712
Promotes educational equity for women and girls

AMERICAN BUSINESS WOMEN'S ASSOCIATION NATIONAL HEADQUARTERS
9100 Ward Parkway, Kansas City, MO 64114; (816) 361-6621
Promotes educational and professional achievement of working women

AMERICAN CIVIL LIBERTIES UNION, WOMEN'S RIGHTS PROJECT
132 W. 43rd St., New York, NY 10036; (212) 944-9800
Seeking constitutional equality through litigation

AMERICAN COUNCIL ON EDUCATION, OFFICE OF WOMEN IN HIGHER EDUCATION
One Dupont Circle, Washington, DC 20036-1193; (202) 939-9390
Seeking to advance talented women in academic administration

AMERICAN FEDERATION OF STATE, COUNTY AND MUNICIPAL EMPLOYEES, WOMEN'S RIGHTS DEPARTMENT
1625 L St. NW, Washington, DC 20036; (202) 429-5090
Represents 1.4 million public employees throughout the United States

AMERICAN PUBLIC HEALTH ASSOCIATION—WOMEN'S CAUCUS
Feminist Women's Health Center, 1469 Humboldt Rd. #200, Chico, CA 95928; (916) 891-1917
Formed to provide a forum for feminist input into the APHA

AMERICAN SOCIETY FOR TRAINING AND DEVELOPMENT—WOMEN'S NETWORK
Box 1443, 1640 King St., Alexandria, VA 22313-2043; (703) 683-8100
Serves as a catalyst and consulting resource for women in the human resources and development profession by providing products and services

AMERICAN WOMEN'S ECONOMIC DEVELOPMENT CORPORATION
71 Vanderbilt Ave., New York, NY 10169; (212) 688-1900
Offers business counseling and training for women currently managing or interested in starting their own businesses

AMNESTY INTERNATIONAL USA—WOMEN AND HUMAN RIGHTS PROJECT
322 8th Ave., New York, NY 10001; (212) 633-4200
Public education to empower people to act to eliminate human rights abuses against women internationally

BUSINESS AND PROFESSIONAL WOMEN'S FOUNDATION
2012 Massachusetts Ave. NW, Washington, DC 20036; (202) 293-1100
Promotes equity and economic self-sufficiency for working women of all races, ethnicities, and political beliefs

CATALYST
250 Park Ave. South, New York, NY 10003; (212) 777-8900
Works with businesses to effect change. Researches pay equity, work and family, and other women's issues

CENTER FOR WOMEN POLICY STUDIES
2000 P St. NW, Ste. 508, Washington, DC 20036; (202) 872-1770
Concentrates on women living with HIV/AIDS, work and family issues, workplace diversity, girls and violence, reproductive rights and health, and welfare reform

CLEARINGHOUSE ON WOMEN'S ISSUES
P.O. Box 70603, Friendship Heights, MD 20813; (202) 362-3789
Exchanges and disseminates educational information and materials on issues related to discrimination on the basis of sex and marital status, with particular emphasis on public policies affecting the economic and educational status of women

COMMITTEE OF 200
625 N. Michigan Ave., Ste. 500, Chicago, IL 60611; (312) 751-3477
Organization of businesswomen dedicated to exchanging business ideas, solutions, and support

COMMUNICATIONS CONSORTIUM
1333 H St. NW, Washington, DC 20005; (202) 682-1270
Helps public interest groups maximize their use of telecommunications systems as tools for education and policy change

DOMESTIC VIOLENCE INSTITUTE
50 South Steele St., Ste. 850, Denver, CO 80209; (303) 322-3444
Focus is to increase the psychological knowledge base

EMILY'S LIST
805 15th St. NW, Ste. 400, Washington, DC 20005; (202) 326-1400
Political network of female candidates running in gubernatorial or congressional races; helps with fund-raising for these candidates

EQUAL RIGHTS ADVOCATES
1663 Mission St., Ste. 550, San Francisco, CA 94103; (415) 621-0672 Hotline: (415) 621-0505
Dedicated to the empowerment of women through the establishment of their economic, social, and political equality

FEDERATION OF ORGANIZATIONS FOR PROFESSIONAL WOMEN
1825 I St. NW, Ste. 400, Washington, DC 20006; (202) 328-1415
Federation of affiliated women's organizations to promote equal status for women in all educational levels and career fields

THE FEMINIST INSTITUTE, INC.
P.O. Box 30563, Bethesda, MD 20824; (301) 805-5611
Nonprofit organization working for feminist social change and celebrating women's suffrage

FINANCIAL WOMEN INTERNATIONAL
200 North Glebe Rd., Ste. 814, Arlington, VA 22203-3728; (301) 657-8288
To empower women in the financial services industry

THE FOUNDATION FOR WOMEN'S RESOURCES
1300 Jefferson, Ste. 210, Austin, TX 78731; (512) 459-1167
Dedicated to improving the personal, economic, and professional status of women through its varied programs and projects

THE FUND FOR THE FEMINIST MAJORITY
1600 Wilson Blvd., Ste. 704, Arlington, VA 22209; (703) 522-2214
Women's rights, abortion rights, advocacy, and research group that works for the empowerment of women in all walks of life

THE GLOBAL FUND FOR WOMEN
2480 Sand Hill Rd., Ste. 100, Menlo Park, CA 94025-6941; (415) 854-0420
A grantmaking organization that provides funds to seed, strengthen, and link groups that are committed to women's well-being internationally and work for their full participation in society

INSTITUTE FOR WOMEN'S POLICY RESEARCH
1400 20th St. NW, Ste. 104, Washington, DC 20036; (202) 785-5100
Organized to meet the needs of women-centered, policy-oriented research

INTERNATIONAL CENTER FOR RESEARCH ON WOMEN
1717 Massachusetts Ave. NW, Ste. 302, Washington, DC 20036; (202) 797-0007
Works to improve the economic, health, and social status of poor women in developing countries worldwide

INTERNATIONAL WOMEN'S HEALTH COALITION
24 East 21st St., New York, NY 10010; (212) 979-8500 (Joan Dunlop)
Dedicated to improving women's reproductive health in the developing world

INTERNATIONAL WOMEN'S TRIBUNE CENTER
777 United Nations Plaza, 3rd Floor, New York, NY 10017; (212) 687-8633
A communications link for 16,000 individuals and groups working on behalf of women in 160 countries

LEAGUE OF WOMEN VOTERS
1730 M St. NW, Ste. 1000, Washington, DC 20036; (202) 429-1965
To encourage informed and active participation of citizens in government and to influence public policy through education and advocacy

MS. FOUNDATION FOR WOMEN
120 Wall St., 33rd Floor, New York, NY 10005; (212) 742-2300
Supports the efforts of women and girls to govern their own lives and influence the world around them

NATIONAL ABORTION RIGHTS ACTION LEAGUE
1156 15th St. NW, Ste. 700, Washington, DC 20005; (202) 828-9300
Political arm of the pro-choice movement

NATIONAL ASSOCIATION FOR FEMALE EXECUTIVES
30 Irving Place, 5th Floor, New York, NY 10003; (212) 477-2200
Provides women with the tools and information to reach their career goals and achieve financial independence

NATIONAL ASSOCIATION OF COMMISSIONS FOR WOMEN
1828 L St., NW, Ste. 250, Washington, DC 20036-5104; (202) 628-5030 or (800) 338-9267
Membership organization composed of regional, state, county, and local commissions created by the government to improve the status of women

NATIONAL ASSOCIATION OF M.B.A. WOMEN
3923 Georgia Ave. NW, Washington, DC 20011; (202) 723-1267
Dedicated to improving career opportunities for women with M.B.A. degrees and to providing networking and scholarship opportunties

NATIONAL ASSOCIATION OF WOMEN BUSINESS OWNERS
200 N. Michigan Ave., Ste. 300, Chicago, IL 60601; (312) 541-1272
Provides a strong voice for female business owners within all economic, social, and political communities

NATIONAL CENTER FOR POLICY ALTERNATIVES
Women's Economic Justice Center, 1875 Connecticut Ave. NW, Ste. 710, Washington, DC 20009; (202) 387-6030
Improves the lives of economically disadvantaged women and their families by creating sound state policies

NATIONAL CENTER FOR WOMEN AND FAMILY LAW
799 Broadway, Room 402, New York, NY 10003; (212) 674-8200
Addresses women's issues in the family law area

NATIONAL CHAMBER OF COMMERCE FOR WOMEN
10 Waterside Plaza, Ste. 6H, New York, NY 10010; (212) 685-3454
Works with local, regional, and state redevelopment agencies to expand business opportunities for women

NATIONAL COALITION AGAINST DOMESTIC VIOLENCE
P.O. Box 34103, Washington, DC 20043-4103; (202) 638-6388
Dedicated to ending violence in the lives of women and children

NATIONAL COMMITTEE ON PAY EQUITY
1126 16th St., Room 422, Washington, DC 20036; (202) 331-7343
Focuses local and national attention to women's salaries, legislation, job evaluation, and the wage gap

NATIONAL CONFERENCE OF STATE LEGISLATORS WOMEN'S NETWORK
1560 Broadway, Ste. 700, Denver, CO 80202; (303) 830-2200
Purpose is to elevate women in positions of power to advance legislation that will support women and their families

NATIONAL COUNCIIL FOR RESEARCH ON WOMEN
530 Broadway, 10th Floor, New York, NY 10012; (212) 274-0730
Provides resources for feminist research, policy analysis, and educational programs for women and girls

NATIONAL LEAGUE OF CITIES
Women in Municipal Government, 1301 Pennsylvania Ave. NW, Washington, DC 20004; (202) 626-3181
To promote issues of interest to women and the status of women in the nation's cities and towns

NATIONAL NETWORK OF WOMEN'S FUNDS
1821 University Ave., Ste. 409N, St. Paul, MN 55104; (612) 641-0742
Dedicated to generating increased resources for programs that benefit women and girls

NATIONAL ORGANIZATION FOR WOMEN
1000 16th St. NW, #700, Washington, DC 20036; (202) 331-0066
Seeks to end prejudice against women through advocacy and lobbying. More than 600 local chapters

NATIONAL WOMEN'S ECONOMIC ALLIANCE FOUNDATION
1440 New York Ave. NW, Ste. 300, Washington, DC 20005; (202) 393-5257
Serves to place more senior-level women on corporate boards and to promote dialogue

NATIONAL WOMEN'S LAW CENTER
1616 P St. NW, Washington, DC 20036; (202) 328-5169
To influence public and private sector policies and practices to better reflect the needs and rights of women and their families

NATIONAL WOMEN'S PARTY
144 Constitution Ave. NE, Washington, DC 20002; (202) 546-1210
Works for passage of the Equal Rights Amendment

9 TO 5 NATIONAL ASSOCIATION OF WORKING WOMEN
238 W. Wisconsin Ave., Milwaukee, WI 53202; (414) 274-0925
Works on issues of job discrimination, sexual harassment, and work-related health problems

OLDER WOMEN'S LEAGUE
666 11th St. NW, Washington, DC 20001; (202) 783-6686 or (800) 825-3695
Advocacy group for midlife and older women. Focuses on pension, health, and job discrimination. 120 chapters nationwide

WIDER OPPORTUNITIES FOR WOMEN
NATIONAL COMMISSION ON WORKING WOMEN
815 15th St. NW, Washington, DC 20005; (202) 638-3143
Works on sexual harassment and access to nontraditional jobs

WOMEN EMPLOYED
22 W. Monroe St., Chicago, IL 60603; (312) 782-3902
Works on career development, access to better jobs for low-income women, and counseling employees on equal opportunity and prevention of sexual harassment

WOMEN'S ECONOMIC AGENDA PROJECT
518 17th St., Oakland, CA 94612; (510) 451-7379
Works to achieve economic justice for low-income women. Helps groups organize and seeks to empower individuals

WOMEN'S LEGAL DEFENSE FUND
1875 Connecticut Ave. NW, Washington, DC 20009; (202) 986-2600
Advocacy, lobbying, public education, and litigation on precedent-setting issues such as family leave, sexual harassment, pay equity, and pregnancy discrimination

WOMEN WORK! THE NATIONAL NETWORK FOR WOMEN'S EMPLOYMENT
1625 K St. NW, Washington, DC 20006; (202) 467-6346
Formerly Displaced Homemakers Network. Helps women return to workplace and provides training and technical assistance to job-training programs for women in transition

APPENDIX FOUR

Governors' Names, Addresses, and Phone Numbers

Alabama Fob James, Jr. (R), State Capitol, 600 Dexter Ave., Montgomery, AL 36104; (334) 242-7100

Alaska Tony Knowles (D), P.O. Box 110001, Juneau, AK 99811-0001; (907) 465-3500

Arizona Fife Symington (R), State Capitol, 1700 W. Washington, Phoenix, AZ 85007; (602) 542-4331

Arkansas Mike Huckabee (R), 250 State Capitol, Little Rock, AR 72201; (501) 682-2345

California Pete Wilson (R), State Capitol, Sacramento, CA 95814; (916) 445-2864

Colorado Roy Romer (D), 136 State Capitol, Denver, CO 80203-1792; (303) 866-2471

Connecticut John G. Rowland (R), 210 Capitol Ave., Hartford, CT 06106; (203) 566-4840

Delaware Tom Carper (D), Tatnall Building, William Penn St., Dover, DE 19901; (302) 739-4101

Florida Lawton Chiles (D), The Capitol, Tallahassee, FL 32399-0001; (904) 488-2272

Georgia Zell Miller (D), 203 State Capitol, Atlanta, GA 30334; (404) 656-1776

Hawaii Benjamin J. Cayetano (D), State Capitol, 235 S. Beretania St., Honolulu, HI 96813; (808) 586-0034

Idaho Philip E. Batt (R), State House, Boise, ID 83720-0034; (208) 334-2100

Illinois Jim Edgar (R), State Capitol, Room 207, Springfield, IL 62706; (217) 782-6830

Indiana Frank O'Bannon (D), State House, Room 206, Indianapolis, IN 46204; (317) 232-4567

Iowa Terry E. Branstad (R), State Capitol, Des Moines, IA 50319-0001; (515) 281-5211

Kansas Bill Graves (R), State Capitol, Second Floor, Topeka, KS 66612-1590; (913) 296-3232

Kentucky Paul Patton (D), State Capitol, 700 Capitol St., Frankfort, KY 40601; (502) 564-2611

Louisiana Mike Foster (R), P.O. Box 94004, Baton Rouge, LA 70804-9004; (504) 342-7015

Maine Angus S. King Jr. (I), State House, Station 1, Augusta, ME 04333; (207) 287-3531

Maryland Parris N. Glendening (D), State House, 100 State Circle, Annapolis, MD 21401; (410) 974-3901

Massachusetts William F. Weld (R), State House, Room 360, Boston, MA 02133; (617) 727-9173

Michigan John Engler (R), P.O. Box 30013, Lansing, MI 48909; (517) 373-3400

Minnesota Arne H. Carlson (R), 130 State Capitol, 75 Constitution Avenue, St. Paul, MN 55155; (612) 296-3391

Mississippi Kirk Fordice (R), P.O. Box 139, Jackson, MS 39205; (601) 359-3150

Missouri Mel Carnahan (D), P.O. Box 720, Jefferson City, MO 65102; (573) 751-3222

Montana Marc Racicot (R), Governor's Office, State Capitol, Helena, MT 59620-0801; (406) 444-3111

Nebraska E. Benjamin Nelson (D), P.O. Box 94848, Lincoln, NB 68509-4848; (402) 471-2244

Nevada Bob Miller (D), State Capitol, Carson City, NV 89710; (702) 687-5670

New Hampshire Jeanne Shaheen (D), Office of the Governor, State House, Room 208, Concord, NH 03301; (603) 271-2121

New Jersey Christine T. Whitman (R), 125 W. State St., CN-001, Trenton, NJ 08625; (609) 292-6000

New Mexico Gary E. Johnson (R), Office of the Governor, State Capitol, Suite 400, Santa Fe, NM 87503; (505) 827-3000

New York George E. Pataki (R), State Capitol, 138 Eagle St., Albany, NY 12224; (518) 474-7516

North Carolina James B. Hunt Jr. (D), State Capitol, 116 West Jones St., Raleigh, NC 27603-8001; (919) 733-4240

North Dakota Edward T. Schafer (R), 600 E. Boulevard Ave., Bismark, ND 58505-0001; (701) 328-2200

Ohio George V. Voinovich (R), 77 South High Street, 30th Floor, Columbus, OH 43266-0601; (614) 466-3555

Oklahoma Frank Keating (R), Suite 212, State Capitol Building, Oklahoma City, OK 73105; (405) 521-2342

Oregon John Kitzhaber (D), 254 State Capitol, Salem, OR 97310; (503) 378-3111

Pennsylvania Tom Ridge (R), Room 225, Main Capitol Building, Harrisburg, Pa 17120; (717) 787-2500

Rhode Island Lincoln Almond (R), 222 State House, Providence, RI 02903; (401) 277-2080

South Carolina David M. Beasley (R), P.O. Box 11369, Columbia, SC 29211; (803) 734-9818

South Dakota William J. Janklow (R), 500 E. Capitol, Pierre, SD 57501; (605) 773-3212

Tennessee Don Sundquist (R), State Capitol, First Floor, Nashville, TN 37423-001; (615) 741-2001

Texas George W. Bush (R), P.O. Box 12428, Austin, TX 78711; (512) 463-2000

Utah Michael O. Leavitt (R), State Capitol, Suite 210, Salt Lake City, UT 84114; (801) 538-1500

Vermont Howard Dean, M.D. (D), Pavillion Office Building, 109 State St., Montpelier, VT 05609; (802) 828-3333

Virginia George Allen (R), State Capitol, Richmond, VA 23219; (804) 786-2211

Washington Gary Locke (D), P.O. Box 40002, Legislative Building, Olympia, WA 98504-0002; (360) 753-6780

West Virginia Cecil Underwood (R), State Capitol Complex, Charleston, WV 25305-0370; (304) 558-2000

Wisconsin Tommy G. Thompson (R), State Capitol, P.O. Box 7863, Madison, WI 53707; (608) 266-1212

Wyoming Jim Geringer (R), Office of the Governor, State Capitol Building, Room 124, Cheyenne, WY 82002; (307) 777-7434

About the Authors

Phyllis Katz is a well-known social psychologist and founder of the Institute for Research on Social Problems. She has been a professor at New York University and CUNY Graduate Center, was the founding editor of the journal *Sex Roles*, served as president of the Society for the Psychological Study of Social Issues, and currently edits the *Journal of Social Issues*. She was recently awarded the Carolyn Sherif prize from the American Psychological Association's Division on Women for her extensive scholarship in the areas of gender roles and intergroup attitude development.

Margaret Katz was an English major at Cornell and received her master's degree in journalism from the University of Colorado. She was on the staff of *Glamour Magazine*; has written for *The Denver Post*, the *Aspen Times*, and The Associated Press; and runs her own writing and editing consulting business. Mother and daughter, they both live in Boulder, Colorado.

Index

When information is found in tables, this is indicated by *t*.

Abbott Laboratories, FEM ratings, 61*t*, 67*t*, 87*t*, 88
  category ratings of companies, 369*t*
  product names by category, 301*t*, 304*t*, 307–311*t*, 325*t*
ABC. *See* Capital Cities/ABC
Abortion
  China, 220
  Ecuador, vi
  Idaho anti-choice bill, 10
Acacia Group, 114
Adidas USA, FEM ratings, 84*t*
Adidas USA, FEM ratings (*cont.*)
  category ratings of companies, 369*t*
  product names by category, 323*t*
Adolph Coors, FEM ratings, 49*t*
  category ratings of companies, 369*t*
  product names by category, 295*t*
  zero percent club, 51*t*
Advance Publications, FEM ratings, 124*t*, 125*t*
  category ratings of companies, 369*t*
  product names by category, 336–337*t*
Advertising, 58–59, 276 n.9

Aetna Life, FEM ratings, 115*t*, 117
category ratings of companies, 369*t*
product names by category, 332*t*
Affirmative Action, 27
AFL-CIO, 382
Africa, FEM ratings, 239*t*, 239–243; *see also specific countries*
African American women, wage gap, 4
Airborne Express
category ratings of companies, 369*t*
FEM ratings, 159*t*, 160, 344*t*
zero percent club, 160*t*
Air fresheners, product names by category, 312*t*
Airlines, FEM ratings, 138–141, 139*t*, 339*t*
zero percent club, 141*t*
See also specific airlines
Alabama, FEM ratings, 173*t*, 181, 193, 215*t*, 218
Alamo Rent A Car, FEM ratings, 145, 145*t*, 340*t*
category ratings of companies, 369*t*
Alaska, FEM ratings, 172*t*, 203–204, 215*t*
Alaska Airlines, FEM ratings, 139*t*, 141, 339*t*
category ratings of companies, 369*t*
Alberto-Culver, FEM ratings, 39*t*, 45, 60*t*, 71, 71*t*, 74
category ratings of companies, 369*t*
product names by category, 282*t*, 287*t*, 293*t*, 300*t*, 315*t*
Albertson's, FEM ratings, 54*t*
category ratings of companies, 369*t*
product names by category, 296*t*
Alcoholic beverage industry, FEM ratings, 49*t*, 51
product names by category, 295–296*t*
Alco Standard, FEM ratings, 154*t*
category ratings of companies, 369*t*
product names by category, 343*t*
Algeria, violence against women, 219–220
Allied Domecq
FEM ratings, 41*t*, 48–49*t*, 118*t*, 119
category ratings of companies, 369*t*
Allied Domecq (*cont.*)
FEM ratings (*cont.*)
product names by category, 286*t*, 294–296*t*
restaurants, rating by category, 333*t*
zero percent club, 46*t*, 51*t*, 120*t*
Allstate (Sears)
FEM ratings, 115*t*
category ratings of companies, 369*t*
product names by category, 332*t*
zero percent club, 117*t*
America, FEM ratings: *See* North and Central America, FEM ratings; South America, FEM ratings
American Airlines
benefits, 13
FEM ratings, 138–139, 139*t*, 339*t*
category ratings of companies, 369*t*
American Association for Higher Education—Women's Caucus, 382
American Association of University Women, 382
American Brands, FEM ratings, 49*t*, 122*t*, 123, 155*t*, 156
category ratings of companies, 369*t*
product names by category, 296*t*, 333*t*, 343*t*
American Business Women's Association, 383
American Civil Liberties Union, Women's Rights Project, 383
American Council on Education, Office of Women in Higher Education, 383
American Cyanamid, FEM ratings
category ratings of companies, 369*t*
product names by category, 307–308*t*, 310–311*t*
American Express, FEM ratings, 110*t*, 111, 112*t*
category ratings of companies, 369*t*
product names by category, 330*t*, 331*t*
American Federation of State, County and Municipal Employees, Women's Rights Department, 383
American General
FEM ratings, 113*t*, 116*t*, 117

American General (*cont.*)
FEM ratings (*cont.*)
category ratings of companies, 369*t*
product names by category, 331*t*
zero percent club, 114*t*, 117*t*
American Home Products, FEM ratings, 40*t*, 61*t*, 67*t*, 87*t*
category ratings of companies, 369*t*
product names by category, 284*t*, 285*t*, 287*t*, 289*t*, 292*t*, 300–301*t*, 303–305*t*, 307*t*, 309*t*, 311*t*, 325*t*
American International Group, FEM ratings, 113*t*, 116*t*, 117
category ratings of companies, 369*t*
product names by category, 331*t*, 332*t*
American Public Health Association—Women's Caucus, 383
American Society for Training and Development—Women's Network, 383
American Stores, FEM ratings, 54*t*, 77*t*, 78, 319*t*
category ratings of companies, 369*t*
product names by category, 297*t*
American Women's Economic Development Corporation, 383
America West Airlines, FEM ratings, 139*t*, 140, 339*t*
category ratings of companies, 369*t*
Amnesty International USA—Women and Human Rights Project, 383
Amoco, FEM ratings, 101*t*
category ratings of companies, 369*t*
product names by category, 328*t*
Anderson Hickey, FEM ratings, 157*t*, 158, 343*t*
category ratings of companies, 369*t*
Andrus, Governor Cecil, 10
Anheuser-Busch, FEM ratings, 40*t*, 49*t*, 50–51
category ratings of companies, 369*t*
product names by category, 282*t*, 287*t*, 292*t*, 295*t*, 339*t*
Ann Taylor, FEM ratings, 80*t*, 81, 82
category ratings of companies, 369*t*
product names by category, 320*t*
Annual reports of companies, 347
Antal, A. B., 222
Antibiotics, product names by category, 308*t*
Apple Computer, FEM ratings, 147*t*, 149, 150
category ratings of companies, 369*t*
product names by category, 341*t*
Appliance industry, FEM ratings, 103–105*t*, 103–106
product names by category, 328–329*t*
zero percent club, 106*t*
See also specific companies
Archer-Daniels-Midland
FEM ratings, 41*t*
category ratings of companies, 369*t*
product names by category, 290*t*
zero percent club, 46*t*
Argentina, FEM ratings, 224*t*, 266
climate toward women, 271*t*
educational and economic opportunity, 271*t*
family rights and roles, 269*t*
health status of women, 269*t*
legal rights for women, 267*t*
Arizona, FEM ratings, 172*t*, 204, 215*t*
Arkansas, FEM ratings, 173*t*, 181–182, 215*t*, 218
Armstrong World Industries, FEM ratings, 73*t*
category ratings of companies, 370*t*
product names by category, 314*t*
Arthritis, product names by category, 308*t*
Arvin Industries
FEM ratings, 100*t*
category ratings of companies, 370*t*
product names by category, 328*t*
zero percent club, 100*t*
Asthma/allergies, product names by category, 308*t*
A.T. Cross
FEM ratings, 154*t*
category ratings of companies, 369*t*
product names by category, 343*t*
zero percent club, 157*t*
Athletic shoes, product names by category, 323*t*
Atlantic Richfield, FEM ratings, 101*t*, 101–102

Atlantic Richfield (*cont.*)
category ratings of companies, 370*t*
product names by category, 328*t*
AT&T, FEM ratings, 107*t*, 108, 120*t*, 121, 146*t*, 149
category ratings of companies, 370*t*
product names by category, 329*t*, 334*t*, 341*t*
Australia
FEM ratings, 224*t*, 225, 225–226, 231, 232, 258, 259
climate toward women, 271*t*
educational and economic opportunity, 270*t*
family rights and roles, 268*t*
health status of women, 269*t*
legal rights for women, 267*t*
spending power of women, 7
Automobile companies, FEM ratings, 96–99, 97*t*
zero percent club, 99*t*
See also specific companies
Automobile suppliers, FEM ratings, 99–100*t*, 99–101
zero percent club, 100–101*t*
See also specific companies
Aveda, FEM ratings, 60*t*, 62, 63
category ratings of companies, 370*t*
product names by category, 300*t*
Avery Dennison, FEM ratings, 154*t*
category ratings of companies, 370*t*
product names by category, 343*t*
Avis
FEM ratings, 145, 145*t*, 340*t*
category ratings of companies, 370*t*
zero percent club, 146*t*
Avon Products, FEM ratings, 60*t*, 62
category ratings of companies, 370*t*
product names by category, 299*t*, 300*t*

Baby products industry, FEM ratings, 86–87*t*, 86–89
product names by categories, 324–327*t*
zero percent club, 89*t*
See also specific companies
Backlash (Faludi), 5
Baking products, product names by category, 282*t*
BankAmerica, FEM ratings, 110*t*, 111
category ratings of companies, 370*t*
product names by category, 330*t*
Bankers Trust New York, FEM ratings, 113*t*
category ratings of companies, 370*t*
product names by category, 331*t*
Bank of Boston, FEM ratings, 110*t*, 111
category ratings of companies, 370*t*
product names by category, 330*t*
Bank of New York, FEM ratings, 111*t*, 112
category ratings of companies, 370*t*
product names by category, 330*t*
Banks, FEM ratings, 110–111*t*, 110–112
product names by categories, 330–331*t*
zero percent club, 112*t*, 137
See also specific banks
Barad, Jill, 91
Barbecue, product names by category, 312*t*
Bar soaps, product names by category, 297*t*
Bathing products/wipes, babies
product names by categories, 324*t*
Batteries, product names by category, 313*t*
Bayer
FEM ratings, 67*t*, 72*t*
category ratings of companies, 370*t*
product names by category, 303–308*t*, 313–314*t*
zero percent club, 70*t*, 75*t*
Bear Stearns, FEM ratings, 113*t*, 114
category ratings of companies, 370*t*
product names by category, 332*t*
Beer, product names by category, 295*t*
Ben & Jerry's Homemade, FEM ratings, 39*t*, 41–42
category ratings of companies, 370*t*
product names by category, 288*t*
Benefits benefiting women, corporate FEMs, 20–21, 36, 349–350, 351–352
category ratings of companies, 368

Benetton (USA), FEM ratings, 80*t*, 81, 82
  category ratings of companies, 370*t*
  product names by category, 320*t*
Bennoune, K. E., 220
Bertelsmann
  FEM ratings, 124*t*, 125*t*, 133*t*
    category ratings of companies, 370*t*
    product names by category, 334–336*t*, 338*t*
  zero percent club, 128*t*, 133*t*
Beverage industry, FEM ratings, 47–51
  alcoholic beverages, 49*t*
    product names by category, 295–296*t*
  nonalcoholic beverages, 48*t*
    product names by category, 293–295*t*
  zero percent club, 51*t*
  See also specific companies
Birkenstock Footprint Sandals, FEM ratings, 81, 84*t*, 84–85
  category ratings of companies, 370*t*
  product names by category, 323*t*
Birth control, product names by category, 303*t*, 309*t*
Black & Decker, FEM ratings, 104*t*, 122*t*
  category ratings of companies, 370*t*
  FEM ratings, 123
  product names by category, 329*t*, 333*t*
Black Enterprise magazine, 126
Bloomingdale's: *See* Federated Department Stores
Body Shop, FEM ratings, 60*t*, 62, 63
  category ratings of companies, 370*t*
  product names by category, 299*t*, 300*t*
Boise Cascade, FEM ratings, 154*t*, 156
  category ratings of companies, 370*t*
  product names by category, 343*t*
Book publishers: *See* Publishing industry, FEM ratings
Borden, FEM ratings, 40*t*, 72*t*
  category ratings of companies, 370*t*
  product names by category, 283*t*, 286*t*, 289*t*, 292*t*, 318*t*
Borland
  FEM ratings, 147*t*
    category ratings of companies, 370*t*
Borland (*cont.*)
  FEM ratings (*cont.*)
    product names by category, 341*t*
  zero percent club, 150*t*
Bottled water, product names by category, 293*t*
Brazil, FEM ratings, 224*t*, 262–263, 266
  climate toward women, 272*t*
  educational and economic opportunity, 270*t*
  family rights and roles, 268*t*
  health status of women, 269*t*
  legal rights for women, 267*t*
Bread, product names by category, 282*t*
Breakfast foods, product names by category, 283*t*
Brio
  FEM ratings, 89*t*, 89–91
    category ratings of companies, 370*t*
    product names by category, 326*t*
  zero percent club, 92*t*
Bristol-Meyers Squibb, FEM ratings, 60*t*, 67*t*, 70, 87*t*, 88
  category ratings of companies, 370*t*
  product names by category, 298–301*t*, 303–306*t*, 307*t*, 309–310*t*, 325*t*
British Petroleum, FEM ratings, 101*t*
  category ratings of companies, 370*t*
  product names by category, 328*t*
Broadcasting industry, FEM ratings, 128–131
  product names by category, 337–338*t*
  zero percent club, 131*t*
  See also specific companies; Television, FEM ratings
Broadway Stores, FEM ratings, 76*t*, 77, 78, 319*t*
  category ratings of companies, 370*t*
Brown Group, FEM ratings, 84*t*, 93*t*, 94
  category ratings of companies, 370*t*
  product names by category, 324*t*, 327*t*
Brunswick, FEM ratings, 122*t*, 122–123
  category ratings of companies, 370*t*
  product names by category, 333*t*
Budget Rent A Car
  FEM ratings, 145, 145*t*, 340*t*
    category ratings of companies, 370*t*
  zero percent club, 146*t*

Bug/pest killers, product names by category, 313*t*
Bush, President George, 167
Business and Professional Women, 385
Business spenders, women as, 137–160

Cadbury Schweppes, FEM ratings, 40*t*, 48*t*
  category ratings of companies, 370*t*
  product names by category, 282–284*t*, 294*t*, 295*t*
California
  affirmative action plans, 168
  FEM ratings, 171*t*, 204–205, 213–214, 215*t*, 217
Calvert Group, FEM ratings, 112*t*, 113
  category ratings of companies, 371*t*
  product names by category, 331*t*
Campbell Soup, FEM ratings, 39*t*, 48*t*
  category ratings of companies, 371*t*
  product names by category, 282–287*t*, 290–292*t*, 294*t*
Canada
  FEM ratings, 224*t*, 253–254, 254, 258, 259, 260
    climate toward women, 271*t*
    educational and economic opportunity, 270*t*
    family rights and roles, 268*t*
    health status of women, 269*t*
    legal rights for women, 267*t*
  spending power of women, 7
Cancer medication, product names by category, 309*t*
Candy and chewing gum, product names by category, 283–284*t*
Canned goods, product names by category, 284*t*
Canon USA
  FEM ratings, 107*t*, 109, 148*t*, 150, 152*t*, 153
    category ratings of companies, 371*t*
    product names by category, 330*t*, 341*t*, 342*t*
  zero percent club, 109*t*, 150*t*, 153*t*
Capital Cities/ABC
  FEM ratings, 124*t*, 125*t*, 126, 129*t*, 130
Capital Cities/ABC (*cont.*)
  FEM ratings (*cont.*)
    category ratings of companies, 371*t*
    product names by category, 336–338*t*
  zero percent club, 128*t*, 131*t*
Carter, President Jimmy, 167
Carter-Wallace
  FEM ratings, 59, 62*t*, 64–65, 68*t*, 70, 73*t*
    category ratings of companies, 371*t*
    product names by category, 298*t*, 301–303*t*, 305–306*t*, 311–312*t*, 317*t*
  zero percent club, 65*t*, 70*t*, 75*t*
Catalogs, product names by category, 321*t*
Catalyst, 277 (n.17), 347
  address, 384
  Award, American Airlines, 139*t*
Categories, generally
  product names by, 281–344*t*
  ranking, used in, 347–350
  See also specific category
CBS, FEM ratings, 129*t*, 338*t*
  category ratings of companies, 371*t*
Celestial Seasonings
  FEM ratings, 48*t*, 49–50
    category ratings of companies, 371*t*
    product names by category, 294*t*
  zero percent club, 51*t*
Center for American Women in Politics, 278 (n.10)
Center for Women Policy Studies, 385
Central America, FEM ratings: *See* North and Central America, FEM ratings
Charitable donations to women's causes, corporate FEMs, 19, 349, 351
Charles Schwab, FEM ratings, 113*t*
  category ratings of companies, 371*t*
  product names by category, 331*t*
Chase Bank, FEM ratings, 110, 110*t*
  category ratings of companies, 371*t*
  product names by category, 330*t*
Cheese/yogurt/dairy/margarine, product names by category, 284–285*t*
Chemical Bank: *See* Chase Bank
Chesapeake
  FEM ratings, 155*t*, 156
    category ratings of companies, 371*t*

Chesapeake (*cont.*)
FEM ratings (*cont.*)
product names by category, 343*t*
zero percent club, 157*t*
Chevron
FEM ratings, 101*t*, 102
category ratings of companies, 371*t*
product names by category, 328*t*
zero percent club, 103*t*
Children's clothes and shoes industry, FEM ratings, 92–93*t*, 92–94
product names by categories, 326–327*t*
baby products, 324*t*
zero percent club, 94*t*
See also specific companies
Children's Defense League, 171
Children's Television Workshop, FEM ratings, 89*t*, 90
category ratings of companies, 371*t*
product names by category, 326*t*
Chile, FEM ratings, 224*t*, 264, 265, 266
climate toward women, 271*t*
educational and economic opportunity, 270*t*
family rights and roles, 269*t*
health status of women, 269*t*
legal rights for women, 267*t*
China
abortion, 220
FEM ratings, 224*t*, 226–227, 231
climate toward women, 271*t*
educational and economic opportunity, 271*t*
family rights and roles, 268*t*
health status of women, 269*t*
legal rights for women, 267*t*
fertility, 280 (n.8)
Chiquita Brands, FEM ratings, 40*t*, 48*t*
category ratings of companies, 371*t*
product names by category, 285*t*, 289*t*, 291*t*, 292*t*, 294*t*
Chrysler, FEM ratings, 96–97, 97*t*, 145, 145*t*, 340*t*
category ratings of companies, 371*t*
product names by category, 327*t*
Chubb, FEM ratings, 115*t*, 117
Chubb (*cont.*)
category ratings of companies, 371*t*
product names by category, 332*t*
Church & Dwight
FEM ratings, 40*t*, 61*t*, 72*t*
category ratings of companies, 371*t*
product names by category, 282*t*, 298*t*, 302*t*, 312*t*, 314–315*t*, 317*t*
zero percent club, 66*t*, 75*t*
Cigarette companies: *See* Tobacco industry, FEM ratings
Cigna, FEM ratings, 115*t*, 116
category ratings of companies, 371*t*
product names by category, 332*t*
Cinemax: *See* Time Warner
Circle K, FEM ratings, 55
Circulation/blood, product names by category, 309*t*
Citgo Petroleum
FEM ratings, 101*t*, 102
category ratings of companies, 371*t*
product names by category, 328*t*
zero percent club, 103*t*
Citicorp, FEM ratings, 110*t*, 110–111
category ratings of companies, 371*t*
product names by category, 330*t*
Civil Rights Act of 1964, Title VII, 167
Civil rights for women
states, FEM ratings for, 24, 170
sources and measures by category, 354, 356
Cleansers, product names by category, 297–298*t*
Clearinghouse on Women's Issues, 384
Climate toward women
countries, FEM ratings for, 25–26, 223, 271–272*t*
sources and measures by category, 361, 363–364
states, FEM ratings for, 24, 170–171
sources and measures by category, 354–357
Clinton, Hillary, 182
Clorox, FEM ratings, 40*t*, 48*t*, 72*t*
category ratings of companies, 371*t*
product names by category, 286*t*, 291*t*, 293*t*, 312–315*t*, 317*t*

Clothing industry, FEM ratings, 79–83, 80–81*t*
product names by categories, 320–322*t*
zero percent club, 83*t*
See also Children's clothes and shoes industry, FEM ratings; *specific companies*
CNN: *See* Turner Broadcasting
Coca-Cola, FEM ratings, 48*t*, 50
category ratings of companies, 371*t*
product names by category, 293–295*t*
Cold/allergy medicines, product names by category, 303*t*
Colgate-Palmolive, FEM ratings, 61*t*, 67*t*, 72*t*, 74, 87*t*
category ratings of companies, 371*t*
product names by category, 297–298*t*, 301–302*t*, 304*t*, 312–314*t*, 315–316*t*, 324*t*
Colorado
anti-gay rights amendment (1992), 164
Denver Metro Convention & Visitor Bureau, 164
FEM ratings, 172*t*, 205–206, 210, 215*t*, 217
violence against women, 219–220
Comcast, FEM ratings, 129*t*
category ratings of companies, 371*t*
product names by category, 338*t*
Committee of 200, 384
Communications Consortium, 384
Community property, rating for states, 356
Companies, generally
category ratings of, 367–368, 369–381*t*
information gathering, 345–347
See also specific companies
Compaq Computer
FEM ratings, 147*t*
category ratings of companies, 371*t*
product names by category, 341*t*
zero percent club, 150*t*
Computer Associates
FEM ratings, 147*t*
category ratings of companies, 371*t*
product names by category, 341*t*
Computer Associates (*cont.*)
zero percent club, 150*t*
Computers, FEM ratings, 146–148*t*, 146–151
product names by category, 340*t*
zero percent club, 150–151*t*
See also specific companies
ConAgra, FEM ratings, 40*t*, 45, 48*t*, 76*t*, 320*t*
category ratings of companies, 371*t*
product names by category, 285–292*t*, 294*t*, 317*t*
Condé Nast, FEM ratings, 125*t*, 127–128
category ratings of companies, 372*t*
product names by category, 334–335*t*
Condé Nast Traveler, 141
Condiments, product names by category, 285–286*t*
Connecticut, FEM ratings, 168, 172*t*, 174, 180, 215*t*, 217
Consolidated Papers
FEM ratings, 154*t*, 156
category ratings of companies, 372*t*
product names by category, 343*t*
zero percent club, 157*t*
Consumerism, women and
corporate FEMs, 16–23
countries, FEM ratings for, 25–26
determination of spending decisions, 8–9
pocketbook prose, 7, 10–13
scoring guide, 15–28
spending power of women, 3–14
states, FEM ratings for, 23–24
Consumer Reports, 8
Continental Airlines, FEM ratings, 139*t*, 339*t*
category ratings of companies, 372*t*
Continental (Insurance), FEM ratings, 115*t*
category ratings of companies, 372*t*
product names by category, 332*t*
Cookies/desserts, product names by category, 286*t*
Cookware, product names by category, 313*t*
Cooper Industries, FEM ratings, 99*t*
category ratings of companies, 372*t*
product names by category, 327*t*

Cooper Tire & Rubber
FEM ratings, 99*t*, 100
category ratings of companies, 372*t*
product names by category, 327*t*
zero percent club, 101*t*
Coors: *See* Adolph Coors
Corning, FEM ratings, 71*t*
category ratings of companies, 372*t*
product names by category, 313*t*
Corporate FEMs, 16–22
benefits benefiting women, 20–21, 36, 349–352
category ratings of companies, 368
charitable donations to women's causes, 19, 349, 351
derivation of ratings, 22–23, 35–38
discrimination in employment, 22
Femlin Factor, 22–23, 36, 276 (n.8), 350, 352
category ratings of companies, 368
glass ceiling effect, 18–19, 35, 350
category ratings of companies, 368
management opportunities, 17–19, 35, 275 (n.1), 348, 350
category ratings of companies, 367
sexual harassment, 21–22
support of women, 19, 35–36, 349, 351
category ratings of companies, 368
Cosmair, FEM ratings, 60*t*, 62, 64
category ratings of companies, 372*t*
product names by category, 297*t*, 299*t*, 300*t*
Cosmetics and personal hygiene industry, FEM ratings, 59–66, 60–62*t*
product names by category, 297–302*t*
zero percent club, 65–66*t*
See also specific companies
Cough drops/lozenges, product names by category, 303*t*
Council of Economic Priorities, 346
Corporate Conscience Award, 10
Digital Equipment, recognition of, 150
Shopping for a Better World, 34
Countries, FEM ratings, 219–273, 224*t*
climate toward women, 25–26, 223, 271–272*t*
Countries, FEM ratings (*cont.*)
climate toward women (*cont.*)
sources and measures by category, 361, 363–364
differences, 221
economic status of women, 222–223, 270–271*t*
sources and measures by category, 360, 363
educational climate for women, 222–223, 270–271*t*
sources and measures by category, 360, 363
family laws affecting women, 25, 222, 268–269*t*
sources and measures by category, 359, 362–363
governance, women in, 25
health status of women, 222, 269–270*t*
sources and measures by category, 360, 363
legal rights for women, 25, 222, 267–268*t*
sources and measures by category, 359, 361–362
sources and measures by category, 358–366
See also specific countries
CPC
FEM ratings, 40*t*
category ratings of companies, 372*t*
product names by category, 282–283*t*, 285–286*t*, 290*t*, 292*t*
zero percent club, 46*t*
Cubin, Representative Barbara, 213

Dairy Queen: *See* International Dairy Queen
Dannon, 44
Danskin, FEM ratings, 80*t*
category ratings of companies, 372*t*
product names by category, 321–322*t*
Dayton Hudson, FEM ratings, 76*t*, 76–77, 318–319*t*
category ratings of companies, 372*t*
DeanWitter, Discover, FEM ratings, 113*t*
category ratings of companies, 372*t*
product names by category, 331*t*

Delaware, FEM ratings, 169, 172*t*, 182–183, 215*t*
Dell Computer, FEM ratings, 147*t*
  category ratings of companies, 372*t*
  product names by category, 341*t*
Delta Air Lines
  FEM ratings, 139*t*, 141, 339*t*
    category ratings of companies, 372*t*
  zero percent club, 141*t*
Denmark, FEM ratings, 224*t*, 244–245, 252, 258, 259
  climate toward women, 271*t*
  educational and economic opportunity, 270*t*
  family rights and roles, 268*t*
  health status of women, 269*t*
  legal rights for women, 267*t*
Deodorants/antiperspirants, product names by category, 298*t*
Department of Labor
  Affirmative Action filings, 27
  awards by, 36, 64
  Digital Equipment, recognition of, 150
  discrimination in employment, 22
  Eve Award, 47
  Glass Ceiling Commission, 62
  IBM, awards to, 148
  managers, survey on incomes of, 32
  noncompliance with, 352
  Westin Hotels and, 144
Department stores, FEM ratings, 318–319*t*
Depression, product names by category, 310*t*
DHL Worldwide Express, FEM ratings, 159*t*, 160, 344*t*
  category ratings of companies, 372*t*
Diabetes, product names by category, 310*t*
Dial, FEM ratings, 41*t*, 61*t*, 64, 72*t*
  category ratings of companies, 372*t*
  product names by category, 289*t*, 291*t*, 297–298*t*, 301*t*, 312*t*, 314–315*t*
Diapers, product names by category, 325*t*
Diet/digestion/ulcers, product names by category, 310*t*
Diet foods, product names by category, 287*t*
Digital Equipment, FEM ratings, 147*t*, 149, 150
  category ratings of companies, 372*t*
  product names by category, 341*t*
Discount stores, FEM ratings, 319*t*
Discover: *See* DeanWitter, Discover
Discrimination in employment
  corporate FEMs and, 22
  United Nations Statement on the Non-Discriminatory Treatment of Women, 26
Dish soap/sponges, product names by category, 313–314*t*
Disney: *See* Walt Disney
Dole Food, FEM ratings, 40*t*, 48*t*
  category ratings of companies, 372*t*
  product names by category, 284*t*, 294*t*
Domestic Violence Institute, 384
Dominican Republic, FEM ratings, 224*t*, 254–255, 260
  climate toward women, 272*t*
  educational and economic opportunity, 270*t*
  family rights and roles, 268*t*
  health status of women, 270*t*
  legal rights for women, 267*t*
Domini Social Index
  described, 352
  Digital Equipment, recognition of, 150
  discrimination in employment, 22
  Gannett Company, mentioned in, 126
  Herman Miller, recognition of, 158
  inclusion in, 36, 64
  Stride-Rite, recognition of, 85
Domino's Pizza, FEM ratings, 118*t*, 119
  category ratings of companies, 333*t*, 372*t*
Dow Chemical, FEM ratings, 61*t*, 67*t*, 72*t*
  category ratings of companies, 372*t*
  product names by category, 299*t*, 301–302*t*, 304*t*, 306–307*t*, 314–315*t*, 317*t*
Dow Jones, FEM ratings, 125*t*
  category ratings of companies, 372*t*
  product names by category, 335*t*, 337*t*

Dr. Pepper/7Up
FEM ratings, 48*t*, 50
category ratings of companies, 372*t*
zero percent club, 51*t*
Dreyer's Grand Ice Cream
FEM ratings, 41*t*, 42, 45
category ratings of companies, 372*t*
product names by category, 288*t*
zero percent club, 46*t*
Dreyfus Service, FEM ratings, 113*t*
category ratings of companies, 372*t*
product names by category, 331*t*
Drug companies, FEM ratings, 66–68*t*, 66–70
zero percent club, 70*t*
See also Over-the-counter drugs, product names by category; Prescription medications, product names by category; *specific companies*
Drugstores, FEM ratings, 319*t*
DuPont: *See also* E.I. DuPont
Dynamics
FEM ratings, 104*t*, 106
category ratings of companies, 369*t*
product names by category, 329*t*
zero percent club, 106*t*

Earnings: *See* Median earnings
Eastman Kodak, FEM ratings, 60*t*, 71*t*, 87*t*, 88, 152*t*, 152–153
category ratings of companies, 372*t*
product names by category, 298–299*t*, 313–314*t*, 316*t*, 324*t*, 342*t*
Economic climate for women, 3–8
countries, FEM ratings for, 222–223, 270–271*t*
sources and measures by category, 360, 363
states, FEM ratings for, 24, 170
sources and measures by category, 353, 355
Ecuador
FEM ratings, 224*t*, 264–265, 266
climate toward women, 272*t*
educational and economic opportunity, 271*t*
family rights and roles, 269*t*
Ecuador (*cont.*)
FEM ratings (*cont.*)
health status of women, 270*t*
legal rights for women, 267*t*
treatment of women, generally, v–vi
Educational attainment
earnings, gender differences, 33
Educational climate for women
countries, FEM ratings for, 222–223, 270–271*t*
sources and measures by category, 360, 363
Educational Insights
FEM ratings, 90*t*
category ratings of companies, 372*t*
product names by category, 326*t*
zero percent club, 92*t*
EEOC: *See* Equal Employment Opportunity Commission (EEOC)
Egypt, FEM ratings, 224*t*, 233, 238–239
climate toward women, 272*t*
educational and economic opportunity, 271*t*
family rights and roles, 269*t*
health status of women, 270*t*
legal rights for women, 267*t*
E.I. DuPont, FEM ratings, 101*t*, 102
category ratings of companies, 372*t*
product names by category, 328*t*
Electrolux
FEM ratings, 104*t*
category ratings of companies, 372*t*
product names by category, 329*t*
zero percent club, 106*t*
Electronic equipment industry, FEM ratings, 107–108*t*, 107–109
product names by category, 329–330*t*
zero percent club, 109*t*
See also specific companies
Electronic office equipment
FEM ratings, 151–152*t*, 151–153
product names by category, 342*t*
zero percent club, 153*t*
See also Electronic equipment industry, FEM ratings; *specific companies*
Eli Lilly, FEM ratings, 68*t*, 70
category ratings of companies, 372*t*

Eli Lilly, FEM ratings (*cont.*)
product names by category, 308*t*, 310–311*t*
Emerson Electric, FEM ratings, 104*t*, 106, 107*t*, 109
category ratings of companies, 372*t*
product names by category, 329*t*, 330*t*
Emily's List, 384
Employer, state as, 24, 170
sources and measures by category, 353, 355
Employment discrimination: *See* Discrimination in employment
Entertainment industry, FEM ratings, 117–135, 333–339*t*
broadcasting, 128–131
publishing industry, 123–128
product names by category, 334–337
record labels, product names by category, 338*t*
restaurants, 118–120
category ratings of companies, 333*t*
sports equipment, 121–123
product names by category, 333–334*t*
systems operators, product names by category, 338*t*
telephone services, 120–121
product names by category, 334*t*
television, product names by category
cable television, 337*t*
network TV, 338*t*
radio, 338*t*
theme parks, product names by category, 339*t*
video stores, 339*t*
zero percent club, 133*t*
Equal Employment Opportunity Commission (EEOC), 167
Affirmative Action filings, 27
airlines, compliance, 138
discrimination in employment and, 22
Eastman Kodak, statistics, 152
EEO-1 form, 276 (n.1), 346
management opportunities, survey, 17, 18
Microsoft, noncompliance, 150
Equal Employment Opportunity Commission (EEOC) (*cont.*)
Pitney Bowes, statistics, 152
USAir Group, lawsuit, 140
Equal Pay Act of 1963, 167
Equal Rights Advocates, 385
Equal Rights Amendment, 166, 278 n.6
Equitable
FEM ratings, 115*t*
category ratings of companies, 372*t*
product names by category, 332*t*
zero percent club, 117*t*
Esprit de Corp, FEM ratings, 80*t*, 81, 82, 92*t*, 93
category ratings of companies, 373*t*
product names by category, 320*t*, 326*t*
Estée Lauder, FEM ratings, 60*t*, 62–63
category ratings of companies, 373*t*
product names by category, 297*t*, 299–300*t*
Europe, FEM ratings, 243–252, 244*t*; *see also specific countries*
E.W. Scripps
FEM ratings, 126*t*, 126–127, 129*t*, 131
category ratings of companies, 372*t*
product names by category, 337–338*t*
zero percent club, 128*t*, 131*t*
Executive Order 11246, 167
Export processing zones, 255
Exxon
FEM ratings, 101*t*
product names by category, 328*t*
zero percent club, 103*t*
Eye drops/contact lenses and solutions, product names by category, 304*t*

Faludi, Susan, 5
Family and Medical Leave Act
corporate FEMs and, 20, 36, 70, 130–131
state differences, 165
Family laws affecting women
countries, FEM ratings for, 25, 222, 268–269*t*
sources and measures by category, 359, 362–363
states, FEM ratings for, 24, 170

Family laws affecting women (*cont.*)
states, FEM ratings for (*cont.*)
sources and measures by category, 354, 356
See also Family and Medical Leave Act
Farberware, FEM ratings
category ratings of companies, 373*t*
FEM ratings, 73*t*
product names by category, 313*t*
Far East/Pacific countries, FEM ratings, 225*t*, 225–232; *see also specific countries*
Fedders
FEM ratings, 104*t*, 106
category ratings of companies, 373*t*
product names by category, 329*t*
zero percent club, 106*t*
Federal Express, FEM ratings, 159, 159*t*, 344*t*
category ratings of companies, 373*t*
Federated Department Stores, FEM ratings, 76*t*, 79, 319*t*
category ratings of companies, 373*t*
Federation of Organizations for Professional Women, 385
Feminist Evaluation Measure (FEM ratings): *See* FEM ratings (Feminist Evaluation Measure)
Feminist Evaluation Method for Measuring Equality (FEMME system), 11, 14
The Feminist Institute, Inc., 385
Femlin Factor, 22–23, 36
category ratings of companies, 368
defined, 276 (n.8)
description of category, 350
measures used in category, 352
FEMME (Feminist Evaluation Method for Measuring Equality) system, 11, 15
FEM ratings (Feminist Evaluation Measure)
category ratings of companies, 367–381
corporate FEMs, 16–22
countries, rating for, 25–26
sources and measures by category, 358–366
FEM ratings (Feminist Evaluation Measure) (*cont.*)
derivation of ratings, 22–23, 35–38
described, 11, 15
states, rating for, 23–24
sources and measures by category, 353–358
See also specific companies and industries
Fidelity Investments, FEM ratings, 113*t*, 114
category ratings of companies, 373*t*
product names by category, 331*t*
15th Amendment, 166
Film, product names by category, 314*t*
Finance industry, FEM ratings, 110–117
banks, 110–112
specific institution by category, 330–331*t*
financial institutions, 112–114
specific institution by category, 331–332*t*
Financial institutions, FEM ratings, 112–113*t*, 112–114
insurance companies, 114–117
specific institution by category, 332*t*
specific institution by category, 330–332*t*
zero percent club, 114*t*
See also Finance industry, FEM ratings; *specific institutions*
Financial Women's International, 385
Finland, FEM ratings, 224*t*, 245, 252, 258
climate toward women, 271*t*
educational and economic opportunity, 270*t*
family rights and roles, 268*t*
health status of women, 269*t*
legal rights for women, 267*t*
First aid, product names by category, 304*t*
First Brands
FEM ratings, 73*t*, 100, 100*t*
category ratings of companies, 373*t*
product names by category, 316–317*t*, 328*t*

First Brands (*cont.*)
zero percent club, 75*t*, 101*t*
First Fidelity BanCorp, FEM ratings, 110*t*, 111–112
category ratings of companies, 373*t*
product names by category, 330*t*
Fisher Price: *See* Mattel
Flagstar, FEM ratings, 118*t*
category ratings of companies, 373*t*
restaurants, rating by category, 333*t*
Florida, FEM ratings, 172*t*, 183, 193, 215*t*
FOIA: *See* Freedom of Information Act (FOIA)
Food industry, FEM ratings, 38–47, 39–41*t*
averages, 39–47
grocery stores, 38–39
product names by category, 282–293*t*
baby food, 325*t*
zero percent club, 46–47*t*
See also specific companies
Food Lion
boards of directors, 13
FEM ratings, 54*t*
category ratings of companies, 373*t*
product names by category, 297*t*
zero percent club, 56*t*
Footwear industry, FEM ratings, 83–86, 84*t*
product names by categories, 323–324*t*
zero percent club, 86*t*
See also Children's clothes and shoes industry, FEM ratings; *specific companies*
Ford Motor, FEM ratings, 96–99, 97*t*
category ratings of companies, 373*t*
product names by category, 327*t*
Foreign/specialty foods, product names by category, 287*t*
Fortune 500 Companies, 277 (n.17)
Foundation Center, 35, 346
The Foundation for Women's Resources, 385
Founders of companies, female, 352
14th Amendment, 165
Fragrances, product names by category, 299*t*
France, FEM ratings, 224*t*, 246, 259
climate toward women, 271*t*
educational and economic opportunity, 270*t*
family rights and roles, 268*t*
health status of women, 269*t*
legal rights for women, 267*t*
Freedom of Information Act (FOIA), 27
Frozen foods, product names by category, 287–288*t*
Fruit of the Loom
FEM ratings, 81*t*, 83, 93*t*, 94
category ratings of companies, 373*t*
product names by category, 321–322*t*, 327*t*
zero percent club, 83*t*, 94*t*
The Fund for the Feminist Majority, 385
Furniture, office: *See* Office furniture, FEM ratings
Furse, Representative Elizabeth, 210

Gandhi, Indira, 229
Gannett, FEM ratings, 125*t*, 126
category ratings of companies, 373*t*
product names by category, 336*t*
The Gap, FEM ratings, 80*t*, 82, 86, 87*t*, 88, 92*t*, 93
category ratings of companies, 373*t*
product names by category, 320*t*, 324*t*, 326*t*
Gasoline companies, FEM ratings, 101*t*, 101–103
product names by category, 328*t*
zero percent club, 103*t*
See also specific companies
GDP: *See* Gross domestic product (GDP)
GE: *See* General Electric
GEICO, FEM ratings, 116*t*
category ratings of companies, 373*t*
product names by category, 332*t*
Gender differences
management opportunities, 33
managers, incomes of, 32
median earnings, 33, 274 (n.1), 275 (n.6)
median income, 4
Gender discrimination, 4
Genentech, FEM ratings, 67*t*

Genentech (*cont.*)
category ratings of companies, 373*t*
product names by category, 309*t*, 311*t*
General Electric, FEM ratings, 71*t*, 74, 103*t*, 105, 106, 112*t*, 115*t*, 116–117, 129*t*, 130, 159
category ratings of companies, 373*t*
product names by category, 316*t*, 328*t*, 331–332*t*, 338*t*
General Mills, FEM ratings, 39*t*, 118*t*
category ratings of companies, 373*t*
product names by category, 282–284*t*, 287–288*t*, 290–292*t*
restaurants, rating by category, 333*t*
General Motors, FEM ratings, 97*t*, 97–99
category ratings of companies, 373*t*
product names by category, 327*t*
Genesco
FEM ratings, 84*t*, 85–86
category ratings of companies, 373*t*
product names by category, 323*t*
zero percent club, 86*t*
George, Phyllis, 31
Georgia, FEM ratings, 172*t*, 183–184, 215*t*
Georgia-Pacific, FEM ratings, 72*t*, 154*t*
category ratings of companies, 373*t*
product names by category, 316*t*, 318*t*, 343*t*
Gerber Products, FEM ratings, 87*t*, 88, 115*t*
category ratings of companies, 373*t*
product names by category, 324–325*t*, 332*t*
Germany
FEM ratings, 224*t*, 246–247, 251, 259
climate toward women, 271*t*
educational and economic opportunity, 270*t*
family rights and roles, 268*t*
health status of women, 269*t*
legal rights for women, 267*t*
spending power of women, 6
Giant Foods, FEM ratings, 54*t*, 55
category ratings of companies, 373*t*
product names by category, 296*t*
Gillette, FEM ratings, 60, 61*t*, 64–65, 72*t*, 104*t*, 154*t*
category ratings of companies, 373*t*
product names by category, 298–302*t*, 316*t*, 318*t*, 329*t*, 343*t*
Gitano Group
FEM ratings, 81*t*, 83
category ratings of companies, 373*t*
product names by category, 322*t*
zero percent club, 83*t*
Glass Ceiling Commission, 62
Glass ceiling effect, 18–19
category ratings of companies, 368
derivation of ratings, 35
described, 32, 348–349
measures used in category, 350
statistics, 12
Glaxo-Wellcome, FEM ratings, 67*t*
category ratings of companies, 373*t*
product names by category, 308*t*, 310*t*
The Global Fund for Women, 385
Globe Business Furniture, FEM ratings, 157*t*, 158, 343*t*
category ratings of companies, 373*t*
Goldman Sachs Group, FEM ratings, 113*t*, 114
category ratings of companies, 373*t*
product names by category, 331*t*
Goodyear Tire & Rubber, FEM ratings, 99*t*, 100
category ratings of companies, 373*t*
product names by category, 327*t*
Governance, women in
countries, FEM ratings for, 25
states, FEM ratings for, 24, 170
sources and measures by category, 354, 355–356
Governor's names, addresses and phone numbers, 392–395
Grand Metropolitan, FEM ratings, 40*t*, 42, 49*t*, 50–51, 76*t*, 118*t*, 119, 320*t*
category ratings of companies, 373*t*
product names by category, 282–284*t*, 287–289*t*, 295–296*t*
restaurants, rating by category, 333*t*
Grand Union
FEM ratings, 54*t*
category ratings of companies, 373*t*

Grand Union (*cont.*)
FEM ratings (*cont.*)
product names by category, 297*t*
zero percent club, 56*t*
Great Atlantic & Pacific Tea, FEM ratings, 54*t*, 55
category ratings of companies, 374*t*
product names by category, 296*t*
Great Western Financial, FEM ratings, 110*t*, 111
category ratings of companies, 374*t*
financial institutions by category, 330*t*
Greece, FEM ratings, 224*t*, 247–248
climate toward women, 271*t*
educational and economic opportunity, 270*t*
family rights and roles, 268*t*
health status of women, 269*t*
legal rights for women, 267*t*
Grocery stores
FEM ratings, 38–39, 53–56
product names by category, 296–297*t*
zero percent club, 56*t*
See also specific companies
Gross domestic product (GDP)
international comparison, 6*t*, 7
United States, 4–5
Guatemala, FEM ratings, 224*t*, 255, 257, 259, 260
climate toward women, 272*t*
educational and economic opportunity, 271*t*
family rights and roles, 269*t*
health status of women, 270*t*
legal rights for women, 267*t*
Guinness
FEM ratings, 49*t*
category ratings of companies, 374*t*
product names by category, 295*t*, 296*t*
zero percent club, 51*t*
Guns, legislation, 166

Hair care, product names by category, 299*t*
Hamilton Beach/Proctor Silex
FEM ratings, 103*t*, 105, 106
Hamilton Beach/Proctor Silex (*cont.*)
FEM ratings (*cont.*)
category ratings of companies, 374*t*
product names by category, 329*t*
zero percent club, 106*t*
Harcourt General
FEM ratings, 124*t*
category ratings of companies, 374*t*
product names by category, 336*t*
zero percent club, 128*t*
Hardware stores, FEM ratings, 319–320*t*
Hasbro, FEM ratings, 87*t*, 90*t*, 91–92
category ratings of companies, 374*t*
product names by category, 325–326*t*
Hawaii, FEM ratings, 172*t*, 207, 210, 214, 215*t*, 217
HBO: *See* Time Warner
Head Sports, FEM ratings, 122*t*, 123
category ratings of companies, 374*t*
product names by category, 333*t*
Health status of women
countries, FEM ratings for, 222, 269–270*t*
sources and measures by category, 360, 363
Health Valley Natural Foods
FEM ratings, 40*t*
category ratings of companies, 374*t*
product names by category, 288*t*, 292*t*
zero percent club, 46*t*
Hearst, FEM ratings, 124*t*, 125*t*, 126*t*, 129*t*, 131
category ratings of companies, 374*t*
product names by category, 334–337*t*
Heart medication, product names by category, 310–311*t*
Heide, Wilma Scott, 137
Heinz: *See* H.J. Heinz
Helene Curtis, FEM ratings, 61*t*
category ratings of companies, 374*t*
product names by category, 300*t*, 301*t*
Hemorrhoid treatments, product names by category, 304*t*
Herman Miller
FEM ratings, 157*t*, 158, 343*t*
category ratings of companies, 374*t*
zero percent club, 158*t*

Hershey Foods, FEM ratings, 39*t*, 41–42, 48*t*, 49–50, 67*t*, 68–69
category ratings of companies, 374*t*
Hertz
FEM ratings, 145*t*, 145–146, 340*t*
category ratings of companies, 374*t*
zero percent club, 146*t*
Hewlett-Packard
Council of Economic Priorities Corporate Conscience Award, 10
FEM ratings, 107*t*, 108–109, 147*t*, 149, 152*t*, 152–153
category ratings of companies, 374*t*
product names by category, 329*t*, 341*t*, 342*t*
Hill, Anita, 4
Hilton Hotels, FEM ratings, 142*t*, 144, 340*t*
category ratings of companies, 374*t*
Hispanic women, wage gap, 4
H.J. Heinz, FEM ratings, 40*t*, 87*t*
category ratings of companies, 374*t*
product names by category, 285*t*, 287–289*t*, 292*t*, 317*t*, 325*t*
Honda of America
FEM ratings, 96–98, 97*t*
category ratings of companies, 374*t*
product names by category, 327*t*
zero percent club, 99*t*
Hon, FEM ratings, 157*t*, 158, 343*t*
category ratings of companies, 374*t*
Horizon Air, 141
Hormel Foods
FEM ratings, 41*t*, 45
category ratings of companies, 374*t*
product names by category, 284*t*, 286–289*t*, 292*t*
zero percent club, 46*t*
Hormones, product names by category, 311*t*
Hosiery, product names by category, 321*t*
HFS
FEM ratings, 142*t*, 143–144, 340*t*
category ratings of companies, 374*t*
zero percent club, 144*t*
Host Marriott, FEM ratings, 142*t*, 340*t*
category ratings of companies, 374*t*
sources and measures by category, 143
Hot beverages, product names by category, 294*t*
Hotels and motels: *See* Lodging, FEM ratings
Houghton Mifflin, FEM ratings, 124*t*
category ratings of companies, 374*t*
product names by category, 336*t*
Household-based consumption, 95–134
Household/bathroom cleaners, product names by category, 314*t*
Household equipment, FEM ratings, 103–109
appliance industry, 103–106
product names by category, 328–329*t*
electronic equipment, 107–109
product names by category, 329–330*t*
product names by category, 328–330*t*
See also specific companies
Household goods companies, FEM ratings, 71–73*t*, 71–75
product names by categories, 312–318*t*
zero percent club, 75*t*
See also specific companies
Huffy, FEM ratings, 86, 87*t*, 90*t*, 122, 122*t*
category ratings of companies, 374*t*
product names by category, 324*t*, 326*t*, 333*t*
Human Rights in China, 220, 227
Hyatt Hotels and Resorts, FEM ratings, 142*t*, 340*t*
category ratings of companies, 374*t*

IBM, FEM ratings, 146, 146*t*, 148, 151*t*, 152
category ratings of companies, 374*t*
product names by category, 340*t*, 342*t*
IBP, FEM ratings, 41*t*
category ratings of companies, 374*t*
product names by category, 289*t*
Ice cream/whipped cream, product names by category, 288*t*
Iced/cold drinks, product names by category, 293*t*
Idaho, FEM ratings, 172*t*, 207–208, 215*t*

IDS Financial, FEM ratings, 113*t*
category ratings of companies, 374*t*
financial institutions by category, 331*t*
IHOP, FEM ratings, 118*t*, 119
category ratings of companies, 333*t*, 374*t*
Illinois, FEM ratings, 172*t*, 194–195, 202, 215*t*, 217
Imasco, FEM ratings, 52*t*, 118*t*, 120
category ratings of companies, 374*t*
product names by category, 296*t*
restaurants, rating by category, 333*t*
Income: *See* Median income
India, FEM ratings, 224*t*, 227–229, 259
climate toward women, 272*t*
educational and economic opportunity, 271*t*
family rights and roles, 268*t*
health status of women, 270*t*
legal rights for women, 267*t*
Indiana, FEM ratings, 173*t*, 195, 202, 215*t*, 217
Institute for Women's Policy Research, 385
Insurance companies, FEM ratings, 114–117, 115–116*t*, 137
product names by category, 332*t*
zero percent club, 75*t*
See also specific companies
Intel, FEM ratings, 147*t*
category ratings of companies, 374*t*
product names by category, 341*t*
International Center for Research on Women, 386
International comparisons: *See* Countries, FEM ratings
International Dairy Queen, FEM ratings, 118*t*, 120
category ratings of companies, 333*t*, 374*t*
International Paper, FEM ratings, 154*t*, 156
category ratings of companies, 375*t*
product names by category, 343*t*
International Planned Parenthood, 255
International Women's Health Coalition, 386
International Women's Tribune Center, 386
Investcorp
FEM ratings, 54*t*, 113*t*
category ratings of companies, 375*t*
product names by category, 297*t*, 331*t*
zero percent club, 56*t*, 114*t*
Iowa, FEM ratings, 172*t*, 195–196, 215*t*
Islamic countries
violence against women, 220
See also specific countries
Israel, FEM ratings, 224*t*, 234, 238
climate toward women, 272*t*
educational and economic opportunity, 271*t*
family rights and roles, 268*t*
health status of women, 269*t*
legal rights for women, 268*t*
Italy
FEM ratings, 224*t*, 248, 259
climate toward women, 271*t*
educational and economic opportunity, 270*t*
family rights and roles, 268*t*
health status of women, 269*t*
legal rights for women, 267*t*
spending power of women, 6
ITT, FEM ratings, 113*t*, 115*t*
category ratings of companies, 375*t*
product names by category, 331*t*, 332*t*
ITT Sheraton Hotels, FEM ratings, 142*t*, 144, 340*t*
category ratings of companies, 375*t*
Izraeli, D. N., 222

Jamaica, FEM ratings, 224*t*, 250, 256, 257
climate toward women, 271*t*
educational and economic opportunity, 270*t*
family rights and roles, 268*t*
health status of women, 269*t*
legal rights for women, 267*t*
James River of Virginia
FEM ratings, 72*t*, 154*t*, 156
category ratings of companies, 375*t*

James River of Virginia (*cont.*)
FEM ratings (*cont.*)
product names by category, 316*t*, 318*t*, 343*t*
zero percent club, 75*t*, 89*t*, 157*t*
Jams and jellies, product names by category, 288–289*t*
Janeway, Elizabeth, 15
Japan
automobile companies, 96
computer companies, 150
FEM ratings, 224*t*, 226, 229–230, 230–232
climate toward women, 271*t*
educational and economic opportunity, 271*t*
family rights and roles, 268*t*
health status of women, 269*t*
legal rights for women, 267*t*
spending power of women, 6
J.C. Penney, FEM ratings, 76*t*, 319*t*
category ratings of companies, 375*t*
Jeans, product names by category, 321*t*
John Hancock, FEM ratings, 115*t*
category ratings of companies, 375*t*
product names by category, 332*t*
John Paul Mitchell Systems, FEM ratings, 61*t*, 64
category ratings of companies, 375*t*
product names by category, 301*t*
Johnson, President Lyndon B., 167
Johnson & Johnson, FEM ratings, 59, 60*t*, 62, 63, 66*t*, 68, 86*t*, 87
category ratings of companies, 375*t*
product names by category, 297–298*t*, 300–312*t*, 324–325*t*
Johnson Products, FEM ratings, 60*t*, 64
category ratings of companies, 375*t*
product names by category, 298*t*, 300*t*
Johnson Publishing, FEM ratings, 61*t*, 64, 124*t*, 125*t*
category ratings of companies, 375*t*
product names by category, 300–301*t*, 334*t*, 336*t*
Johnson & Son: *See* S.C. Johnson & Son
Juice/juice drinks, product names by category, 294*t*

Kansas, FEM ratings, 172*t*, 196–197, 202, 215*t*, 217
Keebler: *See* United Biscuit Holdings, FEM ratings
Kellogg, FEM ratings, 40*t*
category ratings of companies, 375*t*
product names by category, 283*t*, 285*t*
Kennedy, Florynce, 31
Kentucky
FEM ratings, 173*t*, 184, 215*t*
voting rights of women, history, 166
Kenya, FEM ratings, 224*t*, 240–241, 243, 259, 273
climate toward women, 272*t*
educational and economic opportunity, 271*t*
family rights and roles, 269*t*
health status of women, 270*t*
legal rights for women, 267*t*
Kimberly-Clark, FEM ratings, 61*t*, 72*t*, 86, 87*t*, 139*t*, 154*t*, 339*t*
category ratings of companies, 375*t*
product names by category, 302*t*, 316*t*, 318*t*, 324–325*t*, 343*t*
Kinder, Lydenberg, Domini & Co. (KLD), 346–347
Herman Miller, recognition of, 158
IBM, recognition of, 148
rating by, described, 352
KMart, FEM ratings, 76*t*, 78, 154*t*, 319–320*t*
category ratings of companies, 375*t*
product names by category, 343*t*
Knight-Ridder, FEM ratings, 125*t*
category ratings of companies, 375*t*
product names by category, 336*t*
Kohoni, N. K., 221
Krami, Shuida, 220
Kroger
boards of directors, 13
FEM ratings, 53–55, 54*t*
category ratings of companies, 375*t*
product names by category, 296*t*

L.A. Gear, FEM ratings, 84*t*, 93*t*
category ratings of companies, 375*t*
product names by category, 323*t*, 327*t*

Labor, Department of: *See* Department of Labor
Lands' End
FEM ratings, 81*t*, 83, 93*t*, 94
category ratings of companies, 375*t*
product names by category, 321*t*, 327*t*
zero percent club, 83*t*, 94*t*
Lang Communications, FEM ratings, 124*t*, 127
category ratings of companies, 375*t*
product names by category, 334–335*t*
Lanz, FEM ratings, 81*t*
category ratings of companies, 375*t*
Laura Ashley, FEM ratings, 81*t*, 83
category ratings of companies, 375*t*
product names by category, 321*t*
Lawn/garden supplies, product names by category, 315*t*
League of Women Voters, 386
Legal rights for women
countries, FEM ratings for, 25, 222, 267–268*t*
sources and measures by category, 359, 361–362
history of, U.S., 166–168
Saudi Arabia, 9, 26
states, FEM ratings for, 24, 170
sources and measures by category, 354, 356
Switzerland, 47
Lehman Brothers, FEM ratings, 113*t*, 114
category ratings of companies, 375*t*
product names by category, 331*t*
Leslie Fay, FEM ratings, 80*t*
category ratings of companies, 375*t*
product names by category, 322*t*
Levi Strauss, FEM ratings, 80*t*, 82–83
category ratings of companies, 375*t*
product names by category, 321*t*
Life expectancy, 280 (n.4)
Light bulbs, product names by category, 316*t*
The Limited, FEM ratings, 81*t*, 83, 93*t*, 94
category ratings of companies, 375*t*
product names by category, 320–321*t*, 327*t*
Lingerie/underwear, product names by category, 321*t*
Liquor, product names by category, 295–296*t*
Liz Claiborne, FEM ratings, 80*t*, 81, 82
category ratings of companies, 375*t*
product names by category, 322*t*
Lodging, FEM ratings, 141–144, 142*t*, 339*t*
international hotels, 142
zero percent club, 143*t*
See also specific companies
Lotion/moisturizer, product names by category, 299–300*t*
Lotus Development Corp., FEM ratings, 146*t*, 149
category ratings of companies, 375*t*
product names by category, 341*t*
Louisiana, FEM ratings, 172*t*, 184–185, 215*t*
Lucent Technologies. *See* AT&T

Macy's: *See* R.H. Macy
Magazine publishers. *See* Publishing industry, FEM ratings
Mailing services, FEM ratings, 158–160, 159*t*
product names by category, 344*t*
zero percent club, 160*t*
See also specific companies
Maine, FEM ratings, 172*t*, 174–175, 215*t*
Makeup/lips, product names by category, 300*t*
Malls, clothing stores
product names by category, 320–321*t*
Management opportunities
corporate FEMs, 17–19, 35, 275 (n.1)
category ratings of companies, 367
derivation of ratings, 35
description of category, 348
gender differences, 33
glass ceiling effect, 18–19, 35
measures used in category, 350
retail stores, FEM ratings, 75–76
Manor Care, FEM ratings, 142*t*, 143, 340*t*
category ratings of companies, 375*t*
Marriott: *See* Host Marriott

Mars, FEM ratings, 41*t*, 42
  category ratings of companies, 375*t*
  product names by category, 284*t*, 288*t*, 291*t*, 317*t*
Maryland, FEM ratings, 172*t*, 185–186, 193, 210, 215*t*, 217–218
Masco, FEM ratings, 104*t*
  product names by category, 329*t*
Massachusetts
  FEM ratings, 172*t*, 175–176, 203, 215*t*, 217
  governance, women in, 170
  voting rights of women, history, 166
Matsushita Electric
  FEM ratings, 73*t*, 104*t*, 106, 107*t*, 109, 133, 148*t*, 150, 152*t*, 153
    category ratings of companies, 375*t*
    product names by category, 313*t*, 329–330*t*, 338–339*t*, 342*t*
  zero percent club, 75*t*, 106*t*, 109*t*, 133*t*, 151*t*, 153*t*
Mattel, FEM ratings, 89, 89*t*, 91
  category ratings of companies, 375*t*
  product names by category, 326*t*
Maybelline, FEM ratings, 61*t*, 65
  category ratings of companies, 375*t*
  product names by category, 298*t*, 299*t*, 300*t*
May Department Stores, FEM ratings, 76*t*, 84*t*, 319*t*
  category ratings of companies, 376*t*
  product names by category, 324*t*
Maytag, FEM ratings, 104*t*
  category ratings of companies, 376*t*
  product names by category, 329*t*
McCormick, FEM ratings, 40*t*
  category ratings of companies, 376*t*
  product names by category, 286*t*, 291*t*, 293*t*
McDonald's, FEM ratings, 118*t*, 119
  category ratings of companies, 333*t*, 376*t*
McGraw-Hill, FEM ratings, 113*t*, 125*t*, 127, 128
  category ratings of companies, 376*t*
  product names by category, 331*t*, 335–336*t*
MCI Communications, FEM ratings, 120*t*, 121
MCI Communications (*cont.*)
  category ratings of companies, 376*t*
  product names by category, 334*t*
Mead, FEM ratings, 154*t*, 156
  category ratings of companies, 376*t*
  product names by category, 318*t*, 343*t*
Meats, product names by category, 289*t*
Media General
  FEM ratings, 126*t*, 126–127
    category ratings of companies, 376*t*
    product names by category, 337*t*
  zero percent club, 128*t*
Median earnings, gender differences, 33, 274 (n.1), 275 (n.6)
Median income, gender differences, 4
Medicaid, 185
Mellon Bank
  FEM ratings, 111*t*, 112
    category ratings of companies, 376*t*
    product names by category, 330*t*
  zero percent club, 112*t*
Melville, FEM ratings, 76*t*, 77–78, 80*t*, 82, 84*t*, 89, 89*t*, 91, 318–319*t*, 320*t*
  category ratings of companies, 376*t*
  product names by category, 323*t*, 326*t*
Merck, FEM ratings, 67*t*, 68–69
  category ratings of companies, 376*t*
  product names by category, 308*t*, 310*t*
Meredith, FEM ratings, 125*t*, 129*t*
  category ratings of companies, 376*t*
  product names by category, 334–335*t*, 338*t*
Merrill Lynch, FEM ratings, 113*t*
  category ratings of companies, 376*t*
  product names by category, 331*t*
Methodology and scoring, 345–364
Metropolitan Life Insurance, FEM ratings, 115*t*
  category ratings of companies, 376*t*
  product names by category, 332*t*
Mexico, FEM ratings, 224*t*, 250, 256–257, 260
  climate toward women, 271*t*
  educational and economic opportunity, 271*t*
  family rights and roles, 268*t*
  health status of women, 269*t*
  legal rights for women, 267*t*

Michigan, FEM ratings, 172*t*, 197, 215*t*
Microsoft
  FEM ratings, 147*t*, 150
    category ratings of companies, 376*t*
    product names by category, 341*t*
  zero percent club, 151*t*
Middle East/North Africa, FEM ratings, 232*t*, 232–239, 260; *see also specific countries*
Midwestern states, FEM ratings, 193–202, 194*t*, 214*t*, 217; *see also specific states*
Minnesota
  FEM ratings, 172*t*, 197–198, 202, 215*t*
  governance, women in, 170
Mississippi, FEM ratings, 173*t*, 186–187, 193, 215*t*, 218
Missouri, FEM ratings, 172*t*, 198–199, 215*t*
Mitsubishi
  FEM ratings, 96–98, 97*t*
    category ratings of companies, 376*t*
    product names by category, 327*t*
  zero percent club, 99*t*
Mixers, product names by category, 295*t*
Mobil Oil
  FEM ratings, 72*t*, 101*t*, 102
    category ratings of companies, 376*t*
    product names by category, 317*t*, 328*t*
  zero percent club, 75*t*
Money, 64
Monsanto, FEM ratings, 43, 67*t*, 68, 71*t*, 73
  category ratings of companies, 376*t*
  FEM ratings, 39*t*
  product names by category, 287*t*, 308*t*, 310*t*, 315*t*
Montana, FEM ratings, 173*t*, 208, 215*t*, 217
Montgomery Ward Holding
  FEM ratings, 77*t*, 78, 319–320*t*
    category ratings of companies, 376*t*
  zero percent club, 79*t*
Moore, FEM ratings, 154*t*
  category ratings of companies, 376*t*
  product names by category, 343*t*
Morgan Stanley Group
  FEM ratings, 113*t*, 114
Morgan Stanley Group (*cont.*)
  FEM ratings (*cont.*)
    category ratings of companies, 376*t*
    product names by category, 331*t*
  zero percent club, 114*t*
Morocco, FEM ratings, 224*t*, 234–235, 238–239
  climate toward women, 272*t*
  educational and economic opportunity, 270*t*
  family rights and roles, 269*t*
  health status of women, 270*t*
  legal rights for women, 268*t*
Morris, June, 140
Morris Air, 140
Motorola, FEM ratings, 107*t*, 108–109, 149
  category ratings of companies, 376*t*
  FEM ratings, 147*t*
  product names by category, 329*t*, 341*t*
Mousefeathers, FEM ratings, 92*t*, 94
  category ratings of companies, 376*t*
  product names by category, 326*t*
Moving Comfort, FEM ratings, 80*t*, 81
  category ratings of companies, 376*t*
  product names by category, 322*t*
Ms. Foundation for Women, 386
Ms. magazine, 236
MTV: *See* Viacom
Murray, Senator Patty, 212

Nabisco: *See* RJR Nabisco
NARAL Foundation, 182
Nasrin, Tulima, 220
National Abortion Rights Action League, 386
National Association for Female Executives, 386
National Association of Commissions for Women, 386
National Association of M.B.A. Women, 387
National Association of Women Business Owners, 387
National Association of Working Women, 277 (n.14)
National Center for Policy Alternatives, 387

National Center for Women and Family Law, 387
National Chamber of Commerce for Women, 387
National Coalition Against Domestic Violence, 387
National Commission on Working Women, 389
National Committee on Pay Equity, 387
National Conference of State Legislators Women's Network, 387
National Council for Research on Women, 388
National Council of Adoption, 126
National Directory of Women Elected Officials, 46
National League of Cities, 388
National Networks of Women's Funds, 388
National Organization for Women (NOW), 166, 388
National Public Radio, FEM ratings, 129*t*, 338*t*
  category ratings of companies, 376*t*
National Women's Economic Alliance Foundation, 388
National Women's Law Center, 388
National Women's Party, 388
NBC: *See* General Electric
Nebraska, FEM ratings, 173*t*, 199, 202, 215*t*, 217
Neiman Marcus Group
  product names by category, 319*t*
  category ratings of companies, 376*t*
Nestlé
  FEM ratings, 41*t*, 42–43, 45, 48*t*, 68*t*, 70, 87*t*, 88
    category ratings of companies, 376*t*
    product names by category, 282–290*t*, 293–294*t*, 296*t*, 304*t*, 317*t*, 325*t*
  infant formulas, sales to Third World countries, 8
  zero percent club, 46*t*, 51*t*, 70*t*, 89*t*
Neutrogena, FEM ratings, 61*t*
  category ratings of companies, 376*t*
  product names by category, 298*t*, 299*t*, 301*t*
Nevada, FEM ratings, 172*t*, 209, 214, 215*t*, 217
New Balance Athletic Shoes, FEM ratings, 84*t*, 85–86
  category ratings of companies, 376*t*
  product names by category, 323*t*
Newell
  FEM ratings, 61*t*, 71*t*, 154*t*, 156
    category ratings of companies, 376*t*
    product names by category, 299*t*, 313*t*, 316*t*, 342*t*
  zero percent club, 66*t*, 75*t*, 157*t*
New Hampshire, FEM ratings, 172*t*, 176, 180, 215*t*
New Jersey
  FEM ratings, 172*t*, 176–177, 216*t*
  governance, women in, 171
Newman's Own, FEM ratings, 39*t*, 44, 48*t*
  category ratings of companies, 376*t*
  product names by category, 290–293*t*
New Mexico, FEM ratings, 172*t*, 209–210, 216*t*, 217
Newspaper publishers: *See* Publishing industry, FEM ratings
New Ways to Work, 388
New Woman, 150
New York, FEM ratings, 172*t*, 177–178, 180, 216*t*, 217
New York Times, FEM ratings, 125*t*, 126*t*
  category ratings of companies, 377*t*
  product names by category, 335*t*, 337*t*
Nike, FEM ratings, 84*t*
  category ratings of companies, 377*t*
  product names by category, 323*t*
19th Amendment, 166
9 to 5 National Association of Working Women, 389
Nine West Group
  FEM ratings, 84*t*, 85, 86
    category ratings of companies, 377*t*
    product names by category, 322–323*t*
  zero percent club, 86*t*
Nissan
  FEM ratings, 96–98, 97*t*
    category ratings of companies, 377*t*

Nissan (*cont.*)
FEM ratings (*cont.*)
product names by category, 327*t*
zero percent club, 99*t*
Nonalcoholic beverage industry, FEM ratings, 48*t*
product names by category, 293–295*t*
Noncompliance
Department of Labor, 352
Equal Employment Opportunity Commission (EEOC), 150
Nordstrom, FEM ratings, 76*t*, 77, 318*t*
category ratings of companies, 377*t*
North Africa, FEM ratings: *See* Middle East/North Africa, FEM ratings
North and Central America, FEM ratings, 253*t*, 253–260; *see also specific countries*
North Carolina, FEM ratings, 172*t*, 187, 216*t*
North Dakota, FEM ratings, 172*t*, 199–200, 216*t*, 217
Northeastern states, FEM ratings, 173*t*, 173–180, 214*t*, 217; *see also specific states*
Northern Ireland: *See* United Kingdom
Northwest Airlines, FEM ratings, 139*t*, 141, 339*t*
category ratings of companies, 377*t*
Novell, FEM ratings, 147*t*
category ratings of companies, 377*t*
product names by category, 341*t*
NOW: *See* National Organization for Women (NOW)
NPR: *See* National Public Radio, FEM ratings
Nu Skin, FEM ratings, 61*t*, 64
category ratings of companies, 377*t*

Occidental Petroleum, FEM ratings, 101*t*, 102
category ratings of companies, 377*t*
product names by category, 328*t*
OFCCP: *See* Office of Federal Contract Compliance Programs (OFCCP)
Office Depot, FEM ratings, 155*t*
category ratings of companies, 377*t*
product names by category, 343*t*
Office equipment, FEM ratings, 146–158
computers, 146–148*t*, 146–151
product names by category, 340*t*
electronic office equipment, 151–152*t*, 151–153
product names by category, 342*t*
mailing services, 344*t*
office furniture, 157–158
product names by category, 343*t*
office supplies, 153–157
product names by category, 342–343*t*
Office furniture, FEM ratings, 157*t*, 157–158
product names by category, 343*t*
zero percent club, 158*t*
See also specific companies
Office of Federal Contract Compliance Programs (OFCCP), 27, 37, 167, 275 (n.1)
Office products and services, 137–160
mailing services, 158–160
office equipment, 146–158
travel, 138–146
Office supplies, FEM ratings, 153–157, 154–155*t*
product names by category, 342–343*t*
zero percent club, 157*t*
See also specific companies
Ogilvy, David, 3
Ohio, FEM ratings, 172*t*, 200, 216*t*, 217
Oklahoma, FEM ratings, 172*t*, 188, 216*t*
Older Women's League, 389
"The 100 Best Companies for Working Mothers" (*Working Mother*), 22, 36
America West listed in, 140
Ann Taylor listed in, 82
Colgate-Palmolive listed, 74
described, 34, 347, 352
Herman Miller listed in, 158
Host Marriott listed in, 143
Johnson & Johnson listed in, 59
Nordstrom listed in, 77
Pitney Bowes listed in, 152, 159
Opportunity, equality
countries, FEM ratings for, 25
See also Management opportunities
Oracle
FEM ratings, 147*t*

Oracle (*cont.*)
FEM ratings (*cont.*)
category ratings of companies, 377*t*
product names by category, 341*t*
zero percent club, 151*t*
Oregon, FEM ratings, 172*t*, 210–211, 214, 216*t*
Organizations helping women, addresses, 382–389
Oshkosh B'Gosh, FEM ratings, 80*t*, 82, 87*t*, 88, 92*t*, 93
category ratings of companies, 377*t*
product names by category, 326*t*
O'Sullivan
FEM ratings, 157*t*, 158, 343*t*
category ratings of companies, 377*t*
zero percent club, 158*t*
Over-the-counter drugs, product names by category, 303–307*t*; *see also* Drug companies, FEM ratings; *specific companies*

Pacific countries, FEM ratings: *See* Far East/Pacific countries, FEM ratings
PaineWebber Group
FEM ratings, 113*t*
category ratings of companies, 377*t*
product names by category, 331*t*
zero percent club, 114*t*
Painkillers/relaxants, product names by category, 305*t*, 311*t*
Panama, FEM ratings, 224*t*, 250, 257–258, 259, 260
climate toward women, 271*t*
educational and economic opportunity, 270*t*
family rights and roles, 268*t*
health status of women, 269*t*
legal rights for women, 267*t*
Panasonic: *See* Matsushita
Paper products, product names by category, 316*t*
Paramount, FEM ratings, 124*t*, 133
category ratings of companies, 377*t*
product names by category, 336*t*, 339*t*
Pasta/sauce, product names by category, 290*t*
Paul Revere Insurance, FEM ratings, 116*t*
Paul Revere Insurance (*cont.*)
category ratings of companies, 377*t*
product names by category, 332*t*
PBS: *See* Public Broadcasting System (PBS)
Peanut butter, product names by category, 290*t*
Pearson, FEM ratings, 124*t*
product names by category, 336*t*
Penguin Books USA, FEM ratings
category ratings of companies, 377*t*
Pennsylvania, FEM ratings, 172*t*, 178, 180, 216*t*
Pep Boys—Manny, Moe & Jack
FEM ratings, 100*t*
category ratings of companies, 377*t*
product names by category, 328*t*
zero percent club, 101*t*
PepsiCo, FEM ratings, 39*t*, 43–44, 48*t*, 49–50, 118*t*, 119
category ratings of companies, 377*t*
product names by category, 286*t*, 292–295*t*
restaurants, rating by category, 333*t*
Perón, Isabel, 262
Persian Gulf War of 1991, 236
Personal hygiene industry: *See* Cosmetics and personal hygiene industry, FEM ratings
PET, FEM ratings, 40*t*
category ratings of companies, 377*t*
product names by category, 282–287*t*, 289–290*t*, 292–293*t*
Pet foods, product names by category, 316–317*t*
Pet supplies, product names by category, 317*t*
Pfizer, FEM ratings, 61*t*, 67*t*, 87*t*
category ratings of companies, 377*t*
product names by category, 299–302*t*, 304–305*t*, 307–308*t*, 310–311*t*, 324*t*
Pharmaceutical companies: *See* Drug companies
Pharmacia & Upjohn, FEM ratings, 68*t*, 70
category ratings of companies, 377*t*
product names by category, 303–307*t*, 309–312*t*

Philip Morris, FEM ratings, 38, 40*t*, 46, 48–49*t*, 52*t*, 52–53
  category ratings of companies, 377*t*
  product names by category, 282–291*t*, 293–296*t*
Philips Electronics
  FEM ratings, 104*t*, 107*t*, 109, 133
    category ratings of companies, 377*t*
    product names by category, 329*t*, 330*t*, 338*t*
  zero percent club, 109*t*, 133*t*
Phillips Petroleum, FEM ratings, 101*t*, 102
  category ratings of companies, 377*t*
  product names by category, 328*t*
Phillips-Van Heusen, FEM ratings, 80*t*, 84*t*
  category ratings of companies, 377*t*
  product names by category, 320*t*, 322–323*t*
Pilgrim's Pride, FEM ratings, 41*t*
  category ratings of companies, 377*t*
  product names by category, 285*t*, 289*t*
Pioneer Electronics
  FEM ratings, 107*t*, 109, 147*t*
    category ratings of companies, 377*t*
    product names by category, 329*t*, 341*t*
  zero percent club, 109*t*, 151*t*
Pitney Bowes, FEM ratings, 152*t*, 152–153, 159, 159*t*
  category ratings of companies, 377*t*
  product names by category, 342*t*, 344*t*
Plastic wrap/bags/foil, product names by category, 317*t*
Pocketbook prose, 7, 10–13
Pogrebin, Letty Cottin, v–viii
Polaroid, FEM ratings, 71*t*, 74
  category ratings of companies, 377*t*
  product names by category, 314*t*
Population, international, 221
Powers, D., 96
Pregnancy benefits, 20–21, 36
Pregnancy Discrimination Act of 1978, 167
Pregnancy tests, product names by category, 305*t*
Prepared foods, product names by category, 290–291*t*
Prescription medications, product names by category, 308–312*t*; *see also* Drug companies, FEM ratings; *specific companies*
Primerica, FEM ratings
  category ratings of companies, 377*t*
Prince Manufacturing, FEM ratings, 122*t*, 123
  category ratings of companies, 377*t*
  product names by category, 333*t*
Prochnow, Herbert, 15
Procter & Gamble, FEM ratings, 39*t*, 43, 48*t*, 49–50, 59, 60*t*, 62–64, 67*t*, 68, 71*t*, 73–74, 87*t*, 88
  category ratings of companies, 378*t*
  product names by category, 282*t*, 285*t*, 290*t*, 292–294*t*, 297–303*t*, 305–306*t*, 313–316*t*, 318*t*, 325*t*
Proctor Silex/Hamilton Beach. *See* Hamilton Beach/Proctor Silex
Product names by category, 281–344
Promus Hotel, FEM ratings, 142*t*, 340*t*
  category ratings of companies, 378*t*
Prudential Insurance Company of America, FEM ratings, 115*t*, 116
  category ratings of companies, 378*t*
Public Broadcasting System (PBS), FEM ratings, 129*t*, 131, 338*t*
  category ratings of companies, 378*t*
  product names by category, 332*t*
Publishing industry, FEM ratings, 123–128
  book publishers, 123, 124*t*
    product names by category, 336*t*
  magazine publishers, 123, 124–125*t*
    cooking, home, and travel magazines, 334*t*
    entertainment magazines, 335*t*
    news and business magazines, 335*t*
    product names by category, 334–335*t*
    sports and hobbies magazines, 335*t*
    women's and family magazines, 334*t*
  newspaper publishers, 123, 125–126*t*
    product names by category, 336–337*t*
  zero percent club, 128*t*
  See also specific companies

Publix Super Markets, FEM ratings, 54*t*
category ratings of companies, 378*t*
product names by category, 297*t*

Quaker Oats, FEM ratings, 40*t*, 48*t*
category ratings of companies, 378*t*
product names by category, 283–284*t*, 287*t*, 290–291*t*, 293*t*, 316–317*t*
Quaker State, FEM ratings, 99*t*
category ratings of companies, 378*t*
product names by category, 327*t*
QVC Network
FEM ratings, 129*t*, 131
category ratings of companies, 378*t*
product names by category, 337*t*
zero percent club, 131*t*

Radio industry: *See* Broadcasting industry, FEM ratings
Ralston Purina, FEM ratings, 40*t*, 72*t*
category ratings of companies, 378*t*
product names by category, 282*t*, 286*t*, 313*t*, 317*t*
Ramada International Hotels: *See* Renaissance Hotels
Raytheon, FEM ratings, 104*t*
category ratings of companies, 378*t*
product names by category, 329*t*
Reader's Digest Association, FEM ratings, 124*t*, 125*t*, 127
category ratings of companies, 378*t*
product names by category, 334–336*t*
Reagan, President Ronald W., 167
Reckitt & Colman
FEM ratings, 41*t*, 46, 62*t*, 64–65, 73*t*
category ratings of companies, 378*t*
product names by category, 284–285*t*, 301–302*t*, 312–315*t*
zero percent club, 47*t*, 66*t*, 75*t*
Record labels, FEM ratings
product names by category, 338*t*
Red Apple, 53
Redken Laboratories, FEM ratings, 61*t*
category ratings of companies, 378*t*
product names by category, 301*t*
Reebok, FEM ratings, 81*t*, 84*t*, 85, 93*t*
category ratings of companies, 378*t*
product names by category, 322–323*t*, 326*t*
Renaissance Hotels
FEM ratings, 142*t*, 143–144, 340*t*
category ratings of companies, 379*t*
zero percent club, 144*t*
Rental cars, FEM ratings, 145*t*, 145–146
product names by category, 340*t*
zero percent club, 146*t*
See also specific companies
Republican Party, 1996 convention, 168
Restaurants, FEM ratings, 118–119*t*, 118–120, 137
product names by category, 333*t*
zero percent club, 120*t*
See also specific restaurant
Retail stores, FEM ratings, 75–79, 76–77*t*, 318–320*t*
zero percent club, 79*t*
See also specific companies
Revlon
FEM ratings, 59, 62*t*, 65
category ratings of companies, 378*t*
product names by category, 298*t*, 300–301*t*
zero percent club, 66*t*
Reynolds Metals
FEM ratings, 73*t*
category ratings of companies, 378*t*
product names by category, 313*t*, 317*t*
zero percent club, 75*t*
R.H. Macy, FEM ratings
category ratings of companies, 378*t*
FEM ratings, 76*t*, 77, 78–79, 319*t*
Rhode Island, FEM ratings, 172*t*, 178–179, 216*t*
Rhône-Poulenc Rorer
FEM ratings, 67*t*
category ratings of companies, 378*t*
product names by category, 308–309*t*, 311*t*
zero percent club, 70*t*
Ricoh
FEM ratings, 148*t*, 152*t*, 153
category ratings of companies, 378*t*
product names by category, 341*t*, 342*t*
zero percent club, 151*t*, 153*t*

Rights of women: *See* Legal rights for women
Rivers, Joan, 58
RJR Nabisco
FEM ratings, 40*t*, 52*t*, 52–53
category ratings of companies, 378*t*
product names by category, 282–285*t*, 287*t*, 291–292*t*, 296*t*, 317*t*
zero percent club, 47*t*
Roddick, Anita, 63
Roe v. Wade, 166
Romer, Bea, 206
Romer, Governor Roy, 206
Roosevelt, Eleanor, 219
Royal Dutch Petroleum/Shell Oil
FEM ratings, 101*t*
product names by category, 328*t*
zero percent club, 103*t*
RU-486, 10, 212, 275 (n.8)
Rubbermaid, FEM ratings, 71*t*, 74–75, 89*t*, 91, 154*t*, 155–156
category ratings of companies, 378*t*
product names by category, 315–316*t*, 326*t*, 342*t*
Russia
equality laws, 280 (n.9)
FEM ratings, 224*t*, 244, 249–250, 252
climate toward women, 271*t*
educational and economic opportunity, 270*t*
family rights and roles, 268*t*
health status of women, 269*t*
legal rights for women, 267*t*
Rust Belt, 202

Safeway
boards of directors, 13
FEM ratings, 54*t*, 56
category ratings of companies, 378*t*
product names by category, 297*t*
zero percent club, 56*t*
Salad dressing, product names by category, 291*t*
Salomon Holding, FEM ratings, 113*t*, 114
category ratings of companies, 378*t*
product names by category, 331*t*
Sara Lee, FEM ratings, 39*t*, 71, 72*t*, 74, 80*t*, 83
Sara Lee, FEM ratings (*cont.*)
category ratings of companies, 378*t*
product names by category, 286*t*, 289*t*, 314*t*, 316*t*, 320–322*t*
Saudi Arabia
FEM ratings, 224*t*, 235–237, 238–239, 259
climate toward women, 272*t*
educational and economic opportunity, 271*t*
family rights and roles, 269*t*
health status of women, 270*t*
legal rights for women, 268*t*
women's rights, 9, 26
Savvy, 150
S.C. Johnson & Son
FEM ratings, 61*t*, 72*t*
category ratings of companies, 378*t*
product names by category, 301*t*, 312–315*t*
zero percent club, 66*t*, 75*t*
Schering-Plough, FEM ratings, 60*t*, 67*t*, 69–70, 71*t*, 87*t*
category ratings of companies, 378*t*
product names by category, 299*t*, 303–309*t*, 313*t*, 316*t*, 324–325*t*
School supplies, product names by category, 318*t*
Schroeder, Representative Pat, 20
Scott Paper
FEM ratings, 61*t*, 72*t*, 87*t*, 88
category ratings of companies, 378*t*
product names by category, 302*t*, 316*t*, 318*t*, 324*t*
zero percent club, 66*t*, 75*t*, 89*t*
Scripps: *See* E.W. Scripps
Seagram Company, FEM ratings, 48–49*t*
category ratings of companies, 378*t*
product names by category, 294–296*t*
Searle: *See* Monsanto, FEM ratings
Sears (Allstate): *See* Allstate (Sears)
Sears, Roebuck, FEM ratings, 76*t*, 77–78, 92*t*, 93, 103*t*, 105, 106, 318*t*, 320*t*
category ratings of companies, 378*t*
product names by category, 326*t*, 328*t*
Service Merchandise
FEM ratings, 76*t*, 320*t*
category ratings of companies, 378*t*

Service Merchandise (*cont.*)
zero percent club, 79*t*
7-Eleven, FEM ratings, 55
Sex differences: *See* Gender differences
Sexual harassment, 21–22
Shampoo, product names by category, 300–301*t*
Sharp Electronics
FEM ratings, 105*t*, 106, 108*t*, 109, 148*t*, 150, 152*t*, 153
category ratings of companies, 378*t*
product names by category, 329–330*t*, 342*t*
zero percent club, 106*t*, 109*t*, 151*t*, 153*t*
Shaving needs/razors, product names by category, 301*t*
Shell Oil (Royal Dutch), FEM ratings; category ratings of companies, 378*t*
Sheraton Hotels: *See* ITT Sheraton
Shoe companies: *See* Footwear industry
Shopping for a Better World (Council of Economic Priorities), 34
Singapore, FEM ratings, 224*t*, 230–231, 232
climate toward women, 272*t*
educational and economic opportunity, 270*t*
family rights and roles, 268*t*
health status of women, 269*t*
legal rights for women, 268*t*
Single mothers, economic status, 4
Skin, product names by category, 312*t*
Sleeping pills/stimulants, product names by category, 305–306*t*, 312*t*
Smeal, Eleanor, 166
Smithfield Foods
FEM ratings, 41*t*, 45
category ratings of companies, 379*t*
product names by category, 289*t*
zero percent club, 47*t*
SmithKline Beecham Consumer Brands
FEM ratings, 60*t*, 64, 67*t*, 68–69, 71*t*, 74
category ratings of companies, 379*t*
product names by category, 298*t*, 301–303*t*, 305–306*t*, 308–310*t*, 312*t*, 315*t*
zero percent club, 66*t*, 70*t*
Snacks, product names by category
crackers, bars, 291*t*
popcorn, chips, 292*t*
Soda, product names by category, 295*t*
Sony of America
FEM ratings, 107*t*, 109, 133, 147*t*
category ratings of companies, 379*t*
product names by category, 329*t*, 338*t*, 341*t*
zero percent club, 109*t*, 133*t*, 151*t*
Soup, product names by category, 292*t*
South Africa
apartheid policy, 9
FEM ratings, 224*t*, 241–242, 243
climate toward women, 272*t*
educational and economic opportunity, 271*t*
family rights and roles, 268*t*
health status of women, 270*t*
legal rights for women, 268*t*
South America, FEM ratings, 260–266, 261*t*; *see also specific countries*
South Carolina, FEM ratings, 173*t*, 188–189, 216*t*
South Dakota, FEM ratings, 172*t*, 200–201, 202, 216*t*
Southern states, FEM ratings, 180*t*, 180–193, 214*t*, 217–218; *See also specific states*
Southland
FEM ratings, 54*t*, 55
category ratings of companies, 379*t*
product names by category, 297*t*
zero percent club, 56*t*
Southwest Airlines, FEM ratings, 139*t*, 339*t*
category ratings of companies, 379*t*
Southwestern states, FEM ratings, 140
Spain, FEM ratings, 224*t*, 250, 252
climate toward women, 271*t*
educational and economic opportunity, 271*t*
family rights and roles, 268*t*
health status of women, 269*t*
legal rights for women, 267*t*
Spices/seasonings, product names by category, 293*t*

Sports equipment industry, FEM ratings, 121–123, 122*t*
product names by category, 333–334*t*
See also specific companies
Sportswear, product names by category, 322–323*t*
Sprint, FEM ratings, 120*t*, 121
category ratings of companies, 379*t*
product names by category, 334*t*
Standard Register
FEM ratings, 152*t*, 153, 155*t*, 156
category ratings of companies, 379*t*
product names by category, 342–343*t*
zero percent club, 153*t*, 157*t*
Stanton, Elizabeth Cady, 95, 219
Staples Incorporated, FEM ratings, 154*t*, 155, 156
category ratings of companies, 379*t*
product names by category, 342*t*
States, 163–218
civil rights for women, 24, 170
sources and measures by category, 354, 356
climate toward women, generally, 24, 170
sources and measures by category, 354–357
differences, 164–165
economic climate for women, 24, 170
sources and measures by category, 353, 355
family laws affecting women, 24, 170
sources and measures by category, 354, 356
FEM ratings, 23–24, 168–218, 171–173*t*, 215–216*t*
civil rights for women, 24, 170
climate toward women, FEM ratings, 24, 170–171
economic status of women, 24, 170
employer category, 24, 170, 353, 355
governance, women in, 24, 170
sources and measures by category, 353–358
system of rating, 168–73
See also specific states
governance, women in, 24, 170
States (*cont.*)
governance, women in (*cont.*)
sources and measures by category, 354, 355–356
governor's names, addresses and phone numbers, 389–393
legal rights for women, 24, 170
differences between states, 166–168
sources and measures by category, 354, 356
See also specific states
Stomach/indigestion medicine, product names by category, 306*t*
Stop & Shop
FEM ratings, 54*t*
category ratings of companies, 379*t*
product names by category, 297*t*
zero percent club, 56*t*
Stride Rite, FEM ratings, 84*t*, 85, 92*t*, 93
category ratings of companies, 379*t*
product names by category, 323*t*, 326*t*
Sunbeam
FEM ratings, 104*t*, 106
category ratings of companies, 379*t*
zero percent club, 106*t*
Sun (Sunoco), FEM ratings, 101*t*, 102
category ratings of companies, 379*t*
product names by category, 328*t*
Sun Microsystems
FEM ratings, 147*t*
category ratings of companies, 379*t*
product names by category, 341*t*
zero percent club, 151*t*
Sunoco: *See* Sun (Sunoco), FEM ratings
Suntan lotion, product names by category, 306*t*
Support of women
corporate FEMs, 19, 35–36
description of category, 349
measures used in category, 351
Switzerland
FEM ratings, 224*t*, 247, 250–251, 252, 273
climate toward women, 272*t*
educational and economic opportunity, 271*t*
family rights and roles, 268*t*

Switzerland (*cont.*)
FEM ratings (*cont.*)
health status of women, 269*t*
legal rights for women, 268*t*
women's rights, 47
Syntex, FEM ratings, 67*t*
category ratings of companies, 379*t*
product names by category, 308–309*t*, 311*t*

Tambrands, FEM ratings, 61*t*
category ratings of companies, 379*t*
product names by category, 302*t*
Tampons/sanitary pads/incontinence products, product names by category, 302*t*
Tandy
FEM ratings, 155*t*, 156
category ratings of companies, 379*t*
product names by category, 343*t*
zero percent club, 157*t*
Tanzania, FEM ratings, 224*t*, 242–243, 259, 273
climate toward women, 272*t*
educational and economic opportunity, 271*t*
family rights and roles, 269*t*
health status of women, 270*t*
legal rights for women, 268*t*
TCI: *See* Tele-Communications (TCI), FEM ratings
Tehran, violence against women, 220
Tektronix Inc., FEM ratings, 147*t*, 149–150
category ratings of companies, 379*t*
product names by category, 341*t*
Tele-Communications (TCI), FEM ratings, 129*t*
category ratings of companies, 379*t*
product names by category, 337–338*t*
Telephone services, FEM ratings, 120–121
product names by category, 334*t*
See also specific companies
Television, product names by category
cable television, 337*t*
network TV, 338*t*
See also Broadcasting industry, FEM ratings
Tennessee, FEM ratings, 173*t*, 189–190, 216*t*
Texaco, FEM ratings, 101*t*
category ratings of companies, 379*t*
Texas, FEM ratings, 172*t*, 190–191, 216*t*, 218
Texas Instruments
FEM ratings, 107*t*, 147*t*, 152*t*, 153
category ratings of companies, 379*t*
product names by category, 330*t*, 341–342*t*
zero percent club, 109*t*, 151*t*, 153*t*
Thatcher, Margaret, 58, 252
Theme parks, FEM ratings by category, 339*t*; *see also* Walt Disney
Thermador, FEM ratings; category ratings of companies, 379*t*
Thomas, Justice Clarence, 4
3M, FEM ratings, 67*t*, 69, 71*t*, 74, 154*t*, 155
category ratings of companies, 376*t*
product names by category, 306*t*, 313*t*, 315*t*, 318*t*, 342*t*
Timberland
FEM ratings, 81*t*, 83, 84*t*, 86
category ratings of companies, 379*t*
product names by category, 322–323*t*
zero percent club, 83*t*, 86*t*
Times Mirror, FEM ratings, 124*t*, 125*t*, 129*t*
category ratings of companies, 379*t*
product names by category, 335–338*t*
Time Warner, FEM ratings, 124*t*, 127, 129*t*, 130, 132*t*
category ratings of companies, 379*t*
product names by category, 334–339*t*
Title VII (Civil Rights Act of 1964), 167
Toastmaster
FEM ratings, 104*t*, 106
category ratings of companies, 379*t*
product names by category, 329*t*
zero percent club, 106*t*
Tobacco industry, FEM ratings, 51–53
product names by category, 296*t*
See also specific companies
Toilet paper/tissues, product names by category, 318*t*

Toothbrushes/toothpaste/oral hygiene, product names by category, 302*t*
The Topps
FEM ratings, 41*t*, 45
category ratings of companies, 379*t*
zero percent club, 47*t*
Toro, FEM ratings, 104*t*, 105–106
category ratings of companies, 379*t*
product names by category, 329*t*
Toy industry, FEM ratings, 89–90*t*, 89–92
product names by category, 326*t*
zero percent club, 92*t*
See also specific companies
Toyota
FEM ratings, 97*t*, 97–99
category ratings of companies, 379*t*
product names by category, 327*t*
zero percent club, 99*t*
Toys, influence on children, 15–16
Toys "R" Us, FEM ratings, 90*t*, 93*t*, 94
category ratings of companies, 379*t*
product names by category, 326*t*, 327*t*
Transamerica, FEM ratings, 113*t*, 115*t*
category ratings of companies, 379*t*
product names by category, 331–332*t*
Transportation industry, FEM ratings, 96–103
automobile companies, 96–99
product names by category, 327*t*
automobile suppliers, 99–101
product names by category, 327–328*t*
gasoline companies, 101–103
product names by category, 328*t*
product names by category, 327–328*t*
See also specific companies
Trans World Airlines (TWA)
FEM ratings, 139*t*, 141, 339*t*
category ratings of companies, 380*t*
zero percent club, 141*t*
Travel, FEM ratings
airlines, 138–141, 339*t*
business travel, 138–146, 163
countries, variations by, 219–273
lodging, 141–144, 340*t*
rental cars, 145–146
Travel, FEM ratings (*cont.*)
rental cars (*cont.*)
product names by category, 340*t*
states, variations by, 163–218
Travelers (Commercial Credit), FEM ratings, 110*t*
Travelers Group, FEM ratings, 113*t*, 115*t*
category ratings of companies, 379*t*
product names by category, 330*t*, 331*t*, 332*t*
Travel Industry Association of America, 163
Triarc
FEM ratings, 48*t*, 118*t*, 120
category ratings of companies, 380*t*
product names by category, 295*t*
restaurants, rating by category, 333*t*
zero percent club, 51*t*, 120*t*
Turkey, FEM ratings, 224*t*, 237–238, 238–239
climate toward women, 272*t*
educational and economic opportunity, 271*t*
family rights and roles, 268*t*
health status of women, 270*t*
legal rights for women, 267*t*
Turner Broadcasting System, FEM ratings, 129*t*
category ratings of companies, 380*t*
product names by category, 337*t*
TWA: *See* Trans World Airlines (TWA)
Tyco Toys
FEM ratings, 90*t*, 92
category ratings of companies, 380*t*
product names by category, 326*t*
zero percent club, 92*t*
Tyson Foods, FEM ratings, 41*t*, 45
category ratings of companies, 380*t*
product names by category, 288*t*, 289*t*

Unilever
FEM ratings, 41*t*, 46, 48*t*, 61*t*, 64–65, 68*t*, 70, 72*t*
category ratings of companies, 380*t*
product names by category, 283*t*, 285*t*, 288*t*, 290–294*t*, 297–302*t*, 304*t*, 314–315*t*

Unilever (*cont.*)
zero percent club, 47*t*, 51*t*, 66*t*, 70*t*, 75*t*
Unisys, FEM ratings, 148*t*, 150
category ratings of companies, 380*t*
product names by category, 341*t*
United Airlines, FEM ratings, 139*t*, 339*t*
category ratings of companies, 380*t*
United Asset Management, FEM ratings, 113*t*
category ratings of companies, 380*t*
product names by category, 331*t*
United Biscuit Holdings, FEM ratings, 41*t*
category ratings of companies, 380*t*
product names by category, 291–292*t*
United Kingdom
FEM ratings, 224*t*, 251–252, 259
climate toward women, 271*t*
educational and economic opportunity, 270*t*
family rights and roles, 268*t*
health status of women, 269*t*
legal rights for women, 267*t*
spending power of women, 7
United Nations
Convention on the Elimination of All Forms of Discrimination Against Women, 223, 258, 265, 266
Statement on the Non-Discriminatory Treatment of Women, 26
treaty regarding treatment of women, 249–250
women in governance, survey, 25
World Population Conference, 242
United Parcel Service of America
FEM ratings, 159*t*, 160, 344*t*
category ratings of companies, 380*t*
zero percent club, 160*t*
United States, FEM ratings, 224*t*, 231, 253, 254, 258–260
climate toward women, 272*t*
educational and economic opportunity, 270*t*
family rights and roles, 268*t*
health status of women, 269*t*
legal rights for women, 267*t*
United Technologies (Carrier), FEM ratings, 103*t*, 105, 106
category ratings of companies, 371*t*
product names by category, 329*t*
Universal Foods
FEM ratings, 41*t*
category ratings of companies, 380*t*
product names by category, 282*t*
zero percent club, 47*t*
Upjohn: *See* Pharmacia & Upjohn
U.S. Labor Department, 259
U.S. State Department
Brazil, human rights concerns, 262
Dominican Republic, human rights concerns, 254
Kenya, human rights concerns, 240
Panama, discrimination laws, 258
Saudi Arabia, legal rights for women and, 230
Singapore, human rights concerns, 230
South America, human rights concerns, 261
USAir Group, FEM ratings, 139*t*, 140–141, 339*t*
category ratings of companies, 380*t*
US Life
FEM ratings, 116*t*
category ratings of companies, 380*t*
product names by category, 332*t*
zero percent club, 117*t*
Utah, FEM ratings, 169, 173*t*, 211, 214, 216*t*, 217

Venezuela, FEM ratings, 224*t*, 265–266
climate toward women, 271*t*
educational and economic opportunity, 270*t*
family rights and roles, 269*t*
health status of women, 269*t*
legal rights for women, 267*t*
Vermont, FEM ratings, 172*t*, 179, 216*t*
VF, FEM ratings, 80*t*, 93*t*
category ratings of companies, 380*t*
product names by category, 321–322*t*, 326*t*
Viacom, FEM ratings, 129*t*

Viacom, FEM ratings (*cont.*)
category ratings of companies, 380*t*
product names by category, 337–339*t*
Video stores, FEM ratings, 339*t*
Violence against women
condoning of, 219–221
countries, prevention, 362
states, prevention of, 356
Virginia, FEM ratings, 172*t*, 191–192, 216*t*
Vitamins, product names by category, 306–307*t*
Volkswagen/Audi
FEM ratings, 97, 97*t*
category ratings of companies, 380*t*
product names by category, 327*t*
zero percent club, 99*t*
Voting rights, countries, FEM ratings, 361
Voting Rights Bill of 1965, 166
Vucanovich, Barbara, 209

Wage gap, 4
managers, 32
Walgreen, FEM ratings, 76*t*, 319*t*
category ratings of companies, 380*t*
Wal-Mart Stores, FEM ratings, 77*t*, 78–79, 319–320*t*
category ratings of companies, 380*t*
Walt Disney
FEM ratings, 124*t*, 125*t*, 127, 129*t*, 130–131, 142*t*, 143
category ratings of companies, 380*t*
product names by category, 335–340*t*
zero percent club, 128*t*, 131*t*, 132–133, 133*t*, 144*t*
Wang Laboratories
FEM ratings, 147*t*
category ratings of companies, 380*t*
product names by category, 341*t*
zero percent club, 151*t*
Warnaco Group, FEM ratings, 80*t*, 82
category ratings of companies, 380*t*
product names by category, 321–322*t*
Warner-Lambert, FEM ratings, 39*t*, 61*t*, 67*t*
category ratings of companies, 380*t*
Warner-Lambert, FEM ratings (*cont.*)
FEM ratings
product names by category, 301–307*t*
product names by category, 283–284*t*, 299*t*, 309*t*, 311*t*
Wart removal, product names by category, 307*t*
Washington, FEM ratings, 171*t*, 212, 213, 216*t*, 217
Washington Post, FEM ratings, 125*t*, 126*t*
category ratings of companies, 380*t*
product names by category, 335*t*, 337*t*
Welch Foods
FEM ratings, 41*t*
category ratings of companies, 380*t*
product names by category, 289*t*, 294*t*
zero percent club, 47*t*
Wells Fargo, FEM ratings, 110*t*, 111
category ratings of companies, 380*t*
product names by category, 330*t*
Wendy's, FEM ratings, 118*t*, 119
category ratings of companies, 333*t*, 380*t*
Wenner Media, FEM ratings, 125*t*
category ratings of companies, 379*t*
product names by category, 335*t*
Western Publishing Group
FEM ratings, 124*t*
category ratings of companies, 380*t*
product names by category, 336*t*
zero percent club, 128*t*
Western states, FEM ratings, 202*t*, 202–214, 214*t*, 217
Westin Hotel, FEM ratings, 142*t*, 144, 340*t*
category ratings of companies, 380*t*
West Virginia, FEM ratings, 173*t*, 192, 193, 216*t*, 218
Whirlpool
FEM ratings, 104*t*
category ratings of companies, 380*t*
product names by category, 329*t*
zero percent club, 106*t*
Whitman, FEM ratings, 99*t*, 100
category ratings of companies, 380*t*

Whitman, FEM ratings (*cont.*)
product names by category, 327*t*
Wider Opportunities for Women, 389
William Carter, FEM ratings, 87*t*, 93*t*, 94
category ratings of companies, 380*t*
product names by category, 326*t*
Wine/sparkling wine, product names by category, 296*t*
Winn-Dixie Stores, FEM ratings, 54*t*, 56*t*
category ratings of companies, 381*t*
product names by category, 297*t*
Wisconsin, FEM ratings, 172*t*, 201, 216*t*
Wm. Wrigley Jr., FEM ratings, 39*t*
category ratings of companies, 381*t*
product names by category, 284*t*
Women Employed, 389
Women's Bureau, 21
Women's Economic Agenda Project, 389
Women's Legal Defense Fund, 389
Women's rights: *See* Legal rights for women
Women Work! The National Network for Women's Employment, 389
Woolworth, FEM ratings, 76*t*, 84*t*, 93*t*, 319*t*
category ratings of companies, 381*t*
product names by category, 321*t*, 324*t*, 327*t*
Working Assets Funding Service
FEM ratings, 112*t*, 113–114
category ratings of companies, 381*t*
product names by category, 331*t*
Working Mother
American Express, listed in, 111
America West, 140
awards by, 64
California, recognition of, 205
Cigna, mentioned in, 116
Citicorp, listed in, 111
Colorado, recognition of, 206
Gannett, mentioned in, 126
Herman Miller, recognition of, 158
Host Marriott Corp., 143
IBM, recognition of, 148
"The 100 Best Companies for Working Mothers": *See* "The 100 Best Companies for Working Mothers" (*Working Mother*)
Working Mother (*cont.*)
Pitney Bowes, recognition of, 152, 159
publisher of, 127
Stride-Rite, recognition of, 85
Xerox, recognition of, 149
Working Woman, 11
Wyoming
FEM ratings, 172*t*, 213, 214, 216*t*
voting rights of women, history, 166

Xerox, FEM ratings, 146*t*, 149, 151*t*, 152
category ratings of companies, 381*t*
product names by category, 340*t*, 342*t*

Yeast infection medicines, product names by category, 307*t*
York
FEM ratings, 105*t*, 106
category ratings of companies, 381*t*
product names by category, 329*t*
zero percent club, 106*t*

Zenith Electronics, FEM ratings, 107*t*
category ratings of companies, 381*t*
product names by category, 330*t*
Zero percent club
airlines, 141*t*
appliance industry, 106*t*
automobile companies, 99*t*
automobile suppliers, 100–101*t*
baby products industry, 89*t*
banks, 112*t*
beverage industry, 51*t*
broadcasting industry, 131*t*
children's clothes and shoes industry, 94*t*
clothing industry, 83*t*
computers, 150–151*t*
cosmetics and personal hygiene industry, 65–66*t*
definition of zero percenters, 35
drug companies, 70*t*
electronic equipment industry, 109*t*
electronic office equipment, 153*t*
entertainment industry, 133*t*
financial institutions, 114*t*
food industry, 46–47*t*
footwear industry, 86*t*

Zero percent club (*cont.*)
gasoline companies, 103*t*
grocery stores, 56*t*
household goods companies, 75*t*
insurance companies, 117*t*
lodging, 143*t*
mailing services, 160*t*
office furniture, 158*t*
office supplies, 157*t*
publishing industry, 128*t*
rental cars, 146*t*
restaurants, 120*t*
retail stores, 79*t*
toy industry, 92*t*